AF717919

The Modern Clock

A Study of Time Keeping Mechanism;
Its Construction, Regulation
and Repair.

by
Ward L. Goodrich

Arlington Book Company
1984

This edition, published in 1984 by Arlington Book Company, is an unabridged republication of the work originally published by Hazlitt and Walker, Publishers, Chicago, Illinois in 1905 under the title "The Modern Clock: A Study of Time Keeping Mechanism; Its Construction, Regulation and Repair" by Ward L. Goodrich.

Printed in the United States of America by
Arlington Book Company
P.O. Box 327
Arlington, Virginia 22210-0327 USA.

Library of Congress Cataloging in Publication Data.
Goodrich, Ward L.
The Modern Clock:
Reprint. Originally published: Chicago: Hazlitt & Walker, 1905.
Includes indexes.
1. Clocks and watches. 2. Clocks and watches — Repairing and adjusting. I. Title.
TS545.G63 1984 681.1'13'0288 84-24277

ISBN 0-930 163-23-0

FOREWORD

In writing a foreword to the second edition of this practical book on clock repair, I am glad of the opportunity to voice my appreciation for it. For a long time I have been hoping someone would re-publish this work, which was first printed in 1905. As a clockmaker, I have felt a definite need for it in the years it has been out of print. Good, thoroughgoing clock books are scarce. It is a pleasure to welcome back one of the best.

Truly, there is a dearth of good reference works on clock repair. Most of the published volumes treat the subject from a historical point of view. All are excellent in their way—but their way is not that of the craftsman looking for technical help. He still must search far and wide for the information he wants.

This lack of reliable guidance is what makes the present volume so valuable. Few books on clock repair are as complete and authoritative. In scope it goes far beyond the cataloguing of historical types of clocks. The book provides in one handy volume a wealth of technical advice, which all can consult as the need arises.

There is advice to the beginner, counsel for the specialist and information of value to the hobbyist or collector. In this one book we find basic fundamentals, sound business practices and specific helps on the repair of the principal types of clocks. Not much in the field of clockwork is left unmentioned. The book fills a definite gap on the clockmaker's reference shelf and certainly should be placed there by anyone who calls himself a clockmaker.

The volume was originally intended to serve as a guide for the beginner in clock repair. Indeed, he will find it a real and valuable friend. The sound suggestions, so simply stated, will save him many a pitfall that might otherwise trip up the fledgling clockmaker without experience.

Withal that, probably none will treasure it more than the practicing watchmaker and clockmaker. In the light of their experience they can more readily appreciate the clarity of direction that is given in its pages.

Minor changes have been made here and there in the text. Obvious errors in figures have been corrected. But in the main, the book stands as originally conceived by Goodrich, forty-five years ago. There appears to be no need for any major revisions.

Basic principles are stressed throughout and they are unchanged. The clocks and striking trains discussed are the very ones now coming to the bench for repair, so this information also is timely. The increased use of electric clocks has not done away with the need for knowledge of other types since so many people still cherish historical and ornamental models. After all, who can say when such a book is out of date? Thomas Reid wrote his famous treatise at Edinburg, in 1826. Today, one hundred and twenty-four years later, it still is a standard reference work.

I believe Goodrich meant this book to serve as an inspiration for finer craftsmanship and at the same time to give practical help to that end. I feel it will do just that, even better now than before.

J. E. Coleman

Nashville, Tenn.
July, 1950

The Modern Clock

A Study of Time Keeping Mechanism;
Its Construction, Regulation
and Repair.

BY WARD L. GOODRICH

Author of the Watchmaker's Lathe, Its Use and Abuse.

WITH NUMEROUS ILLUSTRATIONS AND DIAGRAMS

CHICAGO
Hazlitt & Walker, Publishers
1905

CONTENTS

CHAPTER I.

THE NECESSITY FOR BETTER SKILL AMONG CLOCKMAKERS

The need for information of an exact and reliable character in regard to the hard worked and much abused clock has, we presume, been felt by every one who entered the trade. This information exists, of course, but it is scattered through such a wide range of publications and is found in them in such a fragmentary form that by the time a workman is sufficiently acquainted with the literature of the trade to know where to look for such information he no longer feels the necessity of acquiring it.

The rapid increase in the use of personal watches as well as the advent of the electric clock have contributed to the neglect of the pendulum timekeepers. Good clock men are becoming scarce as a result of the shift of interest from this type of mechanism. Successive generations of students and apprentices tend to give less and less attention to clock work as they strive to master the intricacies of the many types of personal timepieces that have mushroomed in recent years. Thus, the sands of time, in passage, have filtered out more and more of those whose main interest lies in clock work. In consequence, the number of informed clock men has not kept pace with the increasing number of "watchmakers."

Unquestionably, the development of the electric clock has hastened the demise of the old-time clock repairman. Electric clocks are for the most part repaired at the factory or with factory assemblies. The modern watchmaker has little real contact with them, except perhaps to sell them initially. This leads many watchmakers to neglect clock repair as not worth bothering about, which is unfortunate. Striking clocks still exist in good numbers, for sentimental or ornamental reasons, if not practical need. Their owners revere them and are usually anxious they be kept going, even at some expense. This offers a profitable field of endeavor for those who have accurate knowledge of clock construction and behavior.

But a watchmaker cannot reap the benefit of this trade if he performs a shoddy job for lack of specialized knowledge. The general public today have been made so time conscious that they will not accept careless workmanship. A watchmaker not qualified to repair a striking mechanism who accepts such a job runs the risk of loss of reputation as well as his customer if he fails to turn out a creditable piece of work.

The apprentice, when he first gets an opportunity to examine a clock movement, usually considers it a very mysterious machine. Later on, if he handles many clocks of the simple order, he becomes tolerably familiar with the time train; but he seldom becomes confident of his ability regarding the striking part, the alarm and the escapement, chiefly because the employer and the older workmen get tired of telling him the same things repeatedly, or because they were similarly treated in their youth, and consider clocks a nuisance, any how, never having learned clock work thoroughly, and therefore being unable to appreciate it. In consequence of such treatment, the newcomer makes half-hearted efforts to learn the portions of the business that puzzle him, and then gives it up, and thereafter does as little as possible to clocks, but begs continually to be put on watch work.

We know of a shop where two and sometimes three workmen (the best in the shop, too) are constantly employed upon clocks which country jewelers have failed to repair. If clock work is dull they will go upon watch work (and they do good work, too), but they enjoy the clocks and will do them in preference to watches, claiming that there is greater variety and more interest in the work than can be found in fitting factory made material into watches, which consist of a time train only. Two of these men have become famous, and are frequently sent for to take care of complicated clocks, with musical and mechanical figure attachments, tower, chimes, etc. The third is much younger, but is rapidly perfecting himself, and is already competent to rebuild minute repeaters and other sorts of the finer kinds of French clocks. He now totally neglects watch work, saying that the clocks give him more money and more fun.

We are confident that this would be also the case with many another American youth if he could find some one to patiently instruct him in the few indispensable facts which lie at the bottom of so much that is mysterious and from which he now turns in disgust. The object of these articles is to explain to the apprentice the mysteries of pendulums, escapements, gearing of trains, and the whole technical scheme of these measurers of time, in such a way that hereafter he may be able to answer his own questions, because he will be familiar with the facts on which they depend.

Because authoritative information on clock work has become harder to get as a result of the situation already discussed, frequent demands have been made for a book on clocks. These demands have therefore induced the writer to undertake its compilation. Works on the subject—nominally so, at least—are in existence, but it will generally be found on examination that they are written by outsiders, not by workmen, and that they treat the subject historically, or from the standpoint of the artistic or the curious. Any information regarding the mechanical movements is fragmentary, if found in them at all, and they are better fitted for the amusement of the general public than for the youth or man who wants to know "how and why." These facts have impelled the writer to ignore history and art in considering the subject; to treat the clock as an existing mechanism which must be understood and made to perform its functions correctly; and to consider cases merely as housings of mechanism, regardless of how beautiful, strange or commonplace those housings may be.

We have used the word "compile" advisedly. The writer has no new ideas or theories to put forth, for the reason that the mechanism we are considering has during the last six hundred years had its mathematics reduced to an exact science; its variable factors of material and mechanical movements developed according to the laws of geometry and trigonometry; its defects observed and pointed out; its performances checked and recorded. To gather these facts, illustrate and explain them, arrange them in their proper order, and point out their relative importance in the whole sum of what we call a clock, is therefore all that will be attempted. In doing this free use has been made of the observations of Saunier, Reid, Glasgow, Ferguson, Britten, Riefler and others in Europe and of Jerome, Playtner, Finn, Learned, Ferson, Howard and various other Americans. The work is therefore presented as a compilation, which it is hoped will be of service in the trade.

In our study of so-called American clocks, we use the word American in the sense of ownership rather than origin, the clocks which come to the American workmen to-day have been made in Germany, France, England and America.

The German clocks are generally those of the Schwartzwald (or Black Forest) district, and differ from others in their structure, chiefly in the following particulars: The movement is supported by a horizontal seat-board in the upper portion of the case. The wooden trains of many of the older type instead of being supported by plates are held in position by pillars, and these pillars are held in position by top and bottom boards. In the better class of wooden clocks the pivot holes in the pillars are bushed with brass tubing, while the movement has a brass 'scape wheel, steel wire pivots and lantern pinions of wood, with steel trundles. In all these clocks the front pillars are friction tight, and are the ones to be removed when taking down the trains. Both these and the modern Swartzwald brass movements use a sprocket wheel and chain for the weights and have exposed pendulums and weights.

The French clocks are of two classes, pendules and carriage clocks, and both are liable to develop more hidden crankiness and apparently causeless refusals to go than ever occurred to all the English, German and American clocks ever put together. There are many causes for this, and unless a man is very new at the business he can tell stories of perversity, that would make a timid apprentice want to quit. Yet the French clocks, when they do go, are excellent time-keepers, finely finished, and so artistically designed that they make their neighbors seem very clumsy by comparison. They are found in great variety, time, half-hour and quarter-hour strike, musical and repeating clocks being a few of the general varieties. The pendulums are very short, to accommodate themselves to the artistic needs of the cases, and nearly all have the snail strike instead of the count wheel. The carriage clocks have watch escapements of cylinder or lever form, and the escapement is frequently turned at right angle by means of bevel gears, or contrate wheel and pinion, and placed on top of the movement.

The English clocks found in America are generally of the "Hall" variety, having heavy, well finished movements, with seconds pendulum and frequently with calendar and chime movements. They, like the German, are generally fitted with weights instead of springs. There are a few English carriage clocks, fitted with springs and fuzees, though most of them, like the French, have springs fitted in going barrels.

The American clocks, with which the apprentice will naturally have most to do, may be roughly divided into time,

time alarm, time strike, time strike alarm, time calendar and electric winding. The American factories generally each make about forty sizes and styles of movements, and case them in many hundreds of different ways, so that the workman will frequently find the same movement in a large number of clocks, and he will soon be able to determine from the characteristics of the movement what factory made the clock, and thus be able to at once turn to the proper catalogue if the name of the maker be erased, as frequently happens.

This comparative study of the practice of different factories will prove very interesting, as the movement comes to the student after a period of prolonged and generally severe use, which is calculated to bring out any existing defects in construction or workmanship; and having all makes of clocks constantly passing through his hands, each exhibiting a characteristic defect more frequently than any other, he is in a much better position to ascertain the merits and defects of each maker than he would be in any factory.

Having thus briefly outlined the kinds of machinery used in measuring time, we will now turn our attention to the examination of the theoretical and mechanical construction of the various parts.

The man who starts out to design and build a clock will find himself limited in three particulars: It must run a specified time; the arbor carrying the minute hand must turn once in each hour; the pendulum must be short enough to go in the case. Two of these particulars are changeable according to circumstances; the length of time run may be thirty hours, eight, thirty, sixty or ninety days. The pendulum may be anywhere from four inches to fourteen feet, and the shorter it is the faster it will go. The one definite point in the time train is that the minute hand must turn once in each hour. We build or alter our train from this point both ways, back through changeable intermediate wheels and pinions to the spring or weight forming the source of power, and forward from it through another changeable series of wheels and pinions to the pendulum. Now as the pendulum governs the rate of the clock we will commence with that and consider it independently.

CHAPTER II.

THE NATURAL LAWS GOVERNING PENDULUMS.

LENGTH OF PENDULUM.—A pendulum is a falling body and as such is subject to the laws which govern falling bodies. This statement may not be clear at first, as the pendulum generally moves through such a small arc that it does not appear to be falling. Yet if we take a pendulum and raise the ball by swinging it up until the ball is level with the point of suspension, as in Fig. 1, and then let it go, we

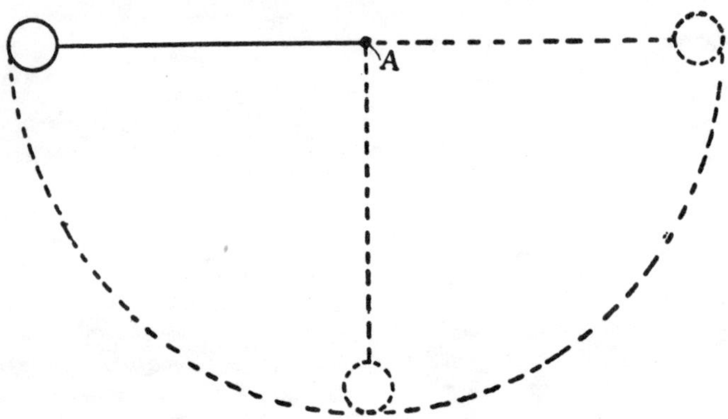

Fig. 1. Dotted lines show path of pendulum.

shall see it fall rapidly until it reaches its lowest point, and then rise until it exhausts the momentum it acquired in falling, when it will again fall and rise again on the other side; this process will be repeated through constantly smaller arcs until the resistance of the air and that of the pendulum spring shall overcome the other forces which operate to keep it in motion and it finally assumes a position of rest at the lowest point (nearest the earth) which the pendulum rod will allow it to assume. When it stops, it will be in line between the center of the earth (center of gravity) and the fixed point from which it is suspended. True, the pendulum bob, when it falls, falls under control of the pendulum rod and has its actions modified by the rod; but it falls just the same, no matter how small its arc of motion may be, and it is this influence of gravity—that force which makes any free body move toward the earth's center—which keeps the pendulum constantly returning to its lowest point and which governs very largely the time taken in moving. Hence, in estimating the length of a pendulum, we must consider gravity as being the prime mover of our pendulum.

The next forces to consider are mass and weight, which, when put in motion, tend to continue that motion indefinitely unless brought to rest by other forces opposing it. This is known as momentum. A heavy bob will swing longer than a light one, because the momentum stored up during its fall will be greater in proportion to the resistance which it encounters from the air and the suspension spring.

As the length of the rod governs the distance through which our bob is allowed to fall, and also controls the direction of its motion, we must consider this motion. Referring again to Fig. 1, we see that the bob moves along the circumference of a circle, with the rod acting as the radius of that circle; this opens up another series of facts. The circumference of a circle equals 3.1416 times its diameter, and the radius is half the diameter (the radius in this case being the pendulum rod). The areas of circles are proportional to the *squares* of their diameters and the circumferences are also proportional to their areas. Hence, the lengths of the paths of bobs moving along these circumferences are in proportion to the *squares* of the lengths of the pendulum rods. This is why a pendulum of half the length will oscillate four times as fast.

Now we will apply these figures to our pendulum. A

body falling in vacuo, in London, accelerates 32.2 feet in one second. Mathematicians designate this force that gravity exerts on a body as g. The circumference of a circle equals 3.1416 times its diameter. Mathematicians represent 3.1416 as π. Now, if we call the time t and the length of a pendulum in feet l, we shall have this formula:

$$t = \pi\sqrt{\frac{l}{g}}$$

Substituting the time, one second, for t, doing the same with the others, and clearing the equation, we shall have:

$$l = \frac{32.2 \text{ feet}}{(3.1416)^2} = 3.2616 \text{ feet}$$

Turning this into its equivalent in inches by multiplying by 12, we shall have 39.1393 inches as the length of a one-second pendulum at London.

Now, as the force of gravity varies somewhat with its distance from the center of the earth, we shall find the value of g in the above formula varying slightly, and this will give us slightly different lengths of pendulum at different places. These values have been found to be as follows:

	Inches.
The Equator is	39
Rio de Janiero	39.01
Madras	39.02
New York	39.1012
Paris	39.13
London	39.14
Edinburgh	39.15
Greenland	39.20
North and South Pole	39.206

Using the above data, we can get the length of any pendulum by multiplying the *length of a one-second pendulum* by the *square* of the *time* required. Thus, to find the length of a pendulum, in London, to beat three seconds:

$3^2 = 9$. $39.1393 \times 9 = 352.2537$ inches $= 29.3544$ feet.

A pendulum beating two-thirds of a second, or 90 beats:

$$(\tfrac{2}{3})^2 = \tfrac{4}{9}. \quad \frac{39.1393 \times 4}{9} = 17.3952 \text{ inches.}$$

A pendulum beating half-seconds or 120 beats:

$$(\tfrac{1}{2})^2 = \tfrac{1}{4}. \quad \frac{39.1393 \times 1}{4} = 9.7848 \text{ inches.}$$

CENTER OF OSCILLATION.—Having now briefly considered the basing facts governing the time of oscillation of the pendulum, let us examine it still further. The pendulum shown in Fig. 1 has all its weight in a mass at its end, but we cannot make a pendulum that way to run a clock, because of physical limitations. We shall have to use a rod stiff enough to transmit power from the clock movement to the pendulum bob and that rod will weigh something. If we use a compensated rod, so as to keep it the same length in varying temperature, it may weigh a good deal in proportion to the bob. How will this affect the pendulum?

If we suspend a rod from its upper end and place alongside of it our ideal pendulum, as in Fig. 2, we shall find that they will not vibrate in equal times if they are of equal lengths. Why not? Because when the rod is swinging (being stiff) a part of its weight rests upon the fixed point of suspension and that part of the rod is consequently not entirely subject to the force of gravity. Now, as the time in which our pendulum will swing depends upon the distance of the *effective* center of its mass from the point of suspension, and as, owing to the difference in construction, the center of mass of one of our pendulums is at the center of its ball, while that of the other is somewhere along the rod, they will naturally swing in different times.

Our other pendulum (the rod) is of the same size all the way up and the center of its *effective* mass would be the center of its weight (gravity) if it were not for the fact which we stated a moment ago that part of the weight is upheld and rendered ineffective by the fixed support of the

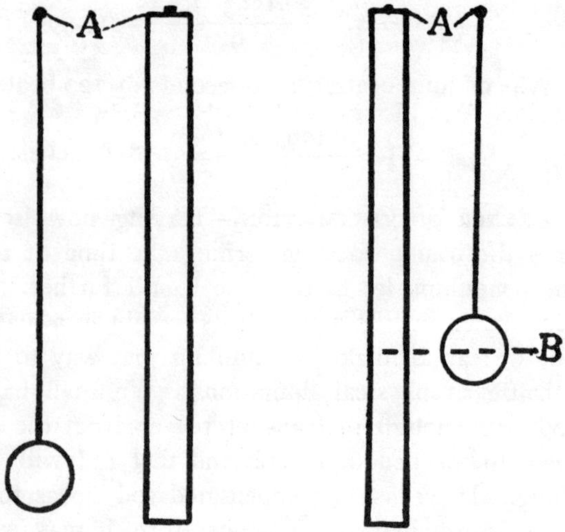

Fig. 2. Two pendulums of equal length but unequal vibration. B, center of oscillation for both pendulums.

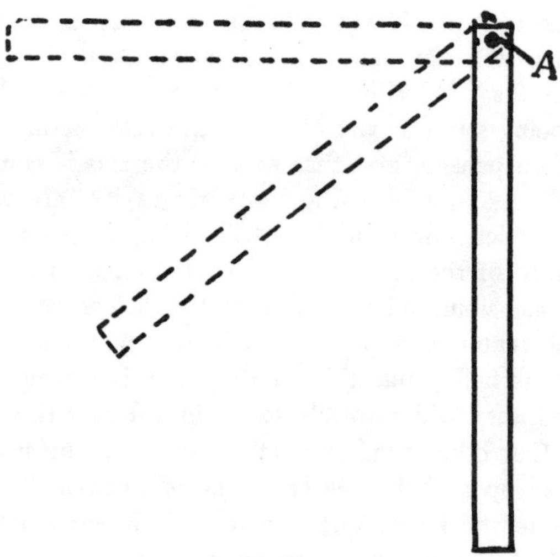

Fig. 3.

pendulum rod, all the while the pendulum is not in a vertical position. If we support the rod in a horizontal position, as in Fig. 3, by holding up the lower end, the point of suspension, A, will support half the weight of the rod; if we hold it at 45 degrees the point of suspension will hold less than half the weight of the rod and more of the rod will be affected by gravity; and so on down until we reach the vertical or up and down position. Thus we see that the force of gravity pulling on our pendulum varies in its effects according to the position of the rod and consequently the *effective* center of its mass also varies with its position and we can only calculate what this mean (or average) position is by a long series of calculations and then taking an average of these results.

We shall find it simpler to measure the time of swing of the rod which we will do by shortening our ball and cord until it will swing in the same time as the rod. This will be at about two-thirds of the length of the rod, so that the *effective* length of our rod is about two-thirds of its real length. This *effective* length, which governs the time of vibration, is called the *theoretical* length of the pendulum and the point at which it is located is called its *center of oscillation.* The distance from the *center* of oscillation to the point of suspension is called the *theoretical* length of the pendulum and is *always* the distance which is given in all tables of lengths of pendulums. This length is the one given for two reasons: First, because, it is the time-keeping length, which is what we are after, and second, because, as we have just seen in Fig. 3, the *real* length of the pendulum increases as more of the weight of the instrument is put into the rod. This explains why the heavy gridiron compensation pendulum beating seconds so common in regulators and which measures from 56 to 60 inches over all, beats in the same time as the wood rod and lead bob measuring 45 inches over all, while one is apparently a third longer than the other

Table Showing the Length of a Simple Pendulum

That performs in one hour any given number of oscillations, from 1 to 20,000, and the variation in this length that will occasion a difference of 1 minute in 24 hours.

Calculated by E. Gourdin.

Number of Oscillations per Hour.	Length in Millimeters.	Variation in Length for One Minute in 24 Hours in Millimeters.	Number of Oscillations per Hour.	Length in Millimeters.	Variation in Length for One Minute in 24 Hours in Millimeters.	Number of Oscillations per Hour.	Length in Millimeters.	Variation in Length for One Minute in 24 Hours in Millimeters.
20,000	32.2	0.04	13,200	73.9	0.10	8,200	191.5	0.26
19,000	35.7	0.05	13,100	75.1	0.10	8,100	196.3	0.27
18,000	39.8	0.05	13,000	76.2	0.10	8,000	201.3	0.27
17,900	40.2	0.06	12,900	77.4	0.11	7,900	206.4	0.28
17,800	40.7	0.06	12,800	78.6	0.11	7,800	211.7	0.29
17,700	41.1	0.06	12,700	79.9	0.11	7,700	217.2	0.30
17,600	41.6	0.06	12,600	81.1	0.11	7,600	223.0	0.30
17,500	42.1	0.06	12,500	82.4	0.11	7,500	229.0	0.31
17,400	42.4	0.06	12,400	83.8	0.11	7,400	235.2	0.32
17,300	43.0	0.06	12,300	85.1	0.12	7,300	241.7	0.33
17,200	43.5	0.06	12,200	86.5	0.12	7,200	248.5	0.34
17,100	44.0	0.06	12,100	88.0	0.12	7,100	255.7	0.35
17,000	44.6	0.06	12,000	89.5	0.12	7,000	262.9	0.36
16,900	45.1	0.06	11,900	91.0	0.12	6,900	270.5	0.37
16,800	45.7	0.06	11,800	92.5	0.13	6,800	278.6	0.38
16,700	46.3	0.06	11,700	94.1	0.13	6,700	286.9	0.39
16,600	46.7	0.07	11,600	95.7	0.13	6,600	295.7	0.40
16,500	47.3	0.07	11,500	97.4	0.13	6,500	304.9	0.41
16,400	47.9	0.07	11,400	99.1	0.13	6,400	314.5	0.43
16,300	48.5	0.07	11,300	100.9	0.14	6,300	324.5	0.44
16,200	49.1	0.07	11,200	102.7	0.14	6,200	335.1	0.46
16,100	49.7	0.07	11,100	104.5	0.14	6,100	346.2	0.47
16,000	50.0	0.07	11,000	106.5	0.14	6,000	357.8	0.48
15,900	51.0	0.07	10,900	108.4	0.15	5,900	370.0	0.50
15,800	51.6	0.07	10,800	110.5	0.15	5,800	382.9	0.52
15,700	52.3	0.07	10,700	112.5	0.15	5,700	396.4	0.54
15,600	52.9	0.07	10,600	114.6	0.16	5,600	410.7	0.56
15,500	53.6	0.07	10,500	116.8	0.16	5,500	425.8	0.58
15,400	54.3	0.08	10,400	119.1	0.16	5,400	440.1	0.60
15,300	55.0	0.08	11,300	111.4	0.17	5,300	458.5	0.62
15,200	55.7	0.08	10,200	123.8	0.17	5,200	476.3	0.65
15,100	56.5	0.08	10,100	126.3	0.17	5,100	495.2	0.67
15,000	57.3	0.08	10,000	128.8	0.18	5,000	515.2	0.70
14,900	58.0	0.08	9,900	131.4	0.18	4,900	536.5	0.73
14,800	58.8	0.08	9,800	134.1	0.18	4,800	559.1	0.76
14,700	59.6	0.08	9,700	136.9	0.19	4,700	583.1	0.79
14,600	60.4	0.08	9,600	139.8	0.19	4,600	608.7	0.83
14,500	61.3	0.08	9,500	142.7	0.19	4,500	636.1	0.86
14,400	62.1	0.09	9,400	145.8	0.20	4,400	665.3	0.90
14,300	63.0	0.09	9,300	148.9	0.20	4,300	696.7	0.95
14,200	63.9	0.09	9,200	152.2	0.21	4,200	730.2	0.99
14,100	64.8	0.09	9,100	155.5	0.21	4,100	766.2	1.04
14,000	65.7	0.09	9,000	159.0	0.22	4,000	805.0	1.09
13,900	66.7	0.09	8,900	162.6	0.22	3,950	825.5	1.12
13,800	67.6	0.09	8,800	166.3	0.23	3,900	846.8	1.15
13,700	68.6	0.09	8,700	170.2	0.23	3,850	869.0	1.18
13,600	69.6	0.09	8,600	173.7	0.24	3,800	892.0	1.21
13,500	70.7	0.09	8,500	178.3	0.24	3,750	915.9	1.25
13,400	71.7	0.10	8,400	182.5	0.25	3,700	940.1	1.28
13,300	72.8	0.10	8,300	187.0	0.25	3,650	966.8	1.31

Table of the Length of a Simple Pendulum,

(CONTINUED.)

Number of Oscillations per Hour.	Length in Meters.	To Produce in 24 Hours 1 Minute.		Number of Oscillations per Hour.	Length in Meters.	To Produce in 24 Hours 1 Minute.	
		Loss, Lengthen by Millimeters.	Gain, Shorten by Millimeters.			Loss, Lengthen by Meters.	Gain, Shorten by Meters.
3 600	0.9939	1.38	1.32	1.900	3.568	0.0050	0.0048
3,550	1.0221	1 42	1.36	1,800	3 975	0 0055	0.0053
3,500	1.0515	1.46	1.40	1,700	4.457	0.0062	0.0059
3.450	1.0822	1.50	1 44	1.600	5.031	0 0070	0.0067
3.400	1.1143	1.55	1.48	1,500	5 725	0.0080	0.0076
3.350	1.1477	1.60	1.53	1.400	6.572	0.0091	0.0087
3,300	1.1828	1.64	1.57	1,300	7.622	0.0106	0.0101
3 250	1.2194	1.69	1.62	1,200	8 945	0 0124	0.0119
3,200	1.2578	1 75	1.67	1,100	10 645	0.0148	0.0142
3,150	1.2981	1.80	1.73	1,000	12.880	0.0179	0.0171
3,100	1.3403	1.86	1 78	900	15 902	0.0221	0.0211
3,050	1.3846	1.93	1.84	800	20.126	0 0280	0.0268
3,000	1.4312	1.99	1 90	700	26.287	0 0365	0.0350
2.900	1.5316	2.13	2.04	600	35 779	0 0497	0.0476
2.800	1 6429	2.28	2 18	500	51 521	0.0716	0.0685
2.700	1 7669	2.46	2 35	400	80 502	0.1119	0.1071
2,600	1 9054	2.65	2 53	300	143 115	0.1989	0.1903
2,500	2 0609	2 87	2.74	200	322 008	0.4476	0.4282
2.400	2.2362	3.11	2 97	100	1,288.034	1.7904	1.7131
2,300	2.4349	3.38	3 24	60	3 577.871	4 9732	4.7586
2.200	2 6612	3.70	3.54	50	5.152.135	7.1613	6.8521
2,100	2 9207	4.06	3 88	1	12,880,337.930	17,903.6700	17,130.8500
2 000	3 2201	4.48	4.28				

In the foregoing tables all dimensions are given in meters and millimeters. If it is desirable to express them in feet and inches, the necessary conversion can be at once effected in any given case by employing the following conversion table, which will prove of considerable value to the watchmaker for various purposes:

Conversion Table of Inches, Millimeters and French Lines.

Inches expressed in Millimeters and French Lines.			Millimeters expressed in Inches and French Lines.			French Lines expressed in Inches and Millimeters.		
Inches.	Equal to		Millimeters.	Equal to		French Lines.	Equal to	
	Millimeters	French Lines.		Inches.	French Lines.		Inches.	Millimeters
1	25 39954	11.25951	1	0.0393708	0.44329	1	0.088414	2.25583
2	50.79908	22.51903	2	0.0787416	0.88659	2	0.177628	4.51166
3	76.19862	33.77854	3	0.1181124	1.32989	3	0 266441	6.76749
4	101.59816	45.03806	4	0.1574832	1.77318	4	0.355255	9.02332
5	126.99771	56.29757	5	0.1968539	2.21648	5	0 444069	11.27915
6	152.39725	67.55709	6	0.2362247	2.65978	6	0.532883	13 53497
7	177.79679	78 81660	7	0 2755955	3.10307	7	0.621697	15 79080
8	203 19633	90 07612	8	0.3149664	3.54637	8	0.710510	18.04663
9	228 59587	101 33563	9	0.3543371	3 98966	9	0 799324	20.30246
10	253.99541	112.59515	10	0.3937079	4.43296	10	0 888138	22.55829
						11	0.976952	24 81412
						12	1 065766	27.06995

CENTER OF GRAVITY.—The watchmaker is concerned only with the theoretical or timekeeping lengths of pendulums, as his pendulum comes to him ready for use; but the clock maker who has to *build* the pendulum to fit not only the movement, but also the case, needs to know more about it, as he must so distribute the weight along its length that it may be given a length of 60 inches or of 44 inches, or anything between them, and still beat seconds, in the case of a regulator. He must also do the same thing in other clocks having pendulums which beat other numbers than 60. Therefore he must know the center of his weights; this is called the center of gravity. This center of gravity is often confused by many with the center of oscillation as its real purpose is not understood. It is simply used as a starting point in *building* pendulums, because there must be a starting point, and this point is chosen because it is always present in every pendulum and it is convenient to work both ways from the center of weight or gravity. In Fig. 2 we have two pendulums, in one of which (the ball and string) the center of gravity is the center of the ball and the center of oscillation is also at the center (practically) of the ball. Such a pendulum is about as short as it can be constructed for any given number of oscillations. The other (the rod) has its center of gravity manifestly at the center of the rod, as the rod is of the same size throughout; yet we found by comparison with the other that its center of oscillation was at two-thirds the length of the rod, measured from the point of suspension, and the *real* length of the pendulum was consequently one-half longer than its *time keeping* length, which is at the center of oscillation. This is farther apart than the center of gravity and oscillation will ever get in actual practice, the most extreme distance in practice being that of the gridiron pendulum previously mentioned. The center of gravity of a pendulum is found at that point at which the pendulum can be balanced horizontally on a knife edge and is marked to measure from when cutting off the rod.

The center of oscillation of a compound pendulum must always be below its center of gravity an amount depending upon the proportions of weight between the rod and the bob. Where the rod is kept as light as it should be in proportion to the bob this difference should come well within the limits of the adjusting screw. In an ordinary plain seconds pendulum, without compensation, with a bob of eighteen or twenty pounds and a rod of six ounces, the difference in the two points is of no practical account, and adjustments for seconds are within the screw of any ordinary pendulum, if the screw is the right length for safety, and the adjusting nut is placed in the middle of the length of the screw threads when the top of the rod is cut off, to place the suspension spring by measurement from the center of gravity as has been already described; also a zinc and iron compensation is within range of the screw if the compensating rods are not made in undue weight to the bob. The whole weight of the compensating parts of a pendulum can be safely made within one and a half pounds or lighter, and carry a bob of twenty-five pounds or over without buckling the rods, and the two points, the center of gravity and the center of oscillation, will be within the range of the screw.

There are still some other forces to be considered as affecting the performance of our pendulum. These are the resistance to its momentum offered by the air and the resistance of the suspension spring.

BAROMETRIC ERROR.—If we adjust a pendulum in a clock with an airtight case so that the pendulum swings a certain number of degrees of arc, as noted on the degree plate in the case at the foot of the pendulum, and then start to pump out the air from the case while the clock is running, we shall find the pendulum swinging over longer arcs as the air becomes less until we reach as perfect a vacuum as we can produce. If we note this point and slowly admit air to the case again we shall find that the arcs of the pendulum's swing will be slowly shortened until the pressure in the case equals that of the surrounding air, when they will be the same as when our experiment was started. If we now pump air into our clock case, the vibrations will become still shorter as the pressure of the air increases, proving conclusively that the resistance of the air has an effect on the swinging of the pendulum.

We are accustomed to measure the pressure of the air as it changes in varying weather by means of the barometer and hence we call the changes in the swing of the pendulum due to varying air pressure the "barometric error." The barometric error of pendulums is only considered in the very finest of clocks for astronomical observatories, master clocks for watch factories, etc., but the resistance of the air is closely considered when we come to shape our bob. This is why bobs are either double-convex or cylindrical in shape, as these two forms offer the least resistance to the air and (which is more important) they offer equal resistance on both sides of the center of the bob and thus tend to keep the pendulum swinging in a straight line back and forth.

THE CIRCULAR ERROR.—As the pendulum swings over a greater arc it will occupy more time in doing it and thus the rate of the clock will be affected, if the barometric changes are very great. This is called the circular error. In ancient times, when it was customary to make pendulums vibrate at least fifteen degrees, this error was of importance

Fig. 4. A, arc of circle. B, cycloid path of pendulum, exaggerated.

and clock makers tried to make the bob take a cycloidal path, as is shown in Fig. 4, greatly exaggerated. This was accomplished by suspending the pendulum by a cord which swung between cycloidal cheeks, but it created so much friction that it was abandoned in favor of the spring as used to-day. It has since been proved that the long and short arcs of the pendulum's vibration are practically isochronous (with a spring of proper length and thickness) up to about six degrees of arc (three degrees each side of zero on the degree plate at the foot of the pendulum) and hence small variations of power in spring-operated clocks and also the barometric error are taken care of, except for greatly increased variations of power, or for too great arcs of vibration. Here we see the reasons for and the amount of swing we can properly give to our pendulum.

TEMPERATURE ERROR.—The temperature error is the greatest which we shall have to consider. It is this which makes the compound pendulum necessary for accurate time, and we shall consequently give it a great amount of space, as the methods of overcoming it should be fully understood.

EXPANSION OF METALS.—The materials commonly used in making pendulums are wood (deal, pine and mahogany), steel, cast iron, zinc, brass and mercury. Wood expands .0004 of its length between 32° and 212° F.; lead, .0028; steel, .0011; mercury, .0180; zinc, .0028; cast iron, .0011; brass, .0020. Now the length of a seconds pendulum, by our tables (3600 beats per hour) is 0.9939 meter; if the rod is brass it will lengthen .002 with such a range of temperature. As this is practically two-thousandths of a meter, this is a gain of two millimeters, which would produce a variation of one minute and forty seconds every twenty-four hours; consequently a brass rod would be a very bad one.

If we take two of these materials, with as wide a difference in expansion ratios as possible, and use the least variable for the rod and the other for the bob, supporting it at the bottom, we can make the expansion of the rod counterbalance the expansion of the bob and thus keep the effective length of our pendulum constant, or nearly so. This is the theory of the compensating pendulum.

CHAPTER III.

COMPENSATING PENDULUMS.

As the pendulum is the means of regulating the time consumed in unwinding the spring or weight cord by means of the escapement, passing one tooth of the escape wheel at each end of its swing, it will readily be seen that lengthening or shortening the pendulum constitutes the means of regulating the clock; this would make the whole subject a very simple affair, were it not that the reverse proposition is also true; viz.; Changing the length of the pendulum will change the rate of the clock and after a proper rate has been obtained further changes are extremely undesirable. This is what makes the temperature error spoken of in the preceding chapter so vexatious where close timing is desired and why as a rule, a well compensated pendulum costs more than the rest of the clock. The sole reason for the business existence of watch and clockmakers lies in the necessity of measuring time, and the accuracy with which it may be done decides in large measure the value of any watchmaker in his community. Hence it is of the utmost importance that he shall provide himself with an accurate means of measuring time, as all his work must be judged finally by it, not only while he is working upon time-measuring devices, but also after they have passed into the possession of the general public.

A good clock is one of the very necessary foundation elements, contributing very largely to equip the skilled mechanic and verify his work. Without some reliable means to get accurate mean time a watchmaker is always at sea—without a compass—and has to trust to his faith and a large amount of guessing, and this is always an embarrassment, no matter how skilled he may be in his craft, or adept in guessing. What I want to call particular attention to is the unreliable and worthless character of the average regulator of the present day. A good clock is not necessarily a high priced instrument and it is within the reach of most watchmakers. A thoroughly good and reliable timekeeper of American make is to be had now in the market for less than one hundred dollars, and the only serious charge that can be made against these clocks is that they cost the consumer too much money. Any of them are thirty-three and a third per cent higher than they should be. About seventy-five dollars will furnish a thoroughly good clock. The average clock to be met with in the watchmakers' shops is the Swiss imitation gridiron pendulum, pin escapement, and these are of the low grades as a rule; the best grades of them rarely ever get into the American market. Almost without exception, the Swiss regulator, as described, is wholly worthless as a standard, as the pendulums are only an imitation of the real compensated pendulum. They are an imitation all through, the bob being hollow and filled with scrap iron, and the brass and steel rods composing the compensating element, along with the cross pieces or binders, are all of the cheapest and poorest description. If one of these pendulums was taken away from the movement and a plain iron bob and wooden rod put to the movement, in its place, the possessor of any such clock would be surprised to find how much better average rate the clock would have the year through, although there would then be no compensating mechanism, or its semblance, in the make up of the pendulum. In brief, the average imitation compensation pendulum of this particular variety is far poorer than the simplest plain pendulum, such as the old style, grandfather clocks were equipped with. A wood rod would be far superior to a steel one, or any metal rod. as may be seen by consulting the expansion data given in the previous chapter

Many other pendulums that are sold as compensating are a delusion in part, as they do not thoroughly compensate, because the elements composing them are not in equilibrium or in due proportion to one another and to the general mechanism.

To all workmen who have a Swiss regulator, I would say that the movement, if put into good condition, will answer very well to maintain the motion of a good pendulum, and that it will pay to overhaul these movements and put to them good pendulums that will pretty nearly compensate. At least a well constructed pendulum will give a very useful and reliable rate with such a motor, and be a great help and satisfaction to any man repairing and rating good watches.

The facts are, that one of the good grade of American adjusted watch movements will keep a much steadier rate when maintained in one position than the average regulator. Without a reliable standard to regulate by, there is very little satisfaction in handling a good movement and then not be able to ascertain its capabilities as to rate. Very many watch carriers are better up in the capabilities of good watches than many of our American repairers are, because a large per cent of such persons have bought a watch of high grade with a published rate, and naturally when it is made to appear to entirely lack a constant rate when compared with the average regulator, they draw the conclusion that the clock is at fault, or that the cleaning and repairing are. Many a fair workman has lost his watch trade, largely on account of a lack of any kind of reliable standard of time in his establishment. There are very few things that a repairer can do in the way of advertising and holding his customers more than to keep a good clock, and furnish good watch owners a means of comparison and thus to confirm their good opinions of their watches.

We have along our railroads throughout the country a standard time system of synchronized clocks, which are an improvement over no standard of comparison; but they cannot be depended upon as a reliable standard, because they are subject to all the uncertainties that affect the telegraph lines—bad service, lack of skill, storms, etc. The clocks furnished by these systems are not reliable in themselves and they are therefore corrected once in twenty-four hours by telegraph, being automatically set to mean time by the mechanism for that purpose, which is operated by a standard or master clock at some designated point in the system.

Now all this is good in a general way; but as a means to regulate a fine watch and use as a standard from day to day, it is not adequate. A standard clock, to be thoroughly serviceable, must always, all through the twenty-four hours, have its seconds hand at the correct point at each minute and hour, or it is unreliable as a standard. The reason is that owing to train defects watches may vary back and forth and these errors cannot be detected with a standard that is right but once a day. No man can compare to a certainty unless his standard is without variation, substantially; and I do not know of any way that this can be obtained so well and satisfactorily as through the means of a thoroughly good pendulum.

Compensating seconds pendulums are, it might be said, the standard time measure. Mechanically such a pendulum is not in any way difficult of execution, yet by far the greater portion of pendulums beating seconds are not at all accurate time measures, as independently of their slight variations in length, any defects in the construction or fitting of their parts are bound to have a direct effect upon the performance of the clock. The average watchmaker as a mechanic has the ability to do the work properly, but he does not fully understand or realize what is necessary, nor appreciate the fact that little things not attended to will render useless all his efforts.

The first consideration in a compensated pendulum is to

maintain the center of oscillation at a fixed distance from the point of suspension and it does not matter how this is accomplished.

So, also, the details of construction are of little consequence, so long as the main points are well looked after—the perfect solidity of all parts, with very few of them, and the free movement of all working surfaces without play, so that the compensating action may be constantly maintained at all times. Where this is not the case the sticking, rattling, binding or cramping of certain parts will give different rates at different times under the same variations of temperature, according as the parts work smoothly and evenly or move only by jerks.

The necessary and useful parts of a pendulum are all that are really admissible in thoroughly good construction. Any and all pieces attached by way of ornament merely are apt to act to the prejudice of the necessary parts and should be avoided. In this chapter we shall give measurements and details of construction for a number of compensated pendulums of various kinds, as that will be the best means of arriving at a thorough understanding of the subject, even if the reader does not desire to construct such a pendulum for his own use.

Principles of Construction.—Compensation pendulums are constructed upon two distinct principles. First, those in which the bob is supported by the bottom, resting on the adjusting screw with its entire height free to expand upward as the rod expands downward from its fixed point of suspension. In this class of pendulums the error of the bob is used to counteract that of the rod and if the bob is made of sufficiently expansible metal it only remains to make the bob of sufficient height in proportion to its expansibility for one error to offset the other. In the second class the attempt is made to leave out of consideration any errors caused by expansion of the bob, by suspending it from the center, so that its expansion downward will exactly balance its expansion upward and hence they will balance each other and may be neglected. Having eliminated the bob from consideration by this means we must necessarily confine our attempt at compensation to the rod in the second method.

The wood rod and lead bob and the mercurial pendulums are examples of the first-class and the wood rod with brass sleeve having a nut at the bottom and reaching to the center of the iron bob and the common gridiron, or compound tubular rod, or compound bar of steel and brass, or steel and zinc, are examples of the second class.

WOOD ROD AND ZINC BOB.—We will suppose that we have one of the Swiss imitation gridiron pendulums which we want to discard, while retaining the case and movement. As these cases are wide and generally fitted with twelve-inch dials, we shall have about twenty inches inside our case and we may therefore use a large bob, lens-shaped, made of cast zinc, polished and lacquered to look like brass.

The bobs in such imitation gridiron pendulums are generally about thirteen inches in diameter and swing about five inches (two and a half inches each side). The pendulums are generally light, convex in front and flattened at the rear, and the entire pendulum measures about 56 inches from the point of suspension to the lower end of the adjusting screw. We will also suppose that we desire to change the appearance of the clock as little as possible, while improving its rate. This will mean that we desire to retain a lens-shaped bob of about the same size as the one we are going to remove.

We shall first need to know the total length of our pendulum, so that we can calculate the expansion of the rod. A seconds pendulum measures 39.2 inches from the point in the suspension spring at the lower edge of the chops to the center of oscillation. With a lens-shaped bob the center of gravity will be practically at the center of the bob, if we use a light wooden rod and a steel adjusting screw and brass nut, as these metal parts, although short, will be heavy enough to nearly balance the suspension spring and that portion of the rod which is above the center. We shall also gain a little in balance if we leave the steel screw long enough to act as an index over the degree plate, in the case, at the bottom of the pendulum, by stripping the thread and turning the end to a taper an inch or so in length.

We shall only be able to use one-half of the expansion upwards of our bob, because the centers of gravity and oscillation will be practically together at the center of the bob. We shall find the center of gravity easily by balancing the pendulum on a knife-edge and thus we will be able to make an exceedingly close guess at the center of oscillation.

Now, looking over our data, we find that we have a suspension spring of steel, then some wood and steel again at the other end. We shall need about one inch of suspension spring. The spring will, of course, be longer than one inch, but we shall hold it in iron chops and the expansion of the chops will equal that of the spring between them, so that only the free part of the spring need be considered. Now from the adjusting screw, where it leaves the last pin through the wood, to the middle position of the rating nut will be about one inch, so we shall have two inches of steel to consider in our figures of expansion.

Now to get the length of the rod. We want to keep our bob about the size of the other, so we will try 14 inches diameter, as half of this is an even number and makes easy figuring in our trials. 39.2 inches, plus 7 (half the diameter of the bob) gives us 46.2 inches; now we have an inch of adjustment in our screw, so we can discard the .2; this leaves us 46 inches of wood and steel for which we must get the expansion.

Wood expands .0004 of its length between 32° and 212° F.
Steel expands .0011 of its length between 32° and 212° F.
Lead expands .0028 of its length between 32° and 212° F.
Brass expands .0020 of its length between 32° and 212° F.
Zinc expands .0028 of its length between 32° and 212° F.
Tin expands .0021 of its length between 32° and 212° F.
Antimony expands .0011 of its length between 32° and 212° F.
Total length of pendulum to adjusting nut 46 inches.
Total length of steel to adjusting nut 2 inches.
Total length of wood to adjusting nut 44 inches.
$.0011 \times 2 = .0022$ inch, expansion of our steel.
$.0004 \times 44 = .0176$ inch, expansion of our wood.

.0198 total expansion of rod.

We have 7 inches as half the diameter of our bob $.0198 \div 7 = .0028\ 2\text{-}7$, which we find from our tables is very close to the expansion of zinc, so we will make the bob of that metal. Now let us check back; the upward expansion of 7 inches of zinc equals $.0028 \times 7 = .0196$ inch, as against .0198 inch downward expansion of the rod. This gives us a total difference of .0002 inch between 32° and 212° or a range of 180° F. This is a difference of .0001 inch for 90° of temperature and is closer than most pendulums ever get.

The above figures are for dry, clear white pine, well baked and shellacked, with steel of average expansion, and zinc of new metal, melted and cast without the admixtures of other metals or the formation of oxide. The presence of tin, lead, antimony and other admixtures in the zinc would of course change the results secured; so also will there be a slight difference in the expansion of the rod if other woods are used. Still the jeweler can from the above get a very close approximation.

Such a bob, 14 inches diameter and 1.5 inches thick, alike on both sides, with an oval hole 1x.5 inches through its center, see Fig. 5, would weigh about 30 to 32 pounds, and

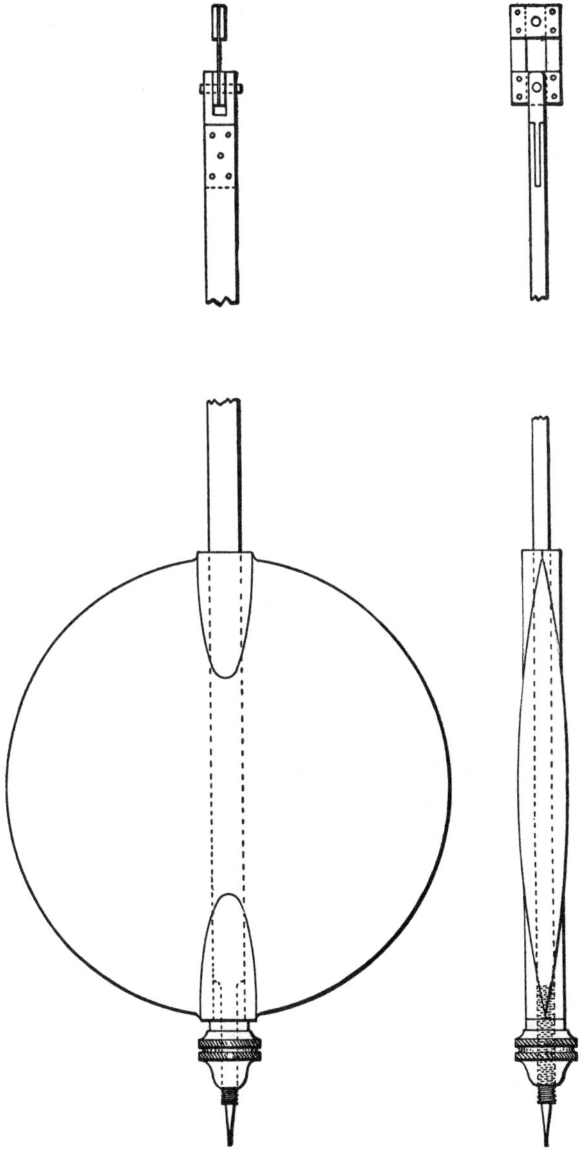

Fig. 5. Zinc bob and wood rod to replace imitation gridiron pendulum.

would have to be hung from a cast iron bracket, Fig. 6, bolted through the clock case to the wall behind it, so as to get a steady rate. It would be nearly constant, as the metal is spread out so as to be quickly affected by temperature; and the shape would hold it well in its plane of oscillation, if both sides were of exactly the same curvature, while the

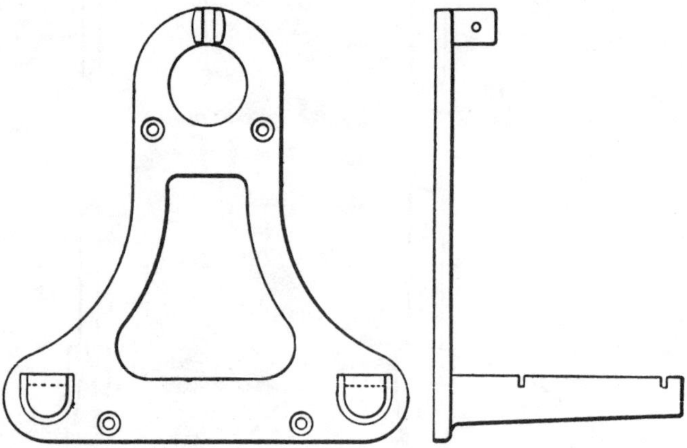

Fig. 6. Cast iron bracket for heavy pendulums and movements.

weight would overcome minor disturbances due to vibration of the building. It would require a little heavier suspension spring, in order to be isochronous in the long and short arcs and this thickening of the spring would need the addition of from one and a half to two pounds more of driving weight.

If so heavy a pendulum is deemed undesirable, the bob would have to be made of cylindrical form, retaining the height, as necessary to compensation, and varying the diameter of the cylinder to suit the weight desired.

Wood Rod and Lead Bob.—The wood should be clear, straight-grained and thoroughly dried, then given several coats of shellac varnish, well baked on. It may be either

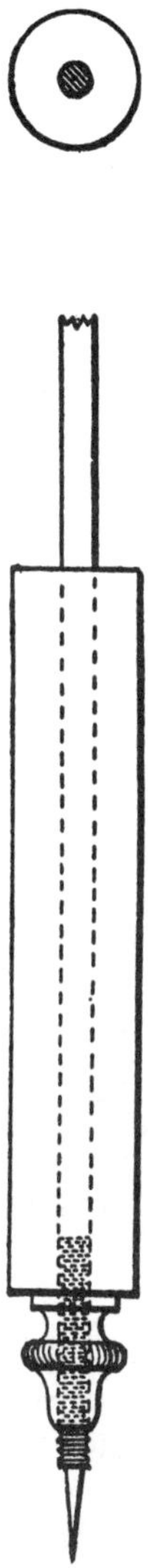

Fig. 7. Wood rod and lead bob.

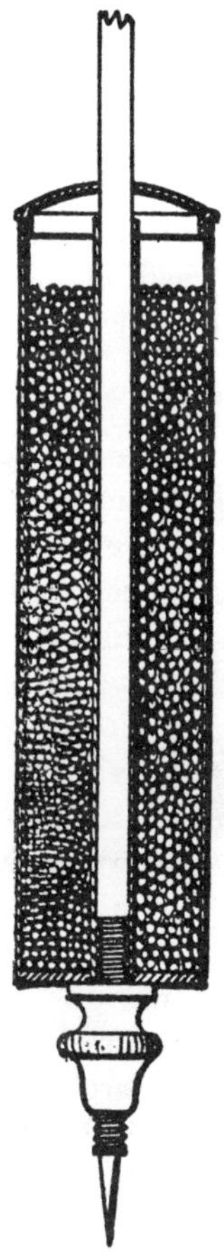

Fig. 8. Bob of metal casing filled with shot.

flat, oval or round in section, but is generally made round because the brass cap at the upper end, the lining for the crutch, and the ferrule for the adjusting screw at the lower end may then be readily made from tubing. For pendulums smaller than one second, the wood is generally hard, as it gives a firmer attachment of the metal parts.

	Inches.
Length, top of suspension spring to bottom of bob....	44.5
Length to bottom of nut....	45.25
Diameter of bob....	2.0
Length of bob....	10.5
Weight of bob, 8 lbs.	
Acting length of suspension spring....	1.0
Width of spring	.45
Thickness	.008
Diameter of rod	.5

The top of the rod should have a brass collar fixed on it by riveting through the rod and it should extend down the rod about three inches, so as to make a firm support for the slit to receive the lower clip of the suspension spring. The lower end should have a slit or a round hole drilled longitudinally three inches up the rod to receive the upper end of the adjusting screw and this should also fit snugly and be well pinned or riveted in place. See Fig. 7. A piece of thin brass tube about one inch in length is fitted over the rod where the crutch works.

In casting zinc and lead bobs, especially those of lens-shapes, the jeweler should not attempt to do the work himself, but should go to a pattern maker, explain carefully just what is wanted and have a pattern made, as such patterns must be larger than the casting in order to take care of the shrinkage due to cooling the molten metal. It will also be better to use an iron core, well coated with graphite when casting, as the core can be made smooth throughout and the exact shape of the pendulum rod, and there will then be no work to be done on the hole when the casting is made. The natural shrinkage of the metal on cooling will free the core, which can be easily driven out when the metal is cold and it will then leave a smooth, well shaped hole to which the rod can be fitted to work easily, but without shake. Lens-shaped bobs, particularly, should be cast flat, with register pins on the flask, so as to get both sides central with the hole, and be cast with a deep riser large enough to put considerable pressure of melted metal on the casting until it is chilled, so as to get a sound casting; it should be allowed to remain in the sand until thoroughly cold, for the same reason, as if cooled quickly the bob will have internal stresses which are liable to adjust themselves sometime after the pendulum is in the clock and thus upset the rate until such interior disturbances have ceased. Cylinders may be cast in a length of steel tubing, using a round steel core and driven out when cold.

If using oval or flat rods of wood, the adjusting screw should be flattened for about three inches at its upper end, wide enough to conform to the width of the rod; then saw a slot in the center of the rod, wide and deep enough to just fit the flattened part of the screw; heat the screw and apply shellac or lathe wax and press it firmly into the slot with the center of the screw in line with the center of the rod; after the wax is cold select a drill of the same size as the rivet wire; drill and rivet snugly through the rod, smooth everything carefully and the job is complete.

If by accident you have got the rod too small for the hole, so that there is any play, give the rod another coat of shellac varnish and after drying thoroughly, sand paper it down until it will fit properly.

Round rods may be treated in the same manner, but it is usual to drill a round hole in such a rod to just fit the wire, then insert and rivet as before after the wax is cold, finishing with a ferrule or cap of brass at the end of the rod.

The slot for the suspension spring is fitted to the upper end of the rod in the same manner.

Pendulum with Shot.—Still another method of making a compensating pendulum, which gives a lighter pendulum, is to make a case of light brass or steel tubing of about three inches diameter, Fig. 8, with a bottom and top of equal weight, so as to keep the center of oscillation about the center of gravity, for convenience in working. The bottom may be turned to a close fit, and soldered, pinned, or riveted into the tube. It is pierced at its center and another tube of the same material as the outer tube, with an internal diameter which closely fits the pendulum rod is soldered or riveted into the center of the bottom, both bottom and top being pierced for its admission and the other parts fitted as previously described.

The length of the case or canister should be about 11.5 inches so as to give room for a column of shot of 10.5 inches (the normal compensating height for lead) and still leave room for correction. Make a tubular case for the driving weight also and then we have a flexible system. If it is necessary to add or subtract weight to obtain the proper arcs of oscillation of the pendulum, it can be readily done by adding to or taking from the shot in the weight case.

Fill the pendulum to 10.5 inches with ordinary sportsmen's shot and try it for rate. If it gains in heat and loses in cold it is over-compensated and shot must be taken from it. If it loses in heat and gains in cold it is under-compensated and shot should be added.

The methods of calculation were given in full in describing the zinc pendulum and hence need not be repeated here, but attention should be called to the fact that there are three materials here, wood, steel or brass and lead and each should be figured separately so that the last two may just counterbalance the first. If the case is made light throughout the effect upon the center of oscillation will be inappreciable as compared with that of the lead, but if made heavier than need be, it will exert a marked influence, particularly if its highest portion (the cover) be heavy, as we then have the effect of a shifting weight high up on the pendulum rod. If made of thin steel throughout and nickel plated, we shall have a light and handsome case for our bob. If this is not practicable, or if the color of brass be preferred, it may be made of that material.

The following table of weights will be of use in making calculations for a pendulum or for clock weights.

Weight of Lead, Zinc and Cast Iron Cylinders One Half Inch Long.

Diameter in Inches.	Weight in Pounds			Diameter in Inches	Weight in Pounds		
	Lead	Zinc	Iron		Lead	Zinc	Iron
.25	.020	012	.012	3 25	3 400	2 098	2.156
.5	.080	.049	.050	3 5	3.944	2 434	2.491
.75	.180	.111	.114	3 75	4 51	2.783	2 865
1	.321	.198	.204	4	5 149	3.177	3 265
1.25	.503	.310	.319	4 25	5 813	3.587	3 686
1.5	.724	.447	.459	4 5	6 519	3 922	4.134
1.75	.984	.607	.624	4.75	7 265	4 483	4.607
2.	1.287	.794	.816	5	8 048	4 966	5.103
2 25	1 630	1.005	1 033	5.25	8 872	5 474	5.626
2.5	2.009	2 239	1 274	5 5	9 737	6.008	5.175
2.75	2.434	1 502	1 544	5.75	10 643	6.567	6.749
3.	2.897	1 788	1 837	6	11.590	7.152	7.350

Example:—Required, the weight of a lead pendulum bob, 3 inches diameter, 9 inches long, which has a hole through it .75 inch in diameter. The weight of a lead cylinder 3 inches diameter in the table is 2 897, which multiplied by 9 (the length given)=26.07 lbs. Then the weight in the table of a cylinder .75 inch diameter is .18 and .18×9=1.62 lbs. And 26.07 - 1.62=24.45, the weight required in lbs.

AUXILIARY WEIGHTS.—If for any reason our pendulum does not turn out with a rating as calculated and we find after getting it to time that it is over compensated, it is a comparatively simple matter to turn off a portion from the bottom of a solid bob. By doing this in very small portions at a time and then testing carefully for heat and cold every time any amount has been removed, we shall in the course of a few weeks arrive at a close approximation to compensation, at least as close as the ordinary standards available to the jeweler will permit. This is a matter of weeks, because if the pendulum is being rated by the standard time which is telegraphed over the country daily at noon, the jeweler, as soon as he gets his pendulum nearly right, will begin to discover variations in the noon signal of from .2 to 5 seconds on successive days. Then it becomes a matter of averages and reasoning, thus: If the pendulum beats to time on the first, second, third, fifth and seventh days, it follows that the signal was incorrect—slow or fast—on the fourth and sixth days.

If the pendulum shows a gain of one second a week on the majority of the days, the observation must be continued without changing the pendulum for another week. If the pendulum shows two seconds gain at the end of this time, we have two things to consider. Is the length right, or is the pendulum not fully compensated? We cannot answer the second query without a record of the temperature variations during the period of observations.

To get the temperature record we shall require a set of maximum and minimum thermometers in our clock case. They consist of mercurial thermometer tubes on the ordinary Fahrenheit scales, but with a marker of colored wood or metal resting on the upper end of the column of mercury in the tube. The tube is not hung vertically, but is placed in an inclined position so that the mark will stay where it is pushed by the column of mercury. Thus if the temperature rises during the day to 84 degrees the mark in the maximum thermometer will be found resting in the tube at 84° whether the mercury is there when the reading is taken or not. Similarly, if the temperature has dropped during the night to 40°, the mark in the minimum thermometer will be found at 40°, although the temperature may be 70° when the reading is taken. After reading, the thermometers are shaken to bring the marks back to the top of the column of mercury and the thermometers are then restored to their positions, ready for another reading on the following day.

These records should be set down on a sheet every day at noon in columns giving date, rate, plus or minus, maximum, minimum, average temperature and remarks as to regulation, etc., and with these data to guide us we shall be in a position to determine whether to move the rating nut or not. If the temperature has been fairly constant we can get a closer rate by moving the nut and continuing the observations. If the temperature has been increasing steadily and our pendulum has been gaining steadily it is probably over-compensated and the bob should be shortened a trifle and the observations renewed.

It is best to "make haste slowly" in such a matter. First bring the pendulum to time in a constant temperature; that will take care of its proper length. Then allow the temperature to vary naturally and note the results.

If the pendulum is under-compensated, so that the bob is too short to take care of the expansion of the rod, auxiliary weights of zinc in the shape of washers (or short cylinders) are placed between the bottom of the bob and the rating nut. This of course makes necessary a new adjustment and another course of observations all around, but it will readily be seen that it places a length of expansible metal between the nut and the center of oscillation and thus makes up for the deficiency of expansion of the bob. Zinc is generally chosen on account of its high rate of expansion, but brass, aluminum and other metals are also used. It is best to use one thick washer, rather than a number of thinner ones, as it is important to keep the construction as solid at this point as possible.

Top Weights.—After bringing the pendulum as close as possible by the compensation and the rating nuts, astronomers and others requiring exact time get a trifle closer rating by the use of top weights. These are generally U-shaped pieces of thin metal which are slipped on the rod above the bob without stopping the pendulum. They raise the center of oscillation by adding to the height of the bob when they are put on, or lower it when they are removed, but they are never resorted to until long after the pendulum is closer to time than the jeweler can get with his limited standards of comparison. They are mentioned here simply that their use may be understood when they may be encountered in cleaning siderial clocks.

Mercurial pendulums also belong to the class of compensation by expansion of the bobs, but they are so numerous and so different that they will be considered separately, later on.

COMPENSATED PENDULUM RODS.—We will now consider the second class, that in which an attempt is made to obtain a pendulum rod of unvarying length.

The oldest form of compensated rod is undoubtedly the gridiron of either nine, five or three rods. As originally made it was an accurate but expensive proposition, as the coefficients of expansion of the brass or zinc and iron or steel had all to be determined individually for each pendulum. Each rod had to be sized accurately, or if this was not done, then each rod had to be fitted carefully to each hole in the cross bars so as to move freely, without shake. The rods were spread out for two purposes, to impress the public and to secure uniform and speedy action in changes of temperature. The weight, which increased rapidly with the increase of diameter of the rod, made a long and large seconds pendulum, some of them measuring as much as sixty-two inches in length, and needing a large bob to look in proportion. Various attempts were made to ornament the great expanse of the gridiron, harps, wreaths and other forms in pierced metal being screwed to the bars. The next advance was in substituting tubes for rods in the gridiron, securing an apparently large rod that was at the same time stiff and light. Then came the era of imitation, in which the rods were made of all brass, the imitation steel portion being nickel plated. With the development of plating they were still further cheapened by being made of steel, with the supposedly brass rods plated with brass and the steel ones with nickel. Thousands of such pendulums are in use to-day; they have the rods riveted to the cross-pieces and are simply steel rods, subject to change of length with every change in temperature. It does no harm to ornament such pendulums, as the rods themselves are merely ornaments, usually all of one metal, plated to change the color.

As three rods were all that were necessary, the clockmaker who desired a pendulum that was compensated soon found his most easily made rod consisted of a zinc bar, wide, thin and flat, placed between two steel parts, like the meat and bread of a sandwich. This gives a flat and apparently solid rod of metal which if polished gives a pleasing appearance, and combines accurate performance with cheapness of construction, so that any watchmaker may make it himself, without expensive tools.

FLAT COMPENSATED ROD.—One of the most easily made zinc and iron compensating pendulums, shown in detail in Fig. 9, is as follows: A lead or iron bob, lens shaped, that is, convex equally on each side, 9 inches diameter and an inch and one-quarter thick at the center. A hole to be made straight through its diameter ½ inch. One-half through the diameter this hole is to be enlarged to ⅝ inch diameter. This will make the hole for half of its length ½ inch and the remaining half ⅝ inch diameter. The ⅝ hole must have a thin tube, just fitting it, and 5 inches long. At one end of this tube is soldered in a nut, with a hole tapped with a tap of thirty-six threads to the inch, and ¼ inch diameter, and at the other end of the tube is

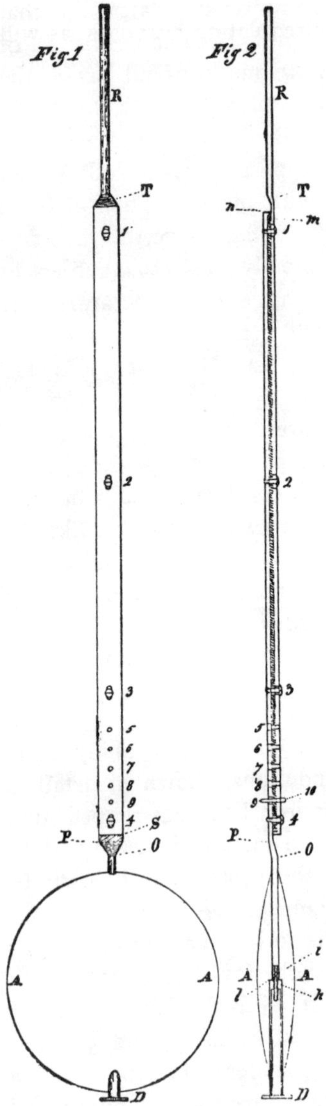

A, the lens-shaped bob; T P, the total length of the compensating part.

R, the upper round part of rod.

The side showing the heads of the screws is the face side and is finished. The screws 1, 2, 3, 4 hold the three pieces from separating, but do not confine the front and middle sections in their lengthwise expansion along the rod, but are screwed into the back iron section, while the holes in the other two sections are slotted smaller than the screw heads.

The holes at the lower extreme of combination 5, 6, 7, 8, 9 are for adjustments in effecting a compensation.

The pin at 10 is the steel adjusting pin, and is only tight in the front bar and zinc bars, being loose in the back bar.

O and P show the angles in the back rod, T shows the angle in the rod at the top, m shows the pin as placed in the iron and zinc sections where they have been soldered as described.

h shows the regulating nut carried by the tube, as described, and terminating in the nut D.

l and i show the screw of 36 threads.

The nut D is to be divided on its edge into 30 divisions.

n is the angle of the back bar to which zinc is soldered.

Fig. 9. Pendulum with compensated rod of steel and zinc.

soldered a collar or disc one inch diameter, which is to be divided into thirty divisions, for regulating purposes, as will be described later on. The whole forms a nut into which the rod screws, and the tube allows the nut to be pushed up to the center of the diameter of the bob, through the large hole, and the nut can be operated then by means of the disc at its lower end. The rod, of flat iron, is in two sections, as follows: That section which enters the bob and terminates in the regulating screw is flat for twenty-six inches, and then rounded to ½ inch for six inches, and a screw cut on its end for two inches, to fit the thread in the nut. The upper end of this section is then to be bent at a right angle, flatwise. This angle piece will be long enough if only 3-16 inch long, so that it covers the thickness of the zinc center rod. The zinc center rod is a bar of the metal, hammered or rolled, 25 inches long, 3-16 inch thick, and ¾ inch wide, and comes up against the angle piece bent on the flat part of the lower section of the rod. Now the upper section of the rod may be an exact duplicate of the lower section, with the flat part only a little longer than the zinc bar, say ½ inch, and the angle turned on the end, as previously described. The balance of the bar may be forged into a rod of 5-16 inch diameter. As has been stated, the zinc bar is placed against the angle piece bent on the upper end of the lower section of the rod, P, n, Fig. 9, and pins must be put through this angle piece into the end of the zinc bar, to hold it in close contact with the iron bar. The upper section of the rod is now to be laid on the opposite side of the zinc bar, with its angle at the other end of the zinc, but not in contact with it, say 1-16 inch left between the angle and the zinc bar. Now all is ready to clamp together—the two flat iron bars with the zinc between them. After clamping, taking care to have the pinned end of the zinc in contact with the angle and the free, or lower end, removed from the other angle about 1-16 inch, three screws should be put through all three bars, with their heads all on the side selected for the front, and one screw may be an inch from the top, another 3 inches from the bottom, and one-half way between the two first mentioned. Now the rod is complete in its composite form, and there is left only the little detail to attend to. Two flat bars, with their ends angled in one case and rounded in the other into rods of given diameter, confining between them, as described, a flat bar of wrought zinc of stated length and of the same thickness and width as the iron bars, comprises the active or compensating elements of the pendulum's rod. The screws that are put through the three bars are each to pass through the front iron bar, without threads in the bar, and only the back iron bar is to have the holes tapped, fitting the screws. All the corresponding holes in the zinc are to be reamed a little larger than the diameter of the screws, and to be freed lengthwise of the bar, to allow of the bar's contracting and expanding without being confined in this action by the screws. At the lower or free end of the zinc bar are to be holes carried clear through all three bars, while the combination is held firmly together by the screws. These holes are to start at ½ inch from the end of the zinc, and each carried straight through all three bars, and then broached true and a steel pin made to accurately fit them from the front side. These holes may be from three to five in number, extending up to a safe distance from the lower screw. The holes in the back bar, after boring, are to be reamed larger than those in the front bar and zinc bar. These holes and the pin serve for adjusting the compensation. The pin holds the front bar and zinc from slipping, or moving past one another at the point pinned, and also allows the back bar to be free of the pin, and not under the influence of the two front bars. The upper end of the second iron section is, as has been mentioned, forged into a round rod about 5-16 inch diameter, and this rod or upper end is to receive the pendulum suspension spring, which may be one single spring, or a compound spring, as preferred.

Now that the pendulum is all ready to balance on the knife edge, proceed as in case of the simple pendulum, and ascertain at what point up the rod the spring must be placed. In this pendulum the rod will be heavier in proportion than the wood rod was to its bob, and the center of gravity of the whole will be found higher up in the bob. However, wherever in the bob the center of gravity is found, that is the starting point to measure from to find the total length of the rod, and the point for the spring. The heavier the rod is in relation to the bob, the higher will the center of gravity of the whole rise in the bob, and the greater will be the total length of the entire pendulum.

In getting up a rod of the kind just described, the main item is to get the parts all so arranged that there will be very little settling of the joints in contact, particularly those which sustain the weight of the bob and the whole dead weight of the pendulum. The nut in the center of the pendulum holds the weight of the bob only, but it should fit against the shoulder formed for the purpose by the juncture of the two holes, and the face of the nut should be turned true and flat, so that there may not be any uneven motion, and only the one imparted by the progressive one of the threads. When this nut is put to its place for the last time, and after all is finished, there should be a little tallow put on to the face of the nut just where it comes to a seat against the shoulder of the bob, as this shoulder being not very well finished, the two surfaces coming in contact, if left dry, might cut and tear each other, and help to make the nut's action slightly unsteady and unreliable. A finished washer can be driven into this lower hole up to the center, friction tight, and serve as a reliable and finished seat for the nut.

In reality, the zinc at the point of contact, where pinned to the angle piece at the top of the lower section, is the point of greatest importance in the whole combination, and if the joint between the angle and the end of the zinc bar is soldered with soft solder, the result will be that of greater certainty in the maintenance of a steady rate. This joint just mentioned can be soldered as follows: File the end of the zinc and the inside surface of the angle until they fit so that no appreciable space is left between them. Then, with a soldering iron, tin the end of the zinc thoroughly and evenly, and then put into the holes already made the two steady pins. Now tin in the same manner the surface of the angle, and see that the holes are free of solder, so that the zinc bar will go to its place easily; then between the zinc and the iron, place a piece of thin writing paper, so that the flat surfaces of the zinc and iron may not become soldered. Set the iron bar upright on a piece of charcoal, and secure it in this position from any danger of falling, and then put the zinc to its place and see that the pins enter and that the paper is between the surfaces, as described. Put the screws into their places, and screw down on the zinc just enough to hold it in contact with the iron bar, but not so tight that the zinc will not readily move down and rest firmly on the angle. Put a little soldering fluid on the tinned joint, and blow with a blow pipe against the iron bar (not touching the zinc with the flame). When the solder in the joint begins to flow, press the zinc down in close contact with the angle, and then cool gradually, and if all the points described have been attended to the joint will be solidly soldered, and the two bars will be as one solid bar bent against itself. The tinning leaves surplus solder on the surfaces sufficient to make a solid joint, and to allow some to flow into the pin holes and also solder the pin to avoid any danger of getting loose in after time, and helps make a much stronger joint. At the time the solder is melted the zinc is sufficiently heated to become quite malleable, and care must be taken not to force it down against the angle in making the joint, or it may be distorted and ruined at the joint. If carefully done the result will be perfect. The paper between the surfaces burns, and is got rid of in washing to remove the soldering fluid. Soda or ammonia will help to remove all traces of the fluid. However, it is best, as a last operation, to put the joint in alcohol for a minute.

This soldering makes the lower section and the zinc practically one piece and without loose joint, and the next joint is that made by the pin pinning the outside bar and the zinc together. This is necessarily formed this way, as in this stage of the operation we do not know just what length the zinc bar will be to exactly compensate for the expansion and contraction of the balance of the pendulum. By the changing of the pin into the different holes, 5, 6, 7, 8, 9, 10, Fig. 9, the zinc is made relatively longer or shorter, and so a compensation is arrived at in time after the clock has been running. After it is definitely settled where the pin will remain to secure the compensation of the rod, then that hole can have a screw put in to match the three upper ones. This screw must be tapped into the front bar and the zinc, and be very free in the back bar to allow of its expansion. It is supposed that in this example given of a zinc and steel compensation seconds pendulum that there has been due allowance made in the lengths of the several bars to allow for adjustment to temperature by the movements of the pin along the course of the several holes described, but the zinc is a very uncertain element, and its ultimate action is largely influenced by its treatment after being cast. Differences of working cast zinc under the hammer or rolls produce wide differences practically, and therefore materially change the results in its combination with iron in their relative expansive action. Wrought zinc can be obtained of any of the brass plate factories, of any dimensions required, and will be found to be satisfactory for the purpose in hand.

The adjusting pin should be well fitted to the holes in the front iron bar, and also fit the corresponding ones in the zinc bar closely, and if the holes are reamed smooth and true with an English clock broach, then the pin will be slightly tapering and fit the iron hole perfectly solid. After one pair of these holes have been reamed, fit the pin and drive it in place perfectly firm, and then with the broach ream all the remaining holes to just the same diameter, and then the pin will move along from one set of holes to another with mechanically accurate results. Otherwise, if poorly fitted, the full effect would not be obtained from the compensating action in making changes in the pin from one set of holes to another. This pin, if made of cast steel, hardened and drawn to a blue, will on the whole be a very good device mechanically.

Many means are used to effect the adjustments for compensation, of more or less value, but whatever the means used, it must be kept in mind that extra care must be taken to have the mechanical execution first class, as on this very much depends the steady rate of the pendulum in after time.

TUBULAR COMPENSATED RODS.—There are tubular pendulums in the market which have a screw sleeve at the top of the zinc element, and by this means the adjustments are effected, and this is thought to be a very accurate mechanism. The most common form of zinc and iron compensation is where the zinc is a tube combined with one iron tube and a central rod, as shown in Figs. 10, 11, 12. The rod is the center piece, the zinc tube next, followed by the iron tube enveloping both. The relative lengths may be the same as those just given in the foregoing example with the compensating elements flat. The relative lengths of the several members will be virtually the same in both combinations.

TUBULAR COMPENSATION WITH ALUMINUM.—The pendulum as seen by an observer appears to him as being a simple single rod pendulum. Figs. 10 and 12 are front and side views; Fig. 11 is an enlarged view of its parts, the upper being a sectional view. Its principal features are: The steel rod S, Fig. 11, 4 mm. in diameter, having at its upper end a hook for fastening to the suspension spring in the usual way; the lower end has a pivot carrying the bushing, T, which solidly connects the steel rod, S, with the aluminum tube, A, the latter being 10 mm. in diameter and its sides 1.5 mm. in thickness of the wall.

The upper end of the aluminum tube is very close to the pendulum hook and is also provided with a bushing, P, Fig. 11. This bushing is permanently connected at the upper end of the aluminum tube with a steel tube, R, 16 mm. in diameter and 1 mm. in thickness. The outer steel tube is the only one that is visible and it supports the bob, the lower part being furnished with a fine thread on which the regulating nut, O, is movable, at the center of the bob.

For securing a central alignment of the steel rod, S, at its lowest part, where it is pivoted, a bushing, M, Fig. 11, is screwed into the steel tube, R. The lower end of the steel tube, R, projects considerably below the lenticular bob (compare Figs. 10 and 12); and is also provided with a thread and regulating weight, G (Figs. 10 and 12), of 100 grammes in weight, which is only used in the fine regulation of small variations from correct time.

The steel tube is open at the bottom and the index at its lower end is fastened to a bridge. Furthermore all three of the bushings, P, T and M, have each three radial cuts, which will permit the surrounding air to act equally and at the same time on the steel rod, S, the aluminum tube, A, and the steel tube, R, and as the steel tube, R, is open at its lower end, and as there is also a certain amount of space between the tubes, the steel rod, and the radial openings in the bushings, there will be a draught of air passing through them, which will allow the thin-walled tubes and thin steel rod to promptly and equally adapt themselves to the temperature of the air.

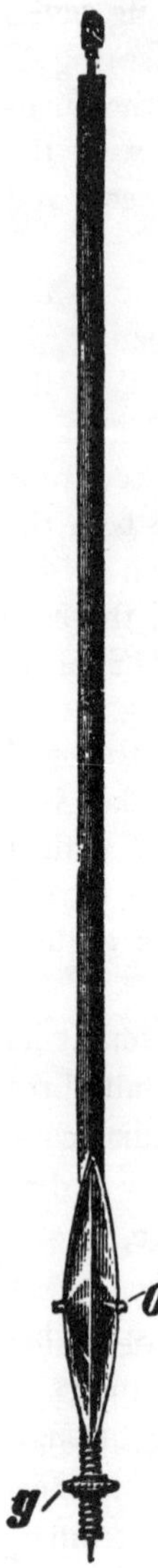

Fig. 10.

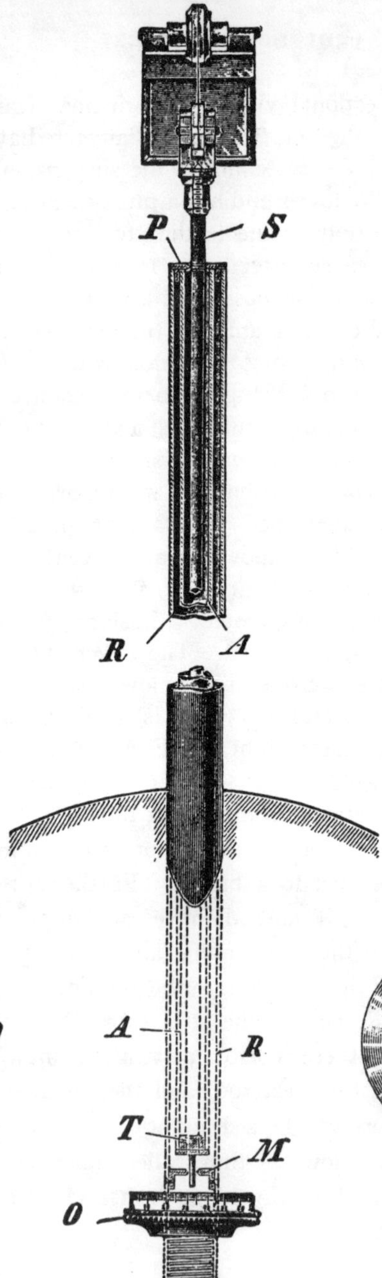

Fig. 11.

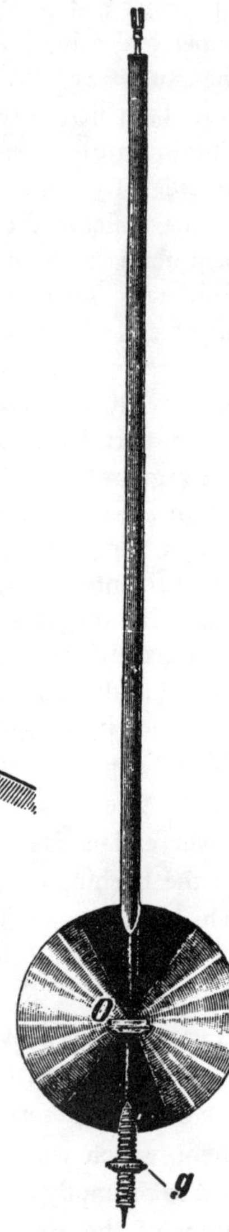

Fig. 12.

The lenticular pendulum bob has a diameter of 24 cm., and is made of red brass. The bob is supported at its center by the regulating nut, O, Figs. 10 and 12. That the bob may not turn on the cylindrical pendulum rod, the latter is provided with a longitudinal groove and working therein are the ends of two shoulder screws which are placed on the back of the bob above and below the regulating nut, O; and thus properly controlling its movements.

From the foregoing description the action of the compensation is readily explained. For the purpose of illustration of its action we will accept the fact that there has been a sudden rise in temperature. The steel rod, S, and the tube, R, will lengthen in a downward direction (including the suspension spring and the pendulum hook), conversely the aluminum tube, A, which is fastened to the steel rod at one end and the steel tube at the other, will lengthen in an upward direction and thus equalize the expansion of the tube, R, and rod, S.

As the coefficients of expansion of steel and aluminum are approximately at the ratio of 1 :2.0313 we find that with such a pendulum construction—accurate calculations presumed—we shall have a complete and exact coincidence in its compensation; in other words, the center of oscillation of the pendulum will be under all conditions at the same distance from the bending point of the suspension spring.

This style of pendulum is made for astronomical clocks in Europe and is furnished in two qualities. In the best quality, the tubes, steel rod, and the bob are all separately and carefully tested as to their expansion, and their coefficients of expansion fully determined in a laboratory; the bushings, P and M, are jeweled, all parts being accurately and finely finished. In the second quality the pendulum is constructed on a general calculation and finished in a more simple manner without impairing its ultimate efficiency.

At the upper part of the steel tube, R, there is a funnel-shaped piece (omitted in the drawing) in which are placed small lead and aluminum balls for the final regulation of the pendulum without stopping it.

The regulation of this pendulum is effected in three ways:

1. The preliminary or coarse regulation by turning the regulating nut, O, and so raising or lowering the bob. 2. The finer regulation by turning the 100 grammes weight, g, having the shape of a nut and turning on the threaded part of the tube, R. 3. The precision regulation is effected by placing small lead or aluminum balls in a small funnel-shaped receptacle attached to the upper part of the tube, R, or by removing them therefrom.

It will readily be seen that this form of pendulum can be used with zinc or brass instead of aluminum, by altering the lengths of the inner rod and the compensating tube to suit the expansion of the metal it is decided to use; also that alterations in length may be made by screwing the bushings in or out, provided that the tube be long enough in the first place. After securing the right position the bushings should have pins driven into them through the tube, in order to prevent further shifting.

CHAPTER IV.

THE CONSTRUCTION OF MERCURIAL PENDULUMS.

Owing to the difficulty of calculating the expansive ratios of metal which (particularly with brass and zinc) vary slightly with differences of manufacture, the manufacture of compensated pendulums from metal rods cannot be reduced to cutting up so many pieces and assembling them from calculations made previously, so that each must be separately built and tested. While this is not a great drawback to the jeweler who wants to make himself a pendulum, it becomes a serious difficulty to a manufacturer, and hence a cheaper combination had to be devised to prevent the cost of compensated pendulums from seriously interfering with their use. The result was the pendulum composed of a steel rod and a quantity of mercury, the latter forming the principal weight for the bob and being contained in steel or glass jars, or jars of cast iron for the heavier pendulums. Other metals will not serve the purpose, as they are corroded by the mercury, become rotten and lose their contents.

Mercury has one deficiency which, however, is not serious, except for the severe conditions of astronomical observatories. It will oxidize after long exposure to the air, when it must be strained and a fresh quantity of metal added and the compensation freshly adjusted. To an astronomer this is a serious objection, as it may interfere with his work for a month, but to the jeweler this is of little moment as the rates he demands will not be seriously affected for about ten years, if the jars are tightly covered.

To construct a reliable gridiron pendulum would cost about fifty dollars while a mercurial pendulum can be well made and compensated for about twenty-five dollars, hence the popularity of the latter form.

Zinc will lengthen under severe variations of temperature as the following will show: Zinc has a decided objectionable quality in its crystalline structure that with temperature changes there is very unequal expansion and contraction, and furthermore, that these changes occur suddenly; this often results in the bending of the zinc rod, causing a binding to take place, which naturally enough prevents the correct working of the compensation.

It is probably not very well known that zinc can change its length at one and the same temperature, and that this peculiar quality must not be overlooked. The U. S. Lake Survey, which has under its charge the triangulation of the great lakes of the United States, has in its possession a steel meter measure, R, 1876; a metallic thermometer composed of a steel and zinc rod, each being one meter in length, marked M. T., 1876s, and M. T. 1876z; and four metallic thermometers, used in connection with the base apparatus, which likewise are made of steel and zinc rods, each of these being four meters in length. All of these rods were made by Repsold, of Hamburg. Comparisons between these different rods show peculiar variations, and which point to the fact that their lengths at the same degree of temperature are not constant. For the purpose of determining these variations accurate investigations were undertaken. The metallic thermometer M. T. 1876 was removed from an observatory room having an equal temperature of about 2° C. and placed for one day in a temperature of +24° C., and also for the same period of time in one of —20° C; it was then replaced in the observatory room, where it remained for twenty-four hours, and comparisons were made during the following three days with the steel thermometer R, 1876, which had been left in the room. From these observations and comparisons the following results were tabulated, which give the mean lengths of the zinc rods of the metallic thermometer. The slight variations of temperature in the observatory room were also taken into consideration in the calculations:

		M. T. 1876s. mm.	M. T. 1876z. mm.	
February	16-24	—0.0006	+0.0152,	previous 7 days at +24°C.
February	25-27	—0.0017	—0.0011,	previous 1 day at —20°C.
March	2-4	+0.0005	+0.0154,	previous 1 day at +24°C.
March	5-8	—0.0058	—0.0022,	previous 1 day at —20°C.

These investigations clearly indicate, without doubt, that the zinc rod at one and the same temperature of about 2° C., is 0.018 mm. longer after having been previously heated to 24° C. than when cooled before to —20° C.

A similar but less complete examination was made with the metallic thermometer four meters in length. These trials were made by that efficient officer, General Comstock, gave the same results, and completely prove that in zinc there are considerable thermal after-effects at work.

To prove that zinc is not an efficient metal for compensation pendulums when employed for the exact measurement of time, a short calculation may be made—using the above conclusions—that a zinc rod one meter in length, after being subjected to a difference of temperature of 44 C. will alter its length 0.018 mm. after having been brought back to its initial degree. For a seconds pendulum with zinc compensation each of the zinc rods would require a length of 64.9 cm. With the above computations we get a difference in length of 0.0117 mm. at the same degree of temperature. Since a lengthening of the zinc rods without a suitable and contemporaneous expansion of the steel rods is synonymous with a shortening of the effectual pendulum length, we have, notwithstanding the compensation, a shortening of the pendulum length of 0.017 mm., which corresponds to a change in the daily rate of about 0.5 seconds.

This will sufficiently prove that zinc is unquestionably not suitable for extremely accurate compensation pendulums, and as neither is permanent under extremes of temperature the advantages of first cost and of correction of error appear to lie with the mercurial form.

The average mercurial compensation pendulums, on sale in the trade are often only partially compensated, as the mercury is nearly always deficient in quantity relatively, and not high enough in the jar to neutralize the action of the rigid metallic elements, composing the structure. The trouble generally is that the mercury forms too small a proportion of the total weight of the pendulum bob. There is a fundamental principle governing these compensating pendulums that has to be kept in mind, and that is that one of the compensating elements is expected to just undo what the other does and so establish through the medium of physical things the condition of the ideal pendulum, without weight or elements outside of the bob. As iron and mercury, for instance, have a pretty fixed relative expansive ratio, then whatever these ratios are after being found, must be maintained in the construction of the pendulum, or the results cannot be satisfactory.

First, there are 39.2 inches of rod of steel to hold the bob between the point of suspension and the center of oscillation, and it has been found that, constructively, in all the ordinary forms of these pendulums, the height of mercury in the bob cannot usually be less than 7.5 inches. Second, that in all seconds pendulums the length of the metal is fixed substantially, while the height of the mercury is a varying one, due to the differing weights of the jars, straps, etc.

Third, the mercury, at its minimum, cannot with jars of ordinary weight be less in height in the jar than 7.5 inches, to effectually counteract what the 39.2 inches of iron does in the way of expanding and contracting under the same exposure.

Whoever observes the great mass of pendulums of this description on sale and in use will find that the height of the mercury in the jar is not up to the amount given above for the least quantity that will serve under the most favorable circumstances of construction. The less weight there is in the rod, jar and frame, the less is the height of mercury which is required; but with most of the pendulums made in the present day for the market, the height given cannot be cut short without impairing the quality and efficiency of the compensation. Any amount less will have the effect of leaving the rigid metal in the ascendancy; or, in other words, the pendulum will be under compensated and leave the pendulum to feel heat and cold by raising and lowering the center of oscillation of the pendulum and hence only partly compensating. A jar with only six inches in height of mercury will in round numbers only correct the temperature error about six-sevenths.

Calculations of Weights.—As to how to calculate the amount of mercury required to compensate a seconds pendulum, the following explanation should make the matter clear to anyone having a fair knowledge of arithmetic only, though there are several points to be considered which render it a rather more complicated process than would appear at first sight.

1st. The expansion in length of steel and cast iron, as given in the tables (these tables differ somewhat in the various books), is respectively .0064 and .0066, while mercury expands .1 in bulk for the same increase of temperature. If the mercury were contained in a jar which itself had no expansion in diameter, then all its expansion would take place in height, and in round numbers it would expand sixteen times more than steel, and we should only require (neglecting at present the allowance to be explained under head 3) to make the height of the mercury—reckoned from the bottom of the jar (inside) to the middle of the column of mercury contained therein—one-sixteenth of the total length of the pendulum measured from the point of suspension to the bottom of the jar, assuming that the rod and the jar are both of steel, and that the center of oscillation is coincident with the center of the column of mercury.

Practically in these pendulums, the center of oscillation is almost identical with the center of the bob.

2d. As we cannot obtain a jar having no expansion in diameter, we must allow for such expansion as follows, and as cast-iron or steel jars of cylindrical shape are undoubtedly the best, we will consider that material and form only.

As above stated, cast iron expands .0066, so that if the original diameter of the jar be represented by 1, its expanded diameter will be 1.0066. Now the area of any circle varies as the square of its diameter, so that before and after its expansion the areas of the jar will be in the ratio of 1^2 to 1.0066^2; that is, in the proportion of 1 to 1.013243; or in round numbers it will be one-seventy-sixth larger in area after expansion than before. It is evident that the mercury will then expand sideways, and that its vertical rise will be diminished to the same extent. Deduct, therefore, the one-seventy-sixth from its expansion in bulk (one-tenth) and we get one-eleventh (or more exactly .086757) remaining. This, then, is the actual vertical rise in the jar, and when compared with the expansion of steel in length it will be found to be about thirteen and a half times greater (more exactly 13.556).

The mercury, therefore (still neglecting head No. 3), must be thirteen and a half times shorter than the length of the pendulum, both being measured as explained above. The pendulum will probably be 43.5 inches long to the bottom of the jar; but as about nine inches of it is cast iron, which has a slightly greater rate of expansion than steel, we will call the length 44 inches, as the half inch added will make it about equivalent to a pendulum entirely of steel. If the height of the mercury be obtained by dividing 44 by 13.5, it will be 3.25 inches high to its center, or 6.5 inches high altogether; and were it not for the following circumstance, the pendulum would be perfectly compensated.

3d. The mercury is the only part of the bob which expands upwards; the jar does not rise, its lower end being carried downward by the expansion of the rod, which supports it. In a well-designed pendulum, the jar, straps, etc., will be from one-fourth to one-third the weight of the mercury. Assume them to be seven pounds and twenty-eight pounds respectively; therefore, the total weight of the bob is thirty-five pounds; but as it is only the mercury (four-fifths) of this total that rises with an increase of temperature, we must increase the weight of the mercury in the proportion of five to four, thus $6.5 \times 5 \div 4 = 8\frac{1}{8}$ inches. Or, what is the same thing, we add one-fourth to the amount of mercury, because the weight of the jar is one-fourth of that of the mercury. Eight and one-eighth inches is, therefore, the ultimate height of the mercury required to compensate the pendulum with that weight of jar. If the jar had been heavier, say one-third the weight of the mercury, then the latter would have to be nearly 8.75 inches high.

If the jar be required to be of glass, then we substitute the expansion of that material in No. 2 and its weight in No. 3.

In the above method of calculating, there are two slight elements of uncertainty: 1st. In assuming that the center of oscillation is coincident with the center of the bob; however, I should suppose that they would never be more than .25 inch apart, and generally much nearer. 2d. The weight of the jar cannot well be exactly known until after it is finished (i. e., bored smooth and parallel inside, and turned outside true with the interior), so that the exact height of the mercury cannot be easily ascertained till then.

I may explain that the reason (in Nos. 1 and 2) we measure the mercury from the bottom to the center of the column, is that it is its center which we wish to raise when an increase of temperature occurs, so that the center may always be exactly the same distance from the point of suspension; and we have seen that 3.25 inches is the necessary quantity to raise it sufficiently. Now that center could not be the center without it had as much mercury over it as it has under it; hence we double the 3.25 and get the 6.5 inches stated.

From the foregoing it will be seen that the average mercury pendulums are better than a plain rod, from the fact that the mercury is free to obey the law of expansion, and so, to a certain degree, does counteract the action of the balance of the metal of the pendulum, and this with a degree of certainty that is not found in the gridiron form, provided always that the height and amount of the mercury are correctly proportional to the total weight of the pendulum.

COMPENSATING MERCURIAL PENDULUMS.—To compensate a pendulum of this kind takes time and study. The first thing to do is to place maximum and minimum thermometers in the clock case, so that you can tell the temperature.

Then get the rate of the clock at a given temperature. For example, say the clock gains two seconds in twenty-four hours, the temperature being at 70°. Then see how much it gains when the temperature is at 80°. We will say it gains two seconds more at 80° than it does when the temperature is at 70°.

In that case we must remove some of the mercury in order to compensate the pendulum. To do this take a syringe and soak the cotton or whatever makes the suction in the syringe with vaseline. The reason for doing this is that mercury is very heavy and the syringe must be air tight before you can take any of the mercury up into it.

You want to remove about two pennyweights of mercury to every second the clock gains in twenty-four hours. Now, after removing the mercury the clock will lose time, because the pendulum is lighter. You must then raise the ball to bring it to time. You then repeat the same operation by getting the rate at 70° and 80° again and see if it gains. When the temperature rises,. if the pendulum still gains, you must remove more mercury; but if it should lose time when the temperature rises you have taken out too much mercury and you must replace some. Continue this operation until the pendulum has the same rate, whether the temperature is high or low, raising the bob when you take out mercury to bring it to time, and lowering the bob when you put mercury in to bring it to time.

To compensate a pendulum takes time and study of the clock, but if you follow out these instructions you will succeed in getting the clock to run regularly in both summer and winter.

Besides the oxidation, which is an admitted fault, there are two theoretical questions which have to do with construction in deciding between the metallic and mercurial forms of compensation. We will present the claims of each side, therefore, with the preliminary statement that (for all except the severest conditions of accuracy) either form, if well made will answer every purpose and that therefore, except in special circumstances, these objections are more theoretical than real.

The advocates of metallic compensation claim that where there are great differences of temperature, the compensated rod, with its long bars will answer more quickly to temperature changes as follows:

The mercurial pendulum, when in an unheated room and not subjected to sudden temperature changes, gives very excellent results, but should the opposite case occur there will then be observed an irregularity in the rate of the clock. The causes which produce these effects are various. As a principal reason for such a condition it may be stated that the compensating mercury occupies only about one-fifth the pendulum length, and it inevitably follows that when the upper strata of the air is warmer than the lower, in which the mercury is placed, the steel pendulum rod will expand at a different ratio than the mercury, as the latter is influenced by a different degree of temperature than the upper part of the pendulum rod. The natural effect will be a lengthening of the pendulum rod, notwithstanding the compensation, and therefore, a loss of time by the clock.

Two thermometers, agreeing perfectly, were placed in the case of a clock, one near the point of suspension, and the other near the middle of the ball, and repeated experiments, showed a difference between these two thermometers of 7° to 10½° F., the lower one indicating less than the higher one. The thermometers were then hung in the room, one at twenty-two inches above the floor, and the other three feet higher, when they showed a difference of 7° between them. The difference of 2.5° more which was found inside the case proceeds from the heat striking the upper part of the case; and the wood, though a bad conductor, gradually increases in temperature, while, on the contrary, the cold rises from the floor and acts on the lower part of the case, The same thermometers at the same height and distance in an unused room, which was never warmed, showed no difference between them; and it would be the same, doubtless, in an observatory.

From the preceding it is very evident that the decrease of rate of the clock since December 13 proceeded from the rod of the pendulum experiencing 7° to 10.5° F. greater heat than the mercury in the bob, thus showing the impossibility of making a mercurial pendulum perfectly compensating in an artificially heated room which varies greatly in temperature. I should remark here that during the entire winter the temperature in the case is never more than 68° F., and during the summer, when the rate of the clock was regular, the thermometer in the case has often indicated 72° to 77° F.

The gridiron pendulum in this case would seem preferable, for if the temperature is higher at the top than at the lower part, the nine compensating rods are equally effected by it. But in its compensating action it is not nearly as regular, and it is very difficult to regulate it, for in any room (artificially heated) it is impossible to obtain a uniform temperature throughout its entire length, and without that all proofs are necessarily inexact.

These facts can also be applied to pendulums situated in heated rooms. In the case of a rapid change in temperature taking place in the observatory rooms, under the domes of observatories, especially during the winter months, and which are of frequent occurrence, a mercurial compensation pendulum, as generally made, is not apt to give a reliable rate. Let us accept the fact, as an example, of a considerable fall in the temperature of the surrounding air; the thin pendulum rod will quickly accept the same temperature, but with the great mass of mercury to be acted upon the responsive effects will only occur after a considerable lapse of time. The result will be a shortening of the pendulum length and a gain in the rate until the mercury has had time to respond, notwithstanding the compensation.

Others who have expressed their views in writing seem to favor the idea that this inequality in the temperature of the atmosphere is unfavorable to the accurate action of the mercurial form of compensation; and however plausible and reasonable this idea may seem at first notice, it will not take a great amount of investigation to show that, instead of being a disadvantage, its existence is beneficial, and an important element in the success of mercurial pendulums.

It appears that the majority of those who have proposed, or have tried to improve Graham's pendulum have overlooked the fact that different substances require different quantities of heat to raise them to the same temperature. In order to warm a certain weight of water, for instance, to the same degree of heat as an equal weight of oil, or an equal weight of mercury, twice as much heat must be given to the water as to the oil, and thirty times as much as to the mercury; while in cooling down again to a given temperature, the oil will cool twice as quick as the water, and the mercury thirty times quicker than the water. This phenomenon is accounted for by the difference in the amount of latent heat that exists in various substances. On the authority of Sir Humphrey Davy, zinc is heated and cooled again ten and three-quarters times quicker than water, brass ten and a half times quicker, steel nine times, glass eight and a half times, and mercury is heated and cooled again thirty times quicker than water.

From the above it will be noticed that the difference in the time steel and mercury takes to rise and fall to a given temperature is as nine to thirty, and also that the difference in the quantity of heat that it takes to raise steel and mercury to a given temperature is in the ratio of nine to thirty.

Now, without entering into minute details on the properties which different substances possess for absorbing or reflecting heat, it is plain that mercury should move in a proportionally different atmosphere from steel in order to be expanded or contracted a given distance in the same length of time; and to obtain this result the amount of difference in the temperature of the atmosphere at the opposite ends of the pendulum must vary a little more or less according to the nature of the material the mercury jars are constructed from.

Differences in the temperature of the atmosphere of a room will generally vary according to its size, the height of the ceiling, and the ventilation of the apartment; and if the difference must continue to exist, it is of importance that the difference should be uniformly regular. We must not lose sight of the fact, however, that clocks having these pendulums, and placed in apartments every way favorable to an equal temperature, and in some instances, the clocks and their pendulums incased in double casing in order to more effectually obtain this result, still the rates of the clock show the same eccentricities as those placed in less favorable position. This clearly shows that many changes in the rates of fine clocks are due to other causes than a change in the temperature of the surounding atmosphere. Still it must be admitted that any change in the condition of the atmosphere that surrounds a pendulum is a most formidable obstacle to be overcome by those who seek to improve compensated pendulums, and it would be of service to them to know all that can possibly be known on the subject.

The differences spoken of above have resulted in some practical improvements, which are: 1st, the division of the mercury into two, three or four jars in order to expose as much surface as possible to the action of the air, so that the expansion of the mercury should not lag behind that of the rod, which it will do if too large amounts of it are kept in one jar. 2nd, the use of very thin steel jars made from tubing, so that the transmission of heat from the air to the mercury may be hastened as much as possible. 3rd, the increase in the number of jars makes a thinner bob than a single jar of the same total weight and hence gives an advantage in decreasing the resistant effect of air friction in dense air, thereby decreasing somewhat the barometric error of the pendulum.

The original form of mercurial pendulums, as made by Graham, and still used in tower and other clocks where extraordinary accuracy is not required, was a single jar which formed the bob and had the pendulum rod extending into the mercury to assist in conducting heat to the variable element of the pendulum. It is shown in section in Fig. 13, which is taken from a working drawing for a tower clock.

The pendulum, Fig. 13, is suspended from the head or cock shown in the figure, and supported by the clock frame itself, instead of being hung on a wall, since the intention is to set the clock in the center of the clockroom, and also because the weight, forty pounds, is not too much for the clock frame to carry. The head, A, forms a revolving thumb-nut, which is divided into sixty parts around the circumference of its lower edge, and the regulating screw, B, is threaded ten to the inch. A very fine adjustment is thus obtained for regulating the time of the pendulum. The lower end of the regulating screw, B, holds the end of the pendulum spring, E, which is riveted between two pieces of steel, C, and a pin, C′, is put through them and the end of the regulating screw, by which to suspend the pendulum.

The cheeks or chops are the pieces D, the lower edges of which form the theoretical point of suspension of the pendulum. These pieces must be perfectly square at their lower edges, otherwise the center of gravity would describe a cylindrical curve. The chops are clamped tightly in place by the setscrews, D′, after the pendulum has been hung.

The lower end of the regulating screw is squared to fit the ways and slotted on one side, sliding on a pin to prevent its turning and therefore twisting the suspension spring when it is raised or lowered.

The spring is three inches long between its points of suspension, one and three-eighths inches wide, and one-sixtieth of an inch thick. Its lower end is riveted between two small blocks of steel, F, and suspended from a pin, F′, in the upper end of the cap, G, of the pendulum rod.

The tubular steel portion of the pendulum rod is seven-eighths of an inch in diameter and one-thirty-second of an inch thickness of the wall. It is enclosed at each end by the solid ends, G and L, and is made as nearly air tight as possible.

The compensation is by mercury inclosed in a cast-iron bob. The mercury, the bob and the rod together weigh forty pounds. The bob of the pendulum is a cast-iron jar, K, three inches in diameter inside, one-quarter inch thick at the sides, and five-sixteenths thick at the bottom, with the cap, J, screwed into its upper end. The cap, J, forms also the socket for the lower end of the pendulum rod, H. The rod, L, one-quarter inch in diameter, screws into the cap, J, and its large end at the same time forms a plug

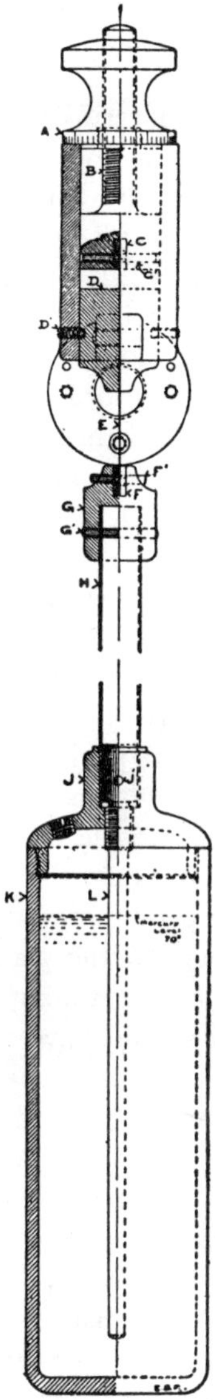

Fig. 13.

for the lower end of the pendulum tube, H. The pin, J′, holds all these parts together. The rod, L, extends nearly to the bottom of the jar, and forms a medium for the transmission of the changes in temperature from the pendulum tube to the mercury. The screw in the cap, J, is for filling or emptying the jar. The jar is finished as smoothly as possible, outside and inside, and should be coated with at least three coats of shellac inside. Of course if one was building an astronomical clock, it would be necessary to boil the mercury in the jar in order to drive off the layer of air between the mercury and the walls of the jar, but with the smooth finish the shellac will give, in addition to the good work of the machinist, the amount of air held by the jar can be ignored.

The cast-iron jar was decided upon because it was safer to handle, can be attached more firmly to the rod with less multiplication of parts, and also on account of the weight as compared with glass, which is the only other thing that should be used, the glass requiring a greater height of jar for equal weight. In making cast iron jars, they should always be carefully turned inside and out in order that the walls of the jar may be of equal thickness throughout; then they will not throw the pendulum out of balance when they are screwed up or down on the pendulum rod in making the coarse regulation before timing by the upper screw. The thread on the rod should have the cover of the jar at about the center of the thread when nearly to time and that portion which extends into the jar should be short enough to permit this.

Ignoring the rod and its parts for the present, and calling the jar one-third of the weight of the mercury, we shall find that thirty pounds of mercury, at .49 pounds per cubic inch, will fill a cylinder which is three inches inside diameter to a height of 8.816 inches, after deducting for the mass of the rod L. when the temperature of the mercury is 60 degrees F. Mercury expands one-tenth in bulk, while cast-iron expands .0066 in diameter: so the sectional area increases as 1.0066^2 or 1.0132 to 1, therefore the mercury will rise .1—.013243, or .086757; then the mercury in our jar will rise .767 of an inch in the ordinary changes of temperature, making a total height of 9.58 inches to provide for; so the jar was made ten inches long.

Pendulums of this pattern as used in the high grade English clocks, are substantially as follows: Rod of steel 5-16 inch diameter; jar about 2.1 inches diameter inside and 8¾ inches deep inside. The jar may be wrought or cast iron and about ⅜ of an inch thick with the cover to screw on with fine thread, making a tight joint. The cover of the jar is to act as a nut to turn on the rod for regulation. The thread cut on the rod should be thirty-six to the inch, and fit into the jar cover easily, so that it may turn without binding. With a thirty-six thread one turn of the jar on the rod changes the rate thirty seconds per day and by laying off on the edge of the cover 30 divisions, a scale is made by which movements for one second per day are obtained.

We will now describe (Fig. 14) the method of making a mercurial pendulum to replace an imitation gridiron pendulum for a Swiss, pin escapement regulator, such as is commonly found in the jewelry stores of the United States, that is, a clock in which the pendulum is supported by the plates of the movement and swings between the front plate and the dial of the movement. In thus changing our pendulum, we shall desire to retain the upper portion of the old rod, as the fittings are already in place and we shall save considerable time and labor by this course. As the pendulum is suspended from the movement, it must be lighter in weight than if it were independently supported by a cast iron bracket, as shown in Fig. 6, so we will make the weight about that of the one we have removed, or about twelve pounds. If it is desired to make the pendulum heavier, four jars of the dimensions given would make it weigh about twenty pounds, or four jars of one inch diameter would make a thinner bob and one weighing about fourteen pounds. As the substitution of a different number or different sizes of jars merely involves changing the lengths of the upper and lower bars of the frame, further drawings will be unnecessary, the jeweler having sufficient mechanical capacity to be able to make them for himself. I might add, however, that the late Edward Howard, in building his astronomical clocks, used four jars containing twenty-eight pounds of mercury for such movements, and the perfection of his trains was such that a seven-ounce driving weight was sufficient to propel the thirty pound pendulum.

The two jars are filled with mercury to a height of 7⅝ inches, are 1⅜ inches in diameter outside and 8⅜ inches in height outside. The caps and foot pieces are screwed on and when the foot pieces are screwed on for the last time the screw threads should be covered with a thick shellac varnish which, when dry, makes the joint perfectly air tight. The jars are best made of the fine, thin tubing, used in bicycles, which can be purchased from any factory, of various sizes and thickness. In the pendulum shown in the illustration, the jar stock is close to 14 wire gauge, or about 2 mm. in thickness. In cutting the threads at the ends of the jars they should be about 36 threads to the inch, the same number as the threads on the lower end of the rod used to carry the regulating nut. A fine thread makes the best job and the tightest joints. The caps to the jars are turned up from cold rolled shafting, it being generally good stock and finishes well. The threads need not be over 3-16 inch, which is ample. Cut the square shoulder so the caps and foot pieces come full up and do not show any thread when screwed home. These jars will hold ten pounds of mercury and this weight is about right for this particular style of pendulum. The jars complete will weigh about seven ounces each.

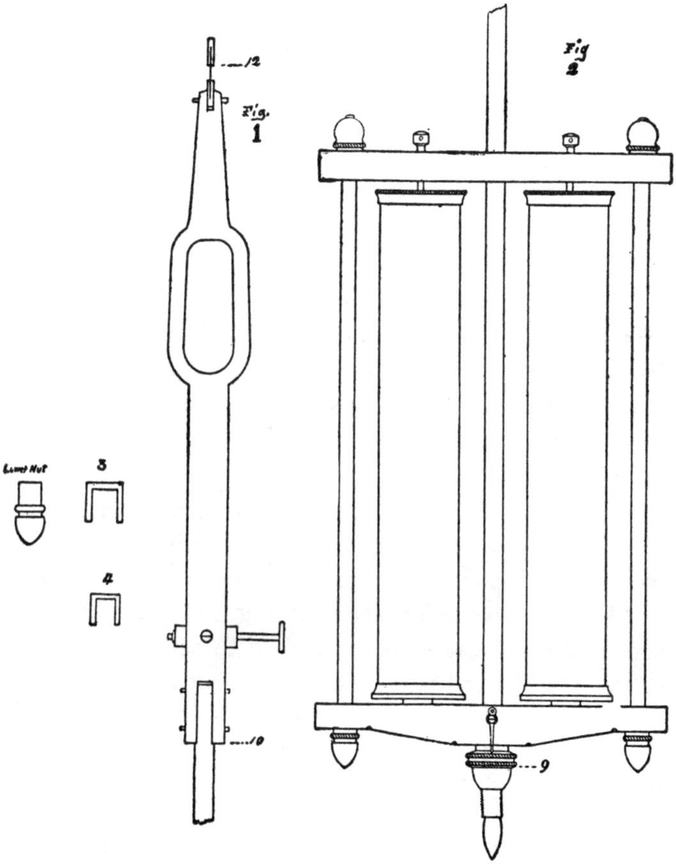

Fig. 14.

The frame is also made of steel and square finished stock is used as far as possible and of the quality used in the caps. The lower bar of the frame is six inches long and ⅝ inch square at the center and tapered, as shown in the illustration. It is made light by being planed away on the under side, an end view being shown at 3. The top bar of the frame, shown at 4, is planed away also and is one-half inch square the whole length and is six inches long. The two side rods are to bind the two bars together, and with the four thumb nuts at the top and bottom make a strong light frame.

The pendulum described is nickel plated and polished, except the jars, which are left half dead; that is, they are frosted with a sand blast and scratch brushed a little. The effect is good and makes a good contrast to the polished parts. The side rods are five inches apart, which leaves one-half inch at the ends outside.

The rod is 5-16 of an inch in diameter and 33 inches long from the bottom of the frame at a point where the regulating nut rests against it to the lower end of the piece of the usual gridiron pendulum shown in Fig. 14 at 10. This piece shown is the usual style and size of those in the majority of these clocks and is the standard adopted by the makers. This piece is 11⅛ inches long from the upper leaf of the suspension spring, which is shown at 12, to the lower end marked 10. By cutting out the lower end of this piece, as showr at 10, and squaring the upper end of the rod, pinning it into the piece as shown, the union can be made easily and any little adjustments for length can be made by drilling another set of holes in the rod and raising the pendulum by so doing to the correct point. A rod whose total length is 37 inches will leave 2 inches for the prolongation below the frame carrying the regulating nut, 9, and for the portion squared at the top, and will then be so long that the rate of the clock will be slow and leave a surplus to be cut off either at the top or bottom, as may seem best.

The screw at the lower end carrying the nut should have 36 threads to the inch and the nut graduated to 30 divisions, each of which is equal in turning the nut to one minute in 24 hours, fast or slow, as the case may be.

The rod should pass through the frame bars snugly and not rattle or bind. It also should have a slot cut so that a pin can be put through the upper bar of the frame to keep the frame from turning on the rod and yet allow it to move up and down about an inch. The thread at the lower end of the rod should be cut about two inches in length and when cutting off the rod for a final length, put the nut in the middle of the run of the thread and shorten the rod at the top. This will be found the most satisfactory method, for when all is adjusted the nut will stand in the middle of its scope and have an equal run for fast or slow adjustment. With the rod of the full length as given, this pendulum had to be cut at the top about one inch to bring to a minute or two in twenty-four hours, and this left all other points below corrected. The pin in the rod should be adjusted the last thing, as this allows the rod to slide on the pin equal distances each way. One inch in the raising or lowering of the frame on the rod will alter the rate for twenty-four hours about eighteen minutes.

Many attempts have been made to combine the good qualities of the various forms of pendulums and thus produce an instrument which would do better work under the severe exactions of astronomical observatories and master clocks controlling large systems. The reader should understand that, just as in watch work, the difficulties increase enormously the nearer we get towards absolute accuracy, and while anybody can make a pendulum which will stay within a minute a month, it takes a very good one to stay within five seconds per month, under the conditions usually found in a store, and such a performance makes it totally unfit for astronomical work, where variations of not over five-thousandths of a second per day are demanded. In order to secure such accuracy every possible aid is given to the pendulum. Barometric errors are avoided by enclosing it in an airtight case, provided with an airpump; the temperature is carefully maintained as nearly constant as possible and its performance is carefully checked against the revolutions of the fixed stars, while various astronomers check their observations against each other by correspondence, so that each can get the rate of his clock by calculations of observations and the law of averages, eliminating personal errors.

One of the successful attempts at such a combination of mercury and metallic pendulums is that of Riefler, as shown in Fig. 15, which illustrates a seconds pendulum one-thirtieth of the actual size.

It consists of a Mannesmann steel tube (rod), bore 16 mm., thickness of metal 1 mm., filled with mercury to about two-thirds of its length, the expansion of the mercury in the tube changing the center of weight an amount sufficient to compensate for the lengthening of the tube by heat, or vice versa. The pendulum, has further, a metal bob weighing several kilograms, and shaped to cut the air. Below the bob are disc shaped weights, attached by screw threads, for correcting the compensation, the number of which may be increased or diminished as appears necessary.

Whereas in the Graham pendulum regulation for temperature is effected by altering the height of the column of

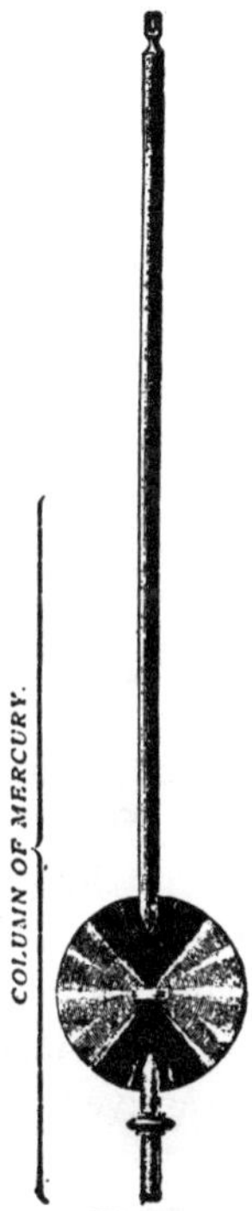

Fig. 15.

mercury, in this pendulum it is effected by changing the position of the center of weight of the pendulum by moving the regulating weights referred to, and thus the height of the column of mercury always remains the same, except as it is influenced by the temperature.

A correction of the compensation should be effected, however, only in case the pendulum is to show sidereal time, instead of mean solar time, for which latter it is calculated. In this case a weight of 110 to 120 grams should be screwed on to correct the compensation.

In order to calculate the effect of the compensation, it is necessary to know precisely the co-efficients of the expansion by heat of the steel rod, the mercury, and the material of which the bob is made.

The last two of these co-efficients of expansion are of subordinate importance, the two adjusting screws for shifting the bob up and down being fixed in the middle of the latter. A slight deviation is, therefore, of no consequence. In the calculation for all these pendulums the co-efficient for the bob is, therefore, fixed at 0.000018, and for the mercury at 0.00018136, being the closest approximation hitherto found for chemically pure mercury, such as that used in these pendulums.

The co-efficient of the expansion of the steel rod is, however, of greater importance. It is therefore, ascertained for every pendulum constructed in Mr. Riefler's factory, by the *physikalisch-technische Reichsanstalt* at Charlottenburg examinations showing, in the case of a large number of similar steel rods, that the co-efficient of expansion lies between 0.00001034 and 0.00001162.

The precision with which the measurements are carried out is so great that the error in compensation resulting from a possible deviation from the true value of the co-efficient of expansion, as ascertained by the Reichsanstalt, does not amount to over ± 0.0017; and, as the precision with which the compensation for each pendulum may be calculated absolutely precludes any error of consequence, Mr. Riefler is in a position to guarantee *that the probable error of compensation in these pendulums will not exceed ± 0.005 seconds per diem and ± 1° variation in temperature.*

A subsequent correction of the compensation is, therefore, superfluous, whereas, with all other pendulums it is necessary, partly because the co-efficients of expansion of the materials used are arbitrarily assumed; and partly because none of the formulæ hitherto employed for calculating the compensation can yield an exact result, for the reason that they neglect to notice certain important influences, in particular that of the weight of the several parts of the pendulum. Such formulæ are based on the assumption that this problem can be solved by simple geometrical calculation, whereas, its exact solution can be arrived at only with the aid of physics.

This is hardly the proper place for details concerning the lengthy and rather complicated calculations required by the method employed. It is intended to publish them later, either in some mathematical journal or in a separate pamphlet. Here I will only say that the object of the whole calculation is to find the allowable or requisite weight of the bob, *i. e.*, the weight proportionate to the co-efficients of expansion of the steel rod, dimensions and weight of the rod and the column of mercury being given in each separate case. To this end the relations of all the parts of the pendulum, both in regard to statics and inertia, have to be ascertained, and for various temperatures.

A considerable number of these pendulums have already been constructed, and are now running in astronomical observatories. One of them is in the observatory of the University of Chicago, and others are in Europe. The precision of this compensation which was discovered by purely theoretical computations, has been thoroughly established by the ascertained records of their running at different temperatures.

The adjustment of the pendulums, which is, of course, almost wholly without influence on the compensation, can be effected in three different ways:

(1.) The rough adjustment, by screwing the bob up or down.

(2.) A finer adjustment, by screwing the correction discs up or down.

(3.) The finest adjustment, by putting on additional weights.

These weights are to be placed on a cup attached to a special part of the rod of the pendulum. Their shape and size is such that they can be readily put on or taken off while the pendulum is swinging. Their weight bears a fixed proportion to the static momentum of the pendulum, so that each additional weight imparts to the pendulum, for twenty-four hours, an acceleration expressed in even seconds and parts of seconds, and marked on each weight.

Each pendulum is accompanied with additional weights of German silver, for a daily acceleration of 1 second each, and ditto of aluminum for an acceleration of 0.5 and 0.1 second respectively.

A metal clasp attached on the rear side of the clock-case, may be pushed up to hold the pendulum in such a way that it can receive no twisting motion during adjustment.

Further, a pointer is attached to the lower end of the pendulum, for reading off the arc of oscillation.

The essential advantages of this pendulum over the mercurial compensation pendulums are the following:

(1.) It follows the changes of temperature more rapidly, because a small amount of mercury is divided over a greater length of pendulum, whereas, in the older ones the entire (and decidedly larger) mass of mercury is situated in a vessel at the lower end of the pendulum rod.

(2.) For this reason differences in the temperature of the air at different levels have no such disturbing influence on this pendulum as on the others.

(3.) This pendulum is not so strongly influenced as the others by changes in the atmospheric pressure, because the principal mass of the pendulum has the shape of a lens, and therefore cuts the air easily.

CHAPTER V.

REGULATIONS, SUSPENSIONS, CRUTCHES AND MINOR POINTS.

REGULATION.—The reader will have noticed that in describing the various forms of seconds pendulums we have specified either eighteen or thirty-six threads to the inch; this is because a revolution of the nut with such a thread gives us a definite proportion of the length of the rod, so that it means an even number of seconds in twenty-four hours.

Moving the bob up or down 1-18 inch makes the clock having a seconds pendulum gain or lose in twenty-four hours one minute, hence the selecting definite numbers of threads has for its reason a philosophical standpoint, and is not a matter of convenience and chance, as seems to be the practice with many clockmakers. With a screw of eighteen threads, we shall get one minute change of the clock's rate in twenty-four hours for every turn of the nut, and if the nut is divided into sixty parts at its edge, each of these divisions will make a change of the clock's rate of one second in twenty-four hours. Thus by using a thread having a definite relation to the length of the rod regulating is made comparatively easy, and a clock can be brought to time without delay. Suppose, after comparing your clock for three or four days with some standard, you find it gains twelve seconds per day, then, turning the nut down twelve divisions will bring the rate down to within one second a day in one operation, if the screw is eighteen threads. With the screw thirty-six threads the nut will require moving just the same number of divisions, only the divisions are twice as long as those with the screw of eighteen threads.

The next thing is the size and weight of the nut. If it is to be placed in the middle of the bob as in Figs. 10, 12 and 15, it should project slightly beyond the surface and its diameter will be governed by the thickness of the bob. If it is an internal nut, worked by means of a sleeve and disc, as in Fig. 9, the disc should be of sufficient diameter to make the divisions long enough to be easily read. If the nut is of the class shown in Fig. 5, 6, 7, a nut is most convenient, 1 inch in diameter, and cut on its edge into thirty equal divisions, each of which is equal to one second in change of rate in twenty-four hours, if the screw has thirty-six threads to the inch. This gives 3.1416 inches of circumference for the thirty divisions, which makes them long enough to be subdivided if we choose, each division being a little over one-tenth of an inch in length, so that quarter-seconds may be measured or estimated.

With some pendulums, Fig. 13, the bob rotates on the rod, and is in the form of a cylinder, say 8½ inches long by 2½ inches in diameter, and the bob then acts on its rod as the nut does, and moves up and down when turned, and in this form of bob the divisions are cut on the outside edge of the cover of the bob, and are so long that each one is subdivided into five or ten smaller divisions, each altering the clock .2 or .1 second per day.

On the top of the bob turn two deep lines, close to the edge, about ⅛-inch apart, and divide the whole diameter into thirty equal divisions, and subdivide each of the thirty into five, and this will give seconds and fifths of seconds for twenty-four hours. Each even seconds division should be marked heavier than the fraction, and should be marked from one to thirty with figures. Just above the cover on the rod should slide a short tube, friction tight, and to this a light index or hand should be fastened, the point of which just reaches the seconds circle on the bob cover, and thus indicates the division, its number and fraction. The tube slides on the rod because the exact place of the hand cannot be settled until it has been settled by experiment. After this it can be fastened permanently, if thought best, though as described it will be all sufficient. While the bob is being raised or lowered to bring the clock to its rate, the bob might get too far away or too near to the index and necessitate its being shifted, and if friction tight this can be readily accomplished, and the hand be brought to just coincide with the divisions and look well and be a means of accomplishing very accurate minute adjustments.

SUSPENSIONS.—Suspensions are of four kinds, cord, wire loop, knife edges and springs. Cords are generally of loosely twisted silk and are seldom found except in the older clocks of French or Swiss construction. They have been entirely displaced in the later makes of European manufactures by a double wire loop, in which the pendulum swings from a central eye in the loop, while the loop rocks upon a round stud by means of an eye at each end of the loop. The eyes should all be in planes parallel to the plane of oscillation of the pendulum, otherwise the bob will take an elliptical path instead of oscillating in a plane. They should also be large enough to roll without friction upon the stud and center of the loop, as any slipping or sliding of either will cause them to soon wear out, besides affecting the rate of the pendulum. Properly constructed loops will give practically no friction and make a very free suspension that will last as long as the clock is capable of keeping time, although it seems to be a very weak and flimsy method of construction at first sight. Care should be taken in such cases to keep the bob from turning when regulating the clock, or the effect upon the pendulum will be the same as if the eyes were not parallel.

Knife-edge suspensions are also rare now, having been displaced by the spring, as it was found the vibrations were too free and any change in power introduced a circular error (See Fig. 4) by making the long swings in longer time.

They are still to be found, however, and in repairing clocks containing them the following points should be observed: The upper surface of the stud on which the pendulum swings should carry the knife edge at its highest point, exactly central with the line of centers of the stud, so that when the pendulum hangs at rest the stud shall taper equally on both sides of the center, thus giving equal freedom to both sides of the swing. Care should be taken that the stud is firmly fixed, with the knife edge exactly at right angles to the movement, and also to the back of the case. The suspension stud and the block on the rod should be long enough to hold the pendulum firmly in line, as the angle in the top of the rod must be the sole means of keeping the pendulum swinging in plane. The student will also perceive the necessity of making the angle occupy the proper position on the rod, especially if the latter be flat. In repairing this suspension it is usual to make the plate, fasten it in place and then drill and file out the hole, as it is easier to get the angles exactly in this way than to complete the plate and then attempt to fasten it in the exact position in which it should be. After fastening the plates in position on the rod, two holes should be drilled, a small one at the apex of the angle (which must be exactly square and true with the rod), and a larger one below it large enough to pass the files easily. The larger hole can then be enlarged to the proper size, filing the angle at the top in such a way that the small hole first drilled forms the groove at the apex of the angle in which the knife edge of the stud shall work when it is completed. Knife-edge suspensions are unfitted for heavy pendulums, as the weight causes the knife edge to work into the groove and cut it, even if the latter be jeweled. Both the edge and groove should be hardened and polished.

Pendulum Suspension Springs.—Next in importance to the pendulum is its suspension spring. This spring should be just stiff enough to make the pendulum swing in all its vibrations in the same time; that is, if the pendulum at one time swung at the bottom of the jar 1¼ inch each side of the center, and at another time it swung only 1 inch each side, that the two should be made in exactly one second. The suspension springs are a point in the construction of a fine pendulum, that there has been very much theorizing on, but the experiments have never thus far exactly corroborated the theories and there are no definite rules to go by, but every maker holds to that plan and construction that gives his particular works the best results. A spring of sufficient strength to materially influence the swing of the pendulum is of course bad, as it necessitates more power to give the pendulum its proper motion and hence there is unnecessary wear on the pallets and escape wheel teeth, and too weak a spring is also bad, as it would not correct any inequalities in the time of swing and would in time break from overloading, as its granular structure would finally change, and rupture of the spring would follow. The office of a spring is to sustain the weight without detriment to strength and elasticity, and if so proportioned to the weight as to be just right, it will make the long and short swings of the pendulum of equal duration. When a pendulum hung by a cord or knife edge instead of a spring is regulated to mean time and swings just two inches at the bottom, any change in the power that swings the pendulum will increase its movement or decrease it, and in either case the rate will change, but with a proper spring the rate will be constant under like conditions. The action of the spring is this: In the long swings the spring, as it bends, lifts the pendulum bob up a little more than the arc of the normal circle in which it swings, and consequently when the bob descends, in going to the center of its swing, it falls a little quicker than it does when held by a cord, and this extra quick drop can be made to neutralize the extra time taken by the bob in making extra long swings. See Fig. 4.

This action is the isochronal action of the spring, the same that is attained in isochronal hair springs in watches.

As with the hairspring, it is quite necessary that the pendulum spring be accurately adjusted to isochronism and my advice to every jeweler is to thoroughly test his regulator, which can easily be done by changing the weight or motive power. If the test should prove the lack of isochronism he can adjust it by following these simple rules. Fig. 16 is the pendulum spring or leaf. If the short arcs should prove the slowest, make the spring a trifle thinner at B; if fastest, reduce the thickness of the spring at A. Continue the test until the long and short arcs are equal. In doing this care must be taken to thin each spring equally, if it is a double spring, and each edge equally, if a single spring, as if one side be left thicker than the other the pendulum will wabble.

The cause of a pendulum wabbling is that there must be something wrong with the suspension spring, or the bridge

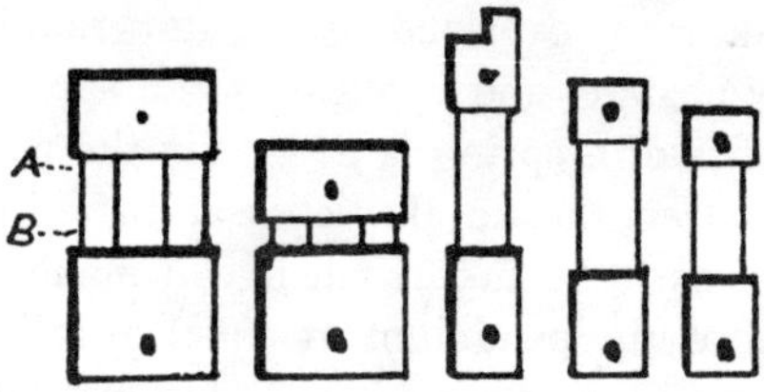

Fig. 16.

that holds the spring. If the suspension spring is bent or kinked, the pendulum will wabble; or if the spring should be of an unequal thickness it will have the same effect on the pendulum; but the main cause of the pendulum wabbling in American clocks is that the slot in the bridge that holds the spring, or the slot in the slide that works up and down on the spring (which is used to regulate the clock) is not parallel. When this slot is not parallel it pinches the spring, front or back, and allows it to vibrate more where it is the freest, causing the pendulum to wabble. We have found that by making these slots parallel the wabbling of the pendulum has ceased in most all cases. If the pallet staff is not at right angles to the crutch, wabbling may be caused by the oblique action of the crutch. This often happens when the movement is not set square in the case.

It occasionally happens in mantel clocks that the pendulum when brought to time is just too long for the case when too thick a spring is used. In such a case thinning the spring will require the bob to be raised a little and also give a better motion. If compelled to make a spring use a piece of mainspring about .007 thick and 3/8 wide for small pendulums and the same spring doubled for heavier pendulums, making the acting part of the spring about 1.5 inches long.

The suspension spring for a rather heavy pendulum is better divided, that is, two springs, held by two sets of clamps, and jointly acting as one spring. The length will be the same as to the acting part, and that part held at each end by the clamps may be 3/4 inch long; total length, 1.5 inches with 3/8 inch at each end held in the clamps. These clamps are best soldered on to the spring with very low flowing solder so as not to draw the temper of the spring, and then two rivets put through the whole, near the lower edge of the clamps. The object of securing the clamps so firmly is so that the spring may not bend beyond the edge of the clamps, as if this should take place the clock will be thrown off of its rate. After a time the rate would settle and become steady, but it only causes an extra period of regulating that does not occur when the clamps hold the spring immovable at this point. About in the center of each of the clamps, when soldered and riveted, is to be a hole bored for a pin, which pins the clamp into the bracket and holds the weight of the pendulum.

The width of this compound spring for a seconds' pendulum of average weight may be .60 inch, from outside to outside, each spring .15 inch wide. This will separate the springs .30 inch in the center. With this form of spring, the lower end of each spring being held in a pair of clamps, the clamps will have to be let into the top of the rod, and held in by a stout pin, or the pendulum finished with a hook which will fit the clamp. In letting the clamp into the rod, the clamp should just go into the mortise and be without side shake, but tilt each way from the center a little on the pin, so that when the pendulum is hung it may hang perpendicular, directly in the center of both springs. Also, the top pair of clamps should fit into a bracket without shake, and tilt a little on a pin, the same as the lower clamps. These two points, each moving a little, helps to take any side twist away, and allows the whole mechanism to swing in line with the center of gravity of the mass from end to end. With the parts well made, as described, the bob will swing in a straight line from side to side, and its path will be without any other motion except the one of slight curvature, due to being suspended by a fixed point at the upper clamp.

PENDULUM SUPPORTS.—Stability in the movement and in the suspension of the pendulum is very necessary in all forms of clocks for accurate time-keeping. The pendulum should be hung on a bracket attached to the back of the case (see Fig. 6), and not be subject to disturbance when the movement is cleaned. Also the movement should rest on two brackets attached to the bracket holding the pendulum and the whole be very firmly secured to the back board of the case. Screws should go through the foot-pieces of the brackets and into a stone or brick wall and be very firmly held against the wall just back of the brackets. Any instability in this part of a clock is very productive of poor rates. The bracket, to be in its best form, is made of cast iron, with a large foot carrying all three separate brackets, well screwed to a strong back-board and the whole secured to the masonry by bolts. Too much firmness cannot be attained, as a lack of it is a very great fault, and many a good clock is a very poor time-keeper, due to a lack of firmness in its supports and fastenings. The late Edward Howard used to make his astronomical clocks with a heavy cast iron back, to which the rest of the case was screwed, so that the pendulum should not swing the case. Any external influence that vibrates a wall or foundation on which a clock is placed, is a disturbing influence, but an instability in a clock's attachment to such supports is a greater one. Many pendulums swing the case in which they hang (from unstable setting up) and never get down to or maintain a satisfactory rate from that cause. This is also aggravated by the habit of placing grandfather clocks on stair landings or other places subject to jarring. The writer knows of several clocks which, after being cleaned, kept stopping until raised off the floor and bolted to the wall, when they at once took an excellent rate. The appearance of resting on the floor may be preserved, if desirable, by raising the clock only half an inch or so, just enough to free it from the floor.

CRUTCHES.—The impulse is transmitted to the pendulum from the pallet staff by means of a wire, or slender rod, fastened at its upper end to the pallet staff and having its lower end terminating in a fork (crutch), loop, or bent at right angles so as to work freely in a slot in the rod. It is also called the verge wire, owing to the fact that older writers and many of the older workmen called the pallet fork the verge, thus continuing the older nomenclature, although of necessity the verge disappeared when the crown wheel was discarded.

In order to avoid friction at this very important point, the centers of both axes of oscillation, that of the pallet arbor and that of the pendulum spring, where it bends, should be in a straight horizontal line. If, for instance, the center of suspension of the pendulum be higher, then the fork and the pendulum describe two different arcs of circles; that of the pendulum will be greater than that of the fork at their meeting point. If, however, the center of suspension of the pendulum be lower than that of the fork, they will also describe two different arcs, and that of the pendulum will be smaller than that of the fork at their point of meeting. This, as can be readily understood, will cause friction in the fork, the pendulum going up and down in it. This is prevented when, as stated before, the center of suspension of the pendulum is in the prolonged straight imaginary line going through the center of the pivots of the fork, which will cause the arcs described by the fork and the pendulum to be the same. It will be well understood from the foregoing that the pendulum should neither be suspended higher nor lower, nor to the left, nor to the right of the fork.

If the centers of motion do not coincide, as is often the case with cheap clocks with recoil escapements, any roughness of the pendulum rod where it slides on the crutch will stop the clock, and repairers should always see to it that this point is made as smooth as possible and be very slightly oiled when setting up. If putting in a new verge wire, the workman can always tell where to bend it to form the loop by noticing where the rod is worn and forming the loop so that it will reach the center of that old crutch or loop mark on the pendulum rod. If the verge wire is too long, it will give too great an arc to the pendulum if the latter is hung below the pallet arbor, as is generally the case with recoil escapements of the cheap clocks, and if it is too short there will not be sufficient power applied to the pendulum when the clock gets dirty and the oil dries, in which case the clock will stop before the spring runs down.

An important thing to look after when repairing is in the verge wire and loop (the slot the pendulum rod goes through). After the clock is set up and oiled, put it on a level shelf; have a special adjusted shelf for this level adjusting, one that is absolutely correct. Have the dial off. If the beat is off on one side, so that it bangs up heavily on one side of the escape wheel, bend the verge wire the same way. That will reverse the action and put it in beat. So far so good—but don't stop now. Just notice whether if that shelf were tipped forward or back, as perhaps your customer's may, that the pendulum should still hang plumb and free. Now if the top of your clock tips forward, the pendulum ball inclines to hang out toward the front. We will suppose you put two small wedges under the back of the case. Now notice in its hanging out whether the pendulum rod pinches or bears in the throat of the verge; or if it tips back, see if the rod hits the other end of the slot. This verge slot should be long enough, with the rod hanging in the middle when adjusted to beat on a level, to admit of the clock pitching forward or back a little without creating a friction on the ends of the slot. This little loop should be open just enough to be nice and free; if open too much, you will notice the pallet fork will make a little jump when carrying the ball over by hand. This is lost motion. If this little bend of wire is not parallel it may be opened enough inside, but if pitched forward a little it will bind in the narrowest part of the V and then the clock will stop. The clock beat and the tipping out or in of the clock case, causing a binding or bearing of the pendulum rod in this verge throat, does more towards stopping clocks just repaired than all other causes.

Putting in Beat.—To put a clock in beat, hang the clock in such a position that when the pendulum is at rest one tooth of the escape wheel will rest on the center of a pallet stone. Screwed on the case of the clock at the bottom of the pendulum there is, or should be, an index marked with degrees. Now, while the escape-wheel tooth is resting on the pallet, as explained above, the index of the pendulum should point to zero on the index. Move the pendulum until the tooth just escapes and note how many degrees beyond zero the pendulum point is. Say it escapes 2° to the left; now move the pendulum until the next tooth escapes—it should escape 2° to the right. But let us suppose it does not escape until the index of the pendulum registers 5° to the right of zero. In this case the rod attached to the pallets must be bent until the escape wheel teeth escape when the pendulum is moved an even number of degrees to the right and left of zero, when the clock will be in beat.

Close Rating with Shot.—Very close rating of a seconds' pendulum, accompanied by records in the book, may be got with the nut alone, but there is the inconvenience of stopping the clock to make an alteration. This may be avoided by having a small cup the size of a thimble or small pill box on the pendulum top. This can be lifted off and put back without disturbing the motion of the pendulum. In using it a number of small shot, selected of equal size, are put in, say 60, and the clock brought as nearly as possible to time by the nut. After a few days the cup may be emptied and put back, when on further trial the value of the 60 shot in seconds a day will be found. This value divided by 60 will give the value of a single shot, by knowing which very small alterations of rate may be made with a definite approach towards accuracy, and in much less time than by putting in or taking out one or more shot at random.

CHAPTER VI.

TORSION PENDULUMS FOR FOUR HUNDRED DAY CLOCKS.

As this pendulum is only found in the 400-day, or annual wind, or anniversary clocks (they are known by all of these names), it is best to describe the pendulum and movement together, as its relations to the work to be done may be more easily perceived.

Rotating pendulums of this kind—that is, in which the bob rotates by the twisting of the suspension rod or spring—will not bear comparison with vibrating pendulums for accurate time keeping. They are only used when a long period between windings is required. Small clocks to go for twelve months with one winding have the torsion pendulum ribbons of flat steel about six inches long, making 15 beats per minute. The time occupied in the beat of such a pendulum depends on the power of the suspending ribbon to resist twisting, and the weight and distance from the center of motion of the bob. In fact, the action of the bob and suspending ribbon is very analogous to that of a balance and balance spring.

In order to get good time from a clock of this character, it should be made with a dead-beat escapement. With such an escapement there is no motion of the escape wheel, after the tooth drops on the locking face of the pallet; the escape wheel is dead and does not move again until it starts to give the pallet impulse. This style of an escapement allows the pendulum as much freedom to vibrate as possible, as the fork in one form of this escapement may leave the pallet pin as soon as the latter strikes the guard pins, as in the ordinary lever escapement of a watch, and it will remain in that position until the return of the fork unlocks the escapement to receive another impulse. B, Fig. 17, represents the escape wheel; C, the pallet; E, pallet staff; D, the pallet pin rivetted on to the pallet staff E, which works in the slot or fork H; this fork is screwed fast to

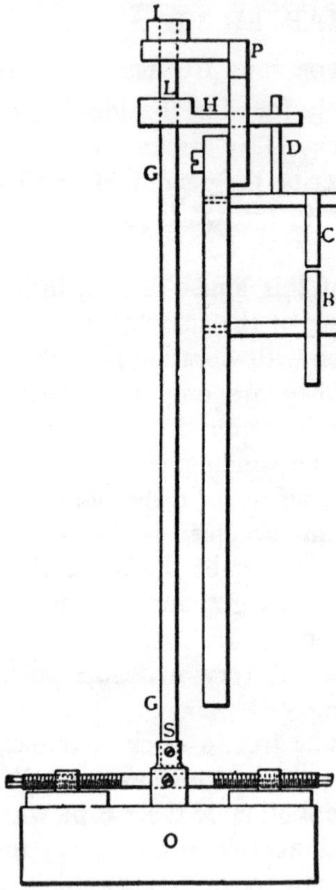

Fig. 17.

the spring. The spring G is made of a piece of flat steel wire and looks like a clock hairspring straightened out. G is fast to the collar I and rests on a seat screwed to the plate of the clock, as shown at P; the spring is also fastened to the pendulum ball O with screws; the ball makes about one and one-half revolutions each beat, which causes the spring to twist. It twists more at the point S than it does at L; as it twists at L it carries the fork with it, so that the latter vibrates from one side to the other, similar to a fork in a watch. This fork H carries the pin D, which is fast to the pallet staff E, far enough to allow the teeth to escape.

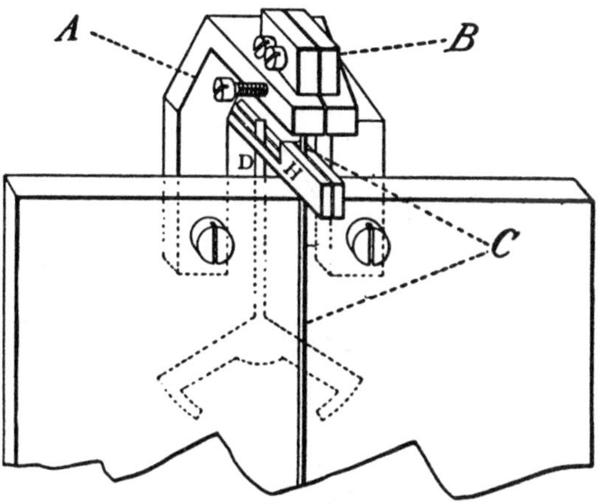

Fig. 18.

In the most common form of this escapement, see Fig. 18, the fork does not allow the pin D to leave the slot H, and the beat pins are absent, the pendulum not being as highly detached as in the form previously mentioned. In this case great care must be taken to have the edges of the slot, which slide on the pallet pin, smooth, parallel and properly beveled, so as not to bind on the pin. The pendulum ball makes from eight to sixteen vibrations a minute. Of course the number depends upon the train of the clock.

In suspending the pendulum it is necessary to verify the drop of the teeth of the escape wheel as follows: The pendulum is suspended and the locking position of the pallets marked, taking as a guiding point the long, regulating screw, which, fixed transversely in the support, serves for adjusting the small suspension block. An impulse of about a third of a turn is given to the pendulum while observing the escapement. If the oscillations of the pendulum, measured on the two sides, taking the locking point as the base, are symmetrical, the drop is also equal, and the rate of the clock regular and exact; but if the teeth of the escape wheel are unlocked sooner on one side than on the other, so that the pendulum in its swing passes beyond the symmetrical

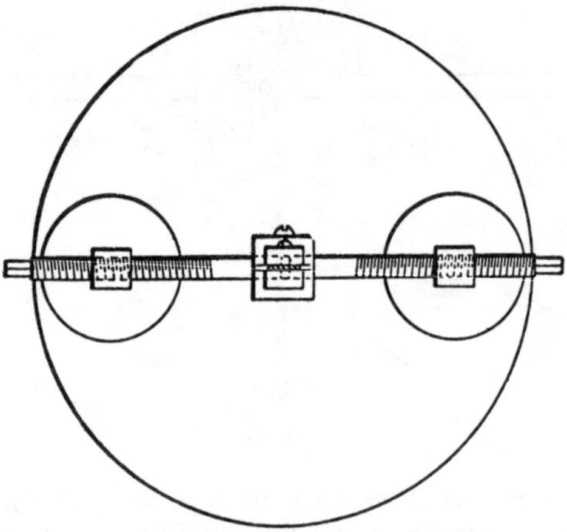

Fig. 19.

point on one of the pallets and does not reach it on the other, it is necessary to correct the unequal drop.

The suspension block B, Fig. 18, between the jaws of which the steel ribbon is pressed by two screws, has a lower cylindrical portion, which is fitted in a hole made in the seat, and is kept immovable by the screw A. If the vibration of the pendulum passes beyond the proper point on the left side, it is necessary to loosen A and turn the suspension block slightly to the right. If the deviation is produced in the opposite direction, it is necessary to turn it to the left. These corrections should be repeated until the drop of the escape wheel teeth on the pallets is exactly equal on the two sides. As the drop is often disturbed by the fact that the long thin steel ribbon has been twisted in cleaning, taking apart or handling by unskilled persons before coming to the watchmaker, it is desirable to test the escapement again, when the clock is put into position on the premises of the buyer.

The timing adjustment of the pendulum is effected with the aid of regulating weights, placed on the ball. By moving these away from the center by means of a right and left hand screw on the center of the disk (see Fig. 19),

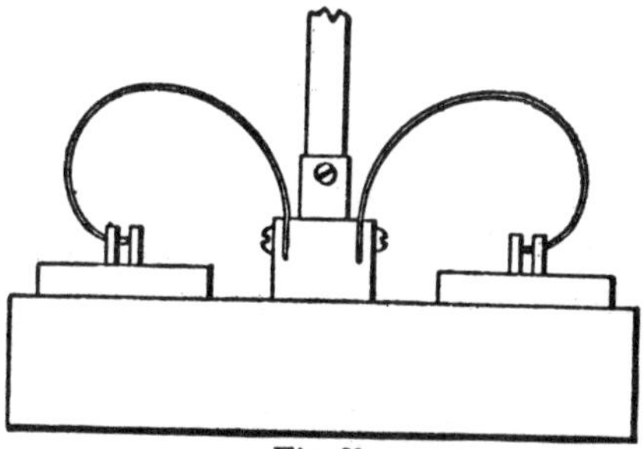

Fig. 20.

the centrifugal force is augmented, the oscillations of the pendulum slackened, and the clock goes slower. The contrary effect is produced if the weights are brought nearer the center. In one form of ball the shifting of the regulating weights is accomplished by a compensating spring of steel and brass like the rim of a watch balance, Fig. 20.

If necessary to replace the pendulum spring, the adjustment is commenced by shortening or lengthening the steel ribbon to a certain extent. For this purpose the end of the spring is allowed to project above the suspension block as a reserve until adjustment has been completed, when it may be cut off. If the space between the ball and the bottom of the case, or the bottom of the movement plates, does not allow of attaining this end, it is necessary to increase or decrease the weight of the disk, adding one or several plates of metal in a depression made in the under side of the ball, and removing the plates screwed to it, which are too light.

There are some peculiarities of the trains of these clocks. The cannon pinion is provided with a re-enforcing spring, serving as guide to the dial work, on which it exercises a sufficient pressure to assure precise working. The pressure of this spring is important, because if the dial work presses too hard on the pinion of the minute wheel, the latter engaging directly with the escape wheel, would transmit to the latter all the force employed in setting the hands. The teeth of the escape wheel would incur damage and the consequent irregularity or even stopping of the clock would naturally follow.

In order that it may run for so long a time, the motive force is transmitted through the train by the intervention of three supplementary wheels between the minute wheel and the barrel, in order to avoid the employment of too large a barrel; the third wheel is omitted; the motion work is geared immediately with the arbor of the escape wheel. It is evident that the system of the three intermediate wheels, of which we have spoken, requires for the motive force a barrel spring much stronger than that of ordinary clocks.

The points which we have noticed are of the most importance with reference to the repair and keeping in order of an annual clock. It very often happens that when the repairer does not understand these clocks, irregularities are sought for where they do not exist. The pivot holes are bushed and the depthings altered, when a more intelligent examination would show that the stopping, or the irregular rate of the clock, proceeds only from the condition of the escapement. Unless, however, they are perfectly adjusted a variation of five minutes a week is a close rate for them, and most of those in use will vary still more.

Annual clocks are enjoying an increased favor with the public; their good qualities allow confidence, the rate being quite regular when in proper order. They are suitable for offices; their silent running recommends them for the sick chamber, and the subdued elegance of their decoration causes the best of them to be valued ornaments in the home.

CHAPTER VII.

PECULIARITIES OF ANGULAR MEASUREMENT—HOW TO READ DRAWINGS.

We now come to a point at which, if we are to keep our pendulum vibrating, we must apply power to it, evenly, accurately and in small doses. In order to do this conveniently we must store up energy by raising a weight or winding a spring and allow the weight to fall or the spring to unwind very slowly, say in thirty hours or in eight days. This brings about the necessity of changing rotary motion to reciprocating motion, and the several devices for doing this are called "escapements" in horology, each being further designated by the names of their inventors, or by some peculiarity of the devices themselves; thus, the Graham is also called the dead beat escapement; Lepaute's is the pin wheel; Dennison's in its various forms is called the gravity; Hooke's is known as the recoil; Brocot's as the visible escapement, etc.

THE MECHANICAL ELEMENTS.—We shall understand this subject more clearly, perhaps, if we first separate these mechanical devices into their component parts and consider them, not as parts of clocks, but as various forms of levers, which they really are. This is perhaps the best place to consider the levers we are using to transmit the energy to the pendulum, as at this point we shall find a greater variety of forms of the lever than in any other place in the clock, and we shall have less difficulty in understanding the methods of calculating for time and power by a thorough preliminary understanding of leverage and the peculiarities of angular or circular motion.

If we take a bar, A, Fig. 21, and place under it a fulcrum, B, then by applying at C a given force, we shall be able to lift at D a weight whose amount will be governed by the relative distances of C and D from the fulcrum B.

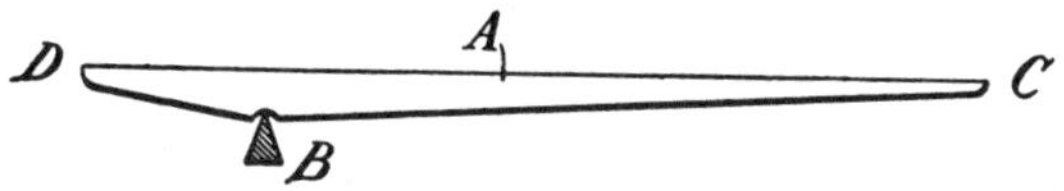

Fig. 21.

If the distance CB is four times that of BD, then a force of 10 pounds at C will lift 40 pounds at D, for one-fourth of the distance through which C moves, minus the power lost by friction. The reverse of this is also true; that is, it will take 40 pounds at D to exert a force of 10 pounds

Fig. 22.

at C and the 10 pounds would be lifted four times as far as the 40 pound weight was depressed.

If instead of a weight we substitute other levers, Fig. 22, the result would be the same, except that we should move the other levers until the ends which were in contact slipped apart.

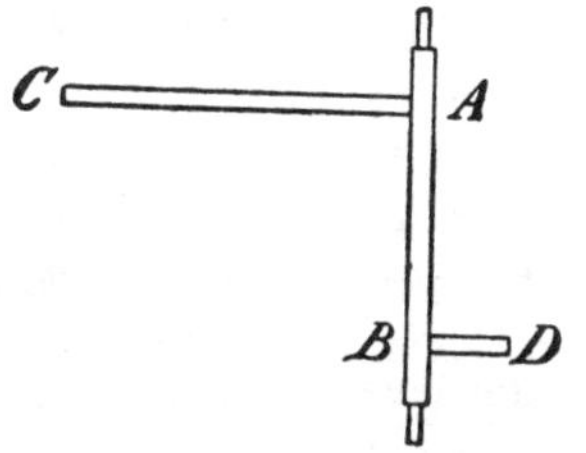

Fig. 23.

If we divide our lever and attach the long end to one portion of an axle, as at A, Fig. 23, and the short end to another part of it at B, the result will be the same as long as the proportions of the lever are not changed. It will still transmit power or impart motion according to the relative lengths of the two parts of the lever. The capacity of our levers, Fig. 22, will be limited by that point at which the ends of the levers will separate, because they are held at the points of the fulcrums and constrained to move in circles by the fulcrums. If we put more levers on the same axles, so spaced that another set will come into action as the first pair are disengaged, we can continue our transmission of power, Fig. 24; and if we follow this with still

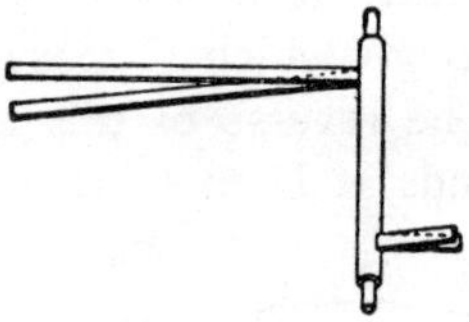

Fig. 24.

others until we can add no more for want of room we shall have wheels and pinions, the collection of short levers forming the pinion and the group of long levers forming the wheel, Fig. 25. Thus every wheel and pinion mounted together on an arbor are simply a collection of levers, each wheel tooth and its corresponding pinion leaf forming one lever. This explains why the force decreases and the motion increases in proportion to the relative lengths of the radii of the wheels and pinions, so that eight or ten turns of the barrel of a clock will run the escape wheel all day.

We now come to the verge or anchor, and here we have the same sort of lever in a different form; the verge wire, which presses on the pendulum rod and keeps it going is the long arm of our lever, but instead of many there is only one. The short arm of our lever is the pallet, and there are two of these. Therefore we have a form of lever in which there is one long arm and two short ones; but as the two are never acting at the same time they do not interfere with each other.

These systems of levers have another advantage, which is that one arm need not be on the opposite side of the ful-

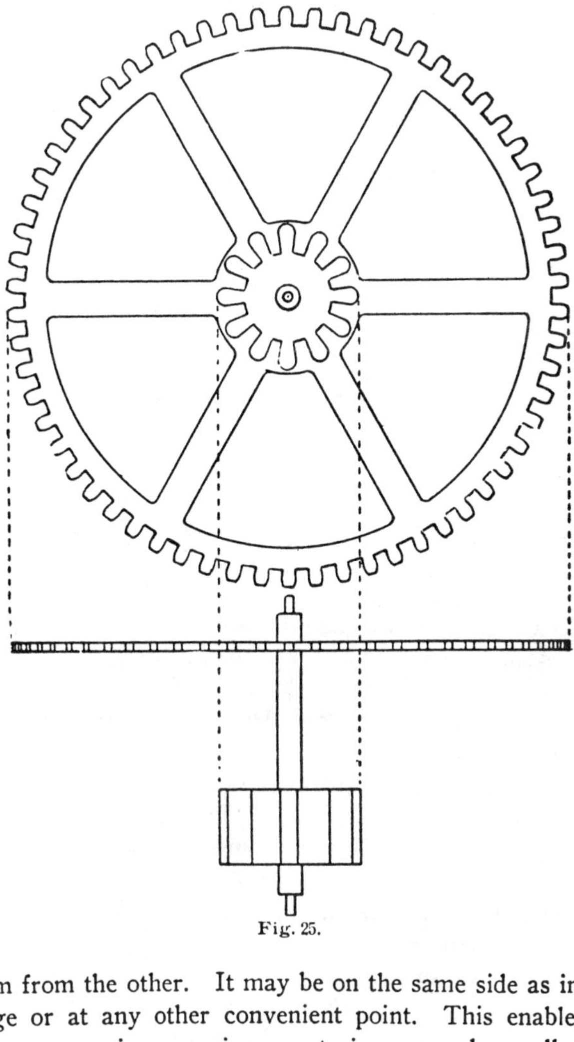

Fig. 25.

crum from the other. It may be on the same side as in the verge or at any other convenient point. This enables us to save space in arranging our trains, as such a collection of wheels and pinions is called, by placing them in any position which, on account of other facts, may seem desirable.

PECULIARITIES OF ANGULAR MOTION.—Now our collections of levers must move in certain directions in order to be serviceable and in order to describe these things properly, we must have names for these movements so that we can convey our thoughts to each other. Let us see how they move. They will not move vertically (up or down) or horizontally (sidewise), because we have taken great pains to prevent them from doing so by confining the central bars of our levers in a fixed position by making pivots on their ends and fitting them carefully into pivot holes in the plates, so that they can move only in one plane, and that movement must be in a circular direction in that predetermined plane. Consequently we must designate any movement in terms of the portions of a circle, *because that is the only way they can move.*

These portions of a circle are called angles, which is a general term meaning always a portion of a circle, measured from its center; this will perhaps be plainer if we consider that whenever we want to be specific in mentioning any particular size of angle we must speak of it in degrees, minutes and seconds, which are the names of the standard parts into which a circle is divided. Now in every circle, large or small, there are 360 degrees, because a degree is 1-360th part of a circle, and this measurement is always *from its center.* Consequently a degree, or any angle composed of a number of degrees, is always the same, because, being *measured from its center,* such measurements of any two circles will coincide as far as they go. If we draw two circles having their centers over each other at A, Fig. 26, and take a tenth part of each, we shall have $360° \div 10 = 36°$, which we shall mark out by drawing radial lines to the circumference of each circle, and we shall find this to be true; the radii of the smaller circle AB and AC will coincide with the radii AD and AE as far as they go. This is because each is the tenth part of its circle, *measured from its center.* Now that portion of the circumference of the circle BC will be smaller than the same portion DE of the larger circle, but each will be a *tenth part of its own circle,* although they are not the same size when measured by a rule on the circumference. This is a point which has bothered so many people when taking up the study of angular measurement that we have tried to make it absurdly

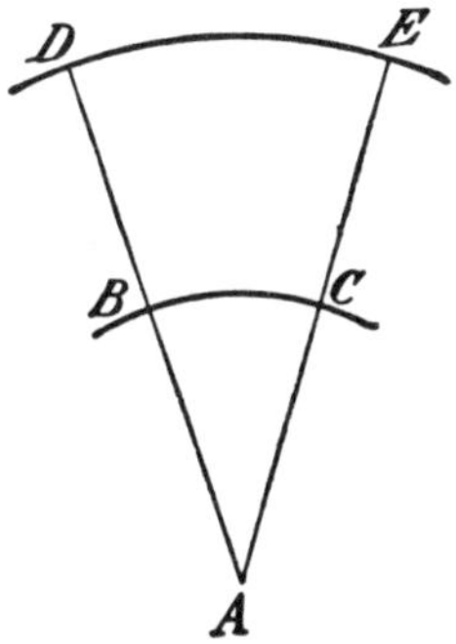

Fig. 26.

clear. An angle *never* means so many feet, inches or millimeters; it *always* means a *portion of a circle, measured from the center.*

There is one feature about these angular (or circular) measurements that is of great convenience, which is that as no definite size is mentioned, but only proportionate sizes, the description of the machine described need not be changed for any size desired, as it will fit all sizes. It thus becomes a flexible term, like the fraction "one-half," changing its size to suit the occasion. Thus, one-half of 300,000 bushels of wheat is 150,000 bushels; one-half of 10 bushels is 5 bushels; one-half of one bushel is two pecks; yet each is *one-half.* It is so with our angles.

There are some other terms which we shall do well to investigate before we leave the subject of angular measurements, which are the relations between the straight and curved lines we shall need to study in our drawings of the various escapements. A radius (plural radii) is a straight line drawn from the center of a circle to its circumference. A tangent is a straight line drawn outside the circumference, touching (but not cutting) it *at right angles* (90 degrees) to a radius drawn to the point of tangency (point where it touches the circumference). A general misunderstanding of this term (tangent) has done much to hinder a proper comprehension of the writers who have attempted to make clear the mysteries of the escapements. Its importance will be seen when we recollect that about the first thing we do in laying out an escapement is to draw tangents to the pitch circle of the escape wheel and plant our pallet center where these tangents intersect on the line of centers. They should always be drawn at *right angles* to the radii which mark the angles we choose for the working portion of our escape wheel. If properly drawn we shall find that the pallet arbor will then *locate itself* at the correct distance from the escape wheel center for any desired angle of escapement. We shall also discover that it will take a different center distance for every different angle and yet each different position will be the correct one for its angle, Fig. 27.

Because an angle is always the same, no matter how far from the center the radii defining it are carried, we are able to work conveniently with large drawing instruments on small drawings. Thus we can use an eight or ten inch protractor in laying off our angles, so as to get the degrees large enough to measure accurately, mark the degrees with dots on our paper and then draw our lines with a straight edge from the center towards the dots, as far as we wish to go. Thus we can lay off the angles on a one-inch escape wheel with a ten-inch protractor more easily and correctly than if we were using a smaller instrument.

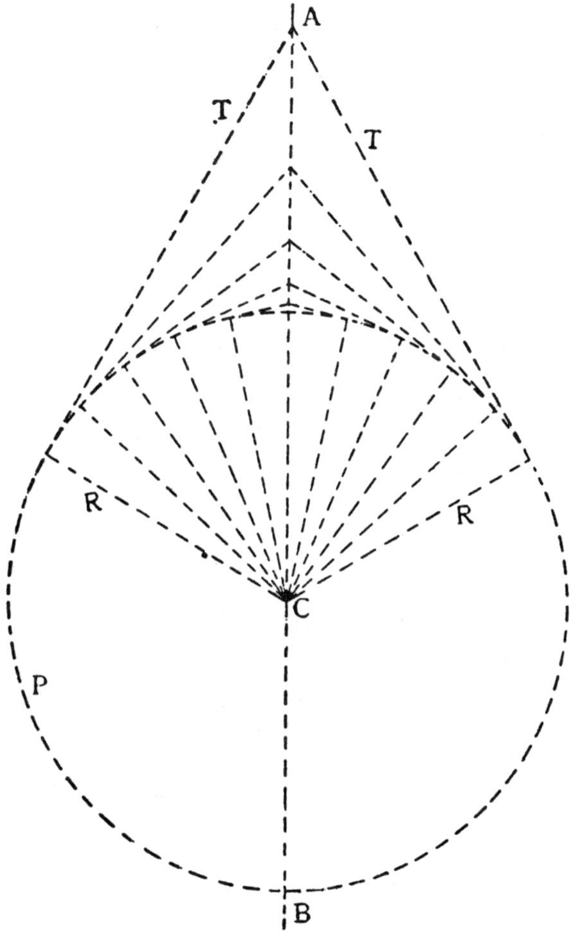

Fig, 27.

Another thing which will help us in understanding these drawings is that the *effective* length of a lever is its distance from the center to the working point, *measured in a straight line.* Thus in a pallet of a clock the distance of the pallets from the center of the pallet arbor is the *effective length* of that arm of the lever, no matter how it may be curved for ornament or for other reasons.

The lines and circles drawn to enable us to take the necessary measurements of angles and center distances are called "construction lines" and are generally dotted on the paper to enable us to distinguish them as lines for measurement only, while the lines which are intended to define the actual shapes of the pieces thus drawn are solid lines. By observing this distinction we are enabled to show the actual shapes of the objects and all their angular measurements clearly on the one drawing.

With these explanations the student should be able to read clearly and correctly the many drawings which follow, and we will now turn our attention to the escapements. In doing this we shall meet with a constant use of certain terms which have a peculiar and special meaning when applied to escapements.

THE LIFT is the amount of angular motion imparted to the verge or anchor by the teeth of the escape wheel pressing against the pallets and pushing first one and then the other out of the way, so that the escape wheel teeth may pass. According as the angular motion is more or less the "lift" is said to be greater or less; as this motion is circular, it must be expressed in degrees. The lifting planes are those surfaces which produce this motion; in clocks with pendulums the lifting planes are generally on the pallets, being those hard and smoothly polished surfaces over which the points of the escape wheel teeth slide in escaping. In lever escapements the lifting planes are frequently on the escape wheel, the pallets being merely round pins. Such an escape wheel is said to have club teeth, as distinguished from the pointed teeth used when the lifting planes are on the pallets. In the cylinder escapement the lifting planes are on the escape wheel; they are curved instead of being straight; and there is but one pallet, which is on the lip of the cylinder. In the forms of lever escapement used in watches and some clocks the lift is divided, part of the lifting planes being also on the pallets; in this case both sets of planes are shorter than if they were entirely on one or the other, but they must be long enough so that combined they will produce the requisite amount of angular motion of the pallets, so as to give the requisite impulse to the pendulum or balance.

THE DROP is the amount of circular motion, measured in degrees, which the escape wheel has from the instant the tooth escapes from one pallet to that point at which it is stopped by the other pallet catching another tooth. During this period the train is running down without imparting any power to the pendulum or balance, hence the drop is entirely lost motion. We must have it, however, as it requires some time for the other pallet to move far enough within the pitch circle of the escape wheel to safely catch and stop the next tooth under all circumstances. It is the freedom and safety of the working plan of our escapement, but it is advisable to keep the drop as small as is possible with safe locking.

THE LOCK is also angular motion and is measured in degrees from the center of the pallet arbor. It is the distance which the pallet has moved inside of the pitch circle of the escape wheel before being struck by the escape wheel tooth. It is measured from the edge of the lifting plane to the point of the tooth where it rests on the locking face of the pallet. A safe lock is necessary in order to prevent the points of the escape wheel teeth butting against the lifting planes, stopping the clock and injuring the teeth. We want to point out that from the instant of escaping to the instant of locking we have the two parts of our escapement propelled by different and entirely separate forces and moving at different speeds. The pallets, after having given impulse to the pendulum, are controlled by the pendulum and moved by it; in the case of a heavy pendulum ball at the end of a 40-inch lever, this control is very steady, powerful and quite slow. The escape wheel, the lightest and fastest in the train, is driven by the weight or spring and moves independently of the pallets during the drop, so that safe locking is important. It should never be too deep, as it would increase the swing of the pendulum too much; this is especially true with short and light pendulums and strong mainsprings.

THE RUN.—After locking the pallet continues to move inward towards the escape wheel center as the pendulum continues its course, and the amount of this motion, measured in degrees from the center of the pallet arbor, is called the run.

When the escapement is properly adjusted the lifting planes are of the same length on both pallets, when they are measured in degrees of motion given to the pallet arbor. They may or may not be equal in *length* when measured by a rule on the faces of the pallets. There should also be an equal and safe lock on each pallet, as measured in degrees of movement of the pallet arbor. The run should also be equal.

The reason why one lifting plane may be longer than the other and still give the same amount of lift is that some escapements are constructed with unequal lockings, so that one radius is longer than the other, and this, as we explained at length in treating of angles, Fig. 26, would make a difference in the length of arc traversed by the longer arm for the same angle of motion.

CHAPTER VIII.

THE GRAHAM OR DEAD BEAT ESCAPEMENT.

This escapement is so called because the escape wheel remains "dead" (motionless) during the periods between the impulses given to the pendulum. It is the original or predecessor of the well known detached lever escapement so common in watches, and it is surprising how many watchmakers who are fairly well posted on the latter form exhibit a surprising ignorance of this escapement as used in clocks. It has like the latter a "lock," "lift" and "run"; the only difference being that it has no "draw," the control by the verge wire rendering the draw unnecessary.

It may be made to embrace any number of teeth of the escape wheel, but, owing to the peculiarities of angular motion referred to in the last chapter, see Fig. 26, B C, D E, the increased arcs traveled as the pallet arms lengthen introduce elements of friction which counterbalance and in some cases exceed the advantage gained by increasing the length of the lever used to propel the pendulum. Similarly, the too short armed escapements were found to cause increased difficulty from faulty fitting of the pivots and their holes, and other errors of workmanship, which errors could not be reduced in the same proportion as the arms were shortened, so that it has been determined by practice that a pallet embracing ninety degrees, or one-fourth of the circumference of the escape wheel, offers perhaps the best escapement of this nature that can be made. Therefore the factories generally now make them in this way. But as many clocks are coming in for repair with greater or less arcs of escapement and the repairers must fix them satisfactorily, we will begin at the beginning by explaining how to make the escapement of any angle whatever, from one tooth up to 140 degrees, or nearly half of the escape wheel.

It is quite a common thing for some workmen to imagine that in making an escapement, the pallets ought to take in a given number of teeth, and that the number which they suppose to be right must not be departed from; but there seems to be no rule that necessarily prescribes any number of teeth to be used arbitrarily. The nearer that the center of motion of the pallets is to the center of the escape wheel, the less will be the number of teeth that will be embraced by the pallets. Fig. 28 is an illustration of the distances between the center of motion of the pallets and the center of the wheel required for 3, 5, 7, 9 and 11 teeth in a wheel of the same size as the circle; but although we have adopted these numbers so as to make a symmetrical diagram, any other numbers that may be desirable can be used with equal propriety. All that is necessary to be done to find the proper center of motion of the pallets is first to determine the number of teeth that are to be embraced, and draw lines (radii) from the points of the outside ones of the number to the center of the wheel, and at right angles to these lines draw other two lines (tangents), and the point where they intersect each other on the line of centers will be the center of motion of the pallets.

It will be seen by the diagram, Fig. 28, that by this method the distance between the centers of motion of the pallets and that of the scape-wheel *takes care of itself* for a given number of teeth and that it is greater when eleven and one-half teeth are to be embraced than for eight or for a less number. These short pallet arms are imagined by some workmen to be objectionable, on the supposition that it will take a heavier weight to drive the clock; but it can easily be shown that this objection is altogether imaginary. Now, bearing in mind the principles of leverage, if the distance between the pallets and escape wheel centers is very

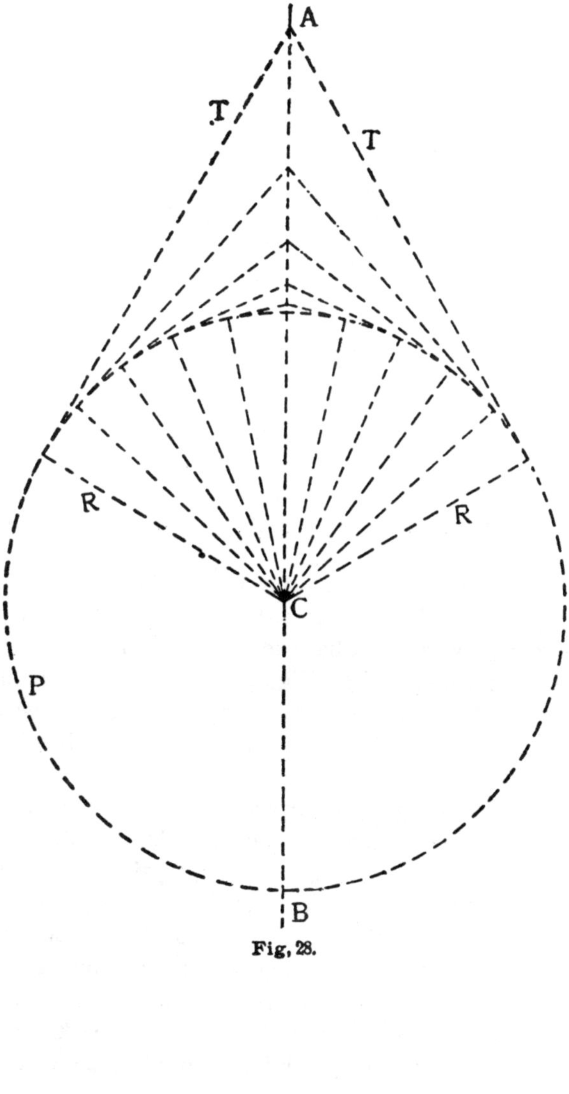

Fig, 28.

long, as in Graham's plan, in which the pallets embraced 138° of the escape wheel, the value of the impulse received from the scape-wheel and communicated through the pallets to the pendulum is no doubt greater with a proper length of verge wire, for, the lifting planes being longer, the leverage is applied to the pendulum for a longer arc of its vibration, yet we must not suppose that from this fact the clock will go

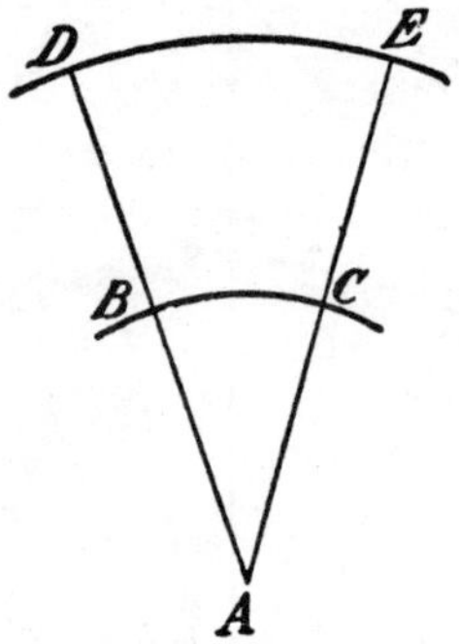

Fig. 29. Note the difference in length of arc for the same angle.

with less weight, for it is easy to see that the longer the pallet-arms are the greater will be the distance the teeth of the escape wheel will have to move (run) on the circular part of the pallets. See Fig. 29. The extra amount of friction, and the consequent extra amount of resistance offered to the pendulum, caused by the extra distance the points of the teeth run on the circular locking planes of the pallets and back again, destroys all the value of the extra amount of impulse given to the pendulum in the first instance by means of the long arms of the pallets. The escape wheel tooth resting on the locking plane of the pallet is quite variable in its effective action, and since it rests on the pallet during a part of each swing of the pendulum and the pendulum is called on to move the pallet back and forth under the tooth, any change in the friction between the tooth and pallet is felt by the pendulum and when the clock gets dirty and the friction between the tooth and pallet is increased, the rate of the clock gets slow, as the friction holds the pendulum from moving as fast as it would without friction. Now, as this friction increases by dirt and thickening of the oil, all these forms of escapements are subject to changes and so change the clock's rate. An increase of the driving weight, or force of the mainspring, of clocks with dead-beat escapements always tends to make their rate slow, from the action mentioned.

It is for this reason that moderately short arms are used in clocks having dead-beat escapements of modern construction. Most of the first-class modern makers of astronomical clocks only embrace seven and one-half teeth, on a 30-tooth wheel, with the centers of motion of the pallets and scape-wheel proportionately nearer, as it can be mathematicallv demonstrated that with the pallets embracing an arc of 90° the application of the power to the pendulum is at right angles to the rod and therefore is most effective.

To Draw the Escapement.—In order to make the matter clearer we show in Fig. 30 the successive stages of drawing an escapement and also the completed work in Figs. 32 and 33 embracing different numbers of teeth. Draw a line, A B, Fig. 30, to serve as a basis for measurements. With a compass draw from some point C on this line a circle to represent the diameter of our escape wheel. Now we shall require to know how many teeth there will be in our escape wheel. There may be 60, 40, 33, 32, 30, or any other number we desire to give it; seconds pendulums generally have 30 teeth in this wheel, because this allows the second hand to be mounted directly on the escape wheel arbor and thus avoids complications. We divide the number of degrees in a circle (360) by the number of teeth we have selected, say 30. $360 \div 30 = 12°$ for each tooth and space. One-fourth of 360° equals 90° and one-fourth of 30 teeth equals seven and one-half teeth; each tooth equaling 12

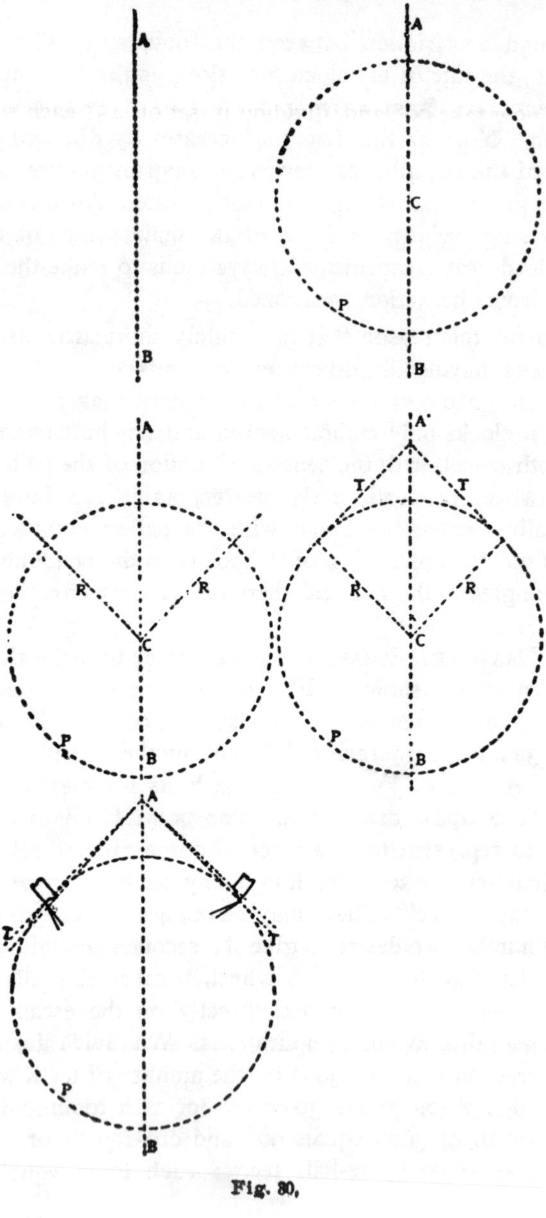

Fig. 80.

degrees, we have $12 \times 7 = 84°$, which gives us six degrees for drop, to ensure the safety of our actions.

We now take 90° and, dividing it, set off 45° each side of our center line and draw radii, R, from the center to the circumference of our circle; this marks the beginnings of our pallets. Now to find our pallet center distance we draw tangents, T (at right angles), from the ends of these radii toward the line of centers. The point where they intersect on the line of centers is the pallet center.

Now we must determine how much motion we are going to give our pendulum, so that we can give the proper lift to our pallets. Four degrees of swing is usual for a seconds pendulum, so we will take four degrees and, dividing it, give two degrees of lift to each pallet. To do this we draw a line two degrees *inside* the tangent, T (towards the escape wheel center), from our pallet center on the entering pallet side and another line from the pallet center two degrees *outside* of the tangent, T, on the exit pallet side. Next, from the pallet center we draw arcs of circles cutting the tangents, T, and the radii, R, where they intersect; this gives us the locking planes on which the teeth of the escape wheel "run" (slide) during the excursions of the pendulum, if the escapement is to have unequal lockings; if the lockings are to be equidistant (if the pallet arms are to be of equal length) the arc for the entering pallet is drawn three degrees below (outside) the radius, R, while that on the exit pallet is drawn three degrees *above* (inside) the exit radius. Finally the lifting planes are drawn from the intersection of the arcs of circles struck from the pallet center with their tangents, T, to the lines, marking the limits of the lift, two degrees away. These lifting planes should be at an angle of 60° from the radii, R, and as a tangent is always at right angles (90°) to its radius, they are consequently at 30° to the tangents running to the pallet center. Thus we can measure these angles from either the escape wheel or the pallet center, as may be most convenient.

When making a new pallet fork, it is most convenient to mark out the lifting planes on the steel at 30° from the tangents, T, as we then do not have to bother with the escape wheel further than to get its center distance and the degrees of arc the lifting planes are to embrace. The workman who is not familiar with this rule is apt to have his ideas upset at first by the angles of inclination toward the center line which the lifting planes will take for different center distances, as owing to the fact that the tangents meet on the center line at different angles for different distances, the lifting planes assume different positions with regard to the center line and he may think that they do not "look right." They are right, however, when drawn at 30° to their tangents. Fig. 31 shows several pallets with different arcs arranged in line for purposes of comparison, each being drawn according to the above rule, as measurements with a protractor will show.

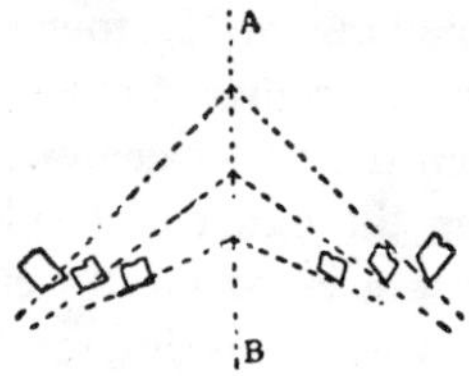

Fig. 31.

We have now arrived at the complete escapement, having finished our pallets. We have, however, nothing to hold them in position; they must be rigidly held in position with regard to each other and the escape wheel, consequently we will make a yoke to connect them to the pallet arbor out of the same steel, giving it any desired shape that will not interfere with the working of the clock. Two of the most usual forms are shown at Figs. 32 and 33.

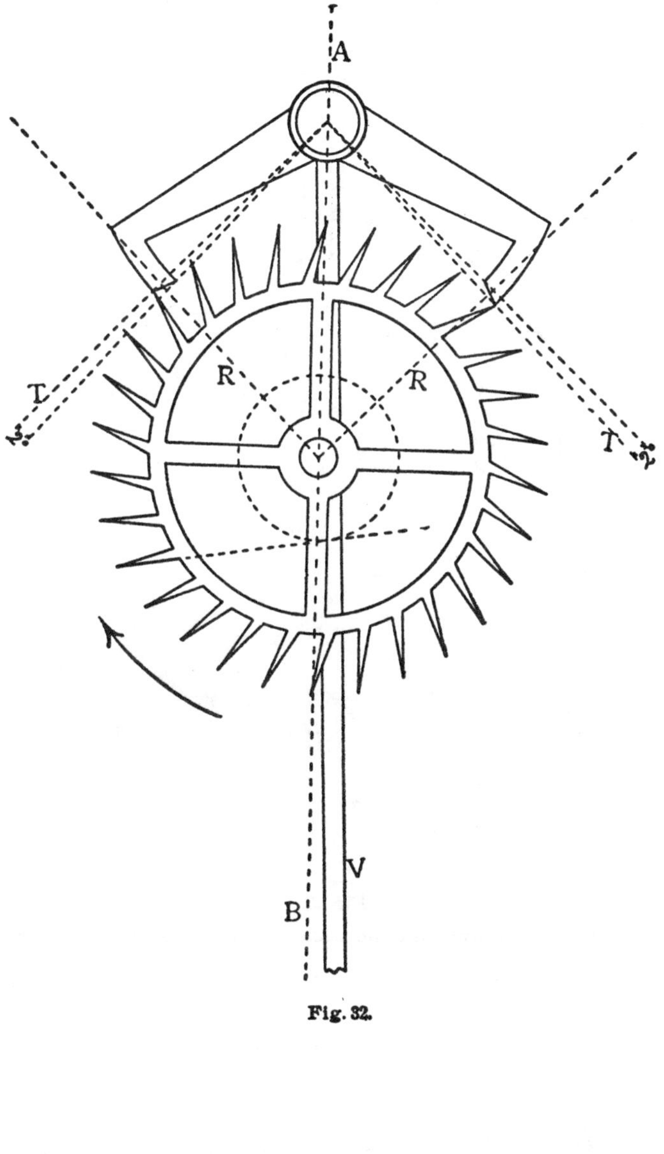

Fig. 32.

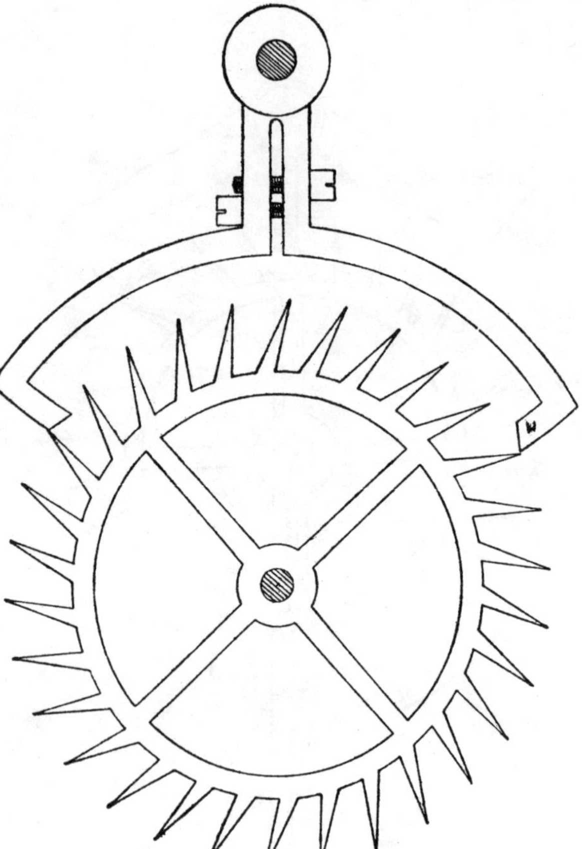

Fig. 33.

Let us see how this rule will work in repairs. Suppose we have a clock brought in with the pallet fork missing, and that the movement is one of those in which the pallet arbor is held by adjustable cocks which have been misplaced or lost, so that we don't know the center distance of the pallet arbor and escape wheel. We shall have to make a new part.

Measure the escape wheel, getting its diameter carefully, take half of this as a radius, and mark out the circle with a fine needle point on some copper, brass or sheet steel, drawing the escapement as detailed in Figs. 30 and 32. Then measure carefully the angles made by the tangents with the center line; take the steel which is to be used in making the pallets and fork; draw on it a center line; lay off the tangents and the lift lines; draw the locking arcs and the lifting planes carefully from the tangents and give the rest of the fork a symmetrical shape. Use needle points to draw with and have your protractor large enough to measure your angles accurately. Then drill or saw out and file to your lines, except on the locking and lifting planes; leave these large enough to stand grinding or polishing after hardening. Harden; draw to a straw color and polish the planes. Your verge will fit if it has not warped in hardening. If this is the case, soften the center, keeping the heat away from the pallets, and bend or twist the arms until the verge will fit the drawing, when laid on top of it. In grinding the pallets the fork should be mounted on its arbor and the latter held between the centers of a rounding up tool while the grinding is done by a lap in the lathe. This insures that the planes will be parallel to the pallet arbor and hence square with the escape wheel teeth, so that they will not create an end thrust on either escape or pallet arbor. It is also the quickest, easiest and most reliable way of doing the job. When clocks come in with the pallets badly cut; soften the center of the fork, place the ends between the jaws of a vise, squeeze enough to bring them

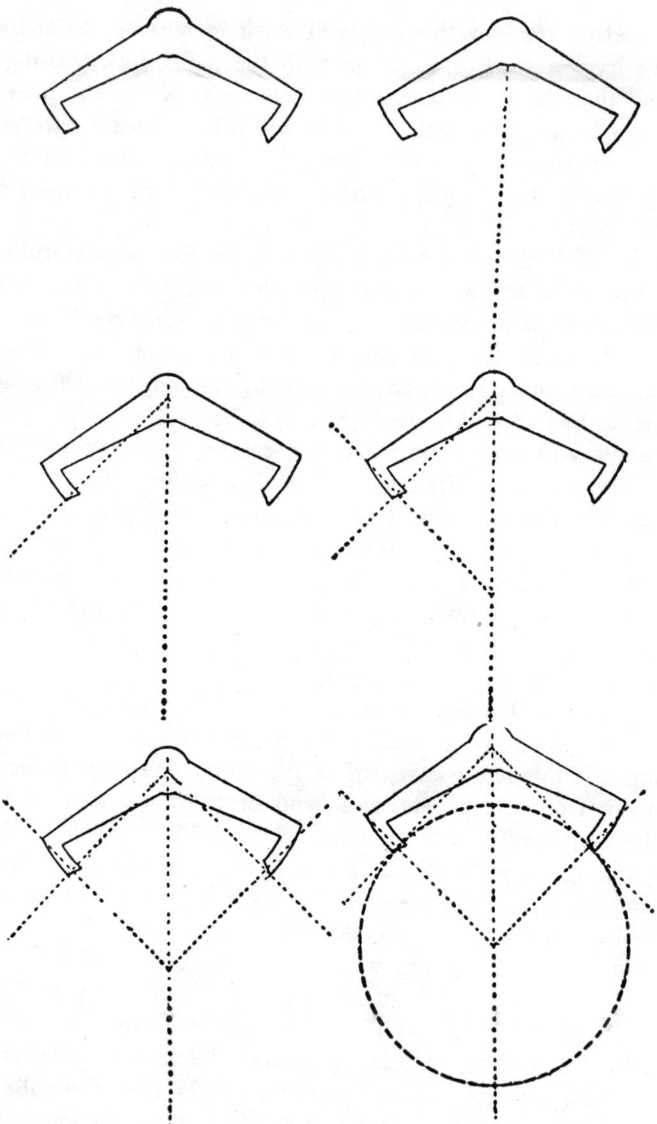

Fig. 34. Drawing escape wheel to fit a tracing from a pallet fork.

closer, mount in the rounding up tool and lap off the cut planes until they are smooth and stand at the proper angle; then polish. This is done quickly.

Can we work the rule backwards? Suppose we get a clock in which we have the pallet arbor adjustable as before, and we have the pallet fork all in good shape, but we have lost the escape wheel, or it has been butchered by somebody before coming to us, so that a new one is required.

Take off the pallet fork; lay it on a sheet of brass and trace around it carefully with a needle point, Fig. 34. Mark the center carefully at the pallet arbor hole and measure carefully the distance between the pallets and mark that center. Draw a center line cutting these centers and extending beyond. Now draw the tangent from the beginning of the entering pallet (as shown by the tracing on our brass) to the pallet center; do the same with the exit pallet. Now take a metal square and place it on one of the tangents exactly, with the end at the beginning of the entering pallet; trace a line cutting the line of centers and we have the radius of our escape wheel. Trace a circle from the intersection of the radius and the center line and we have the circumference of our escape wheel. This circle should also cut the intersection of the tangent and radius on the other side if it is drawn correctly; if it does not do this an error has been made in the drawing.

Having found the diameter and circumference of our escape wheel it may be sawed out and mounted for wheel cutting; or, if we have no wheel cutter and must make the wheel, we must draw it on the brass by hand with a fine needle point before proceeding to saw it out by hand, Fig. 35. Say that the wheel is to have thirty-two teeth, which is a common number; then $360° \div 32 = 11\frac{1}{4}°$ as the space between the points of our teeth. Take a large protractor, one with the degrees large enough to be divided (I use a ten-inch); place its center on the center of our escape wheel, set off $11\frac{1}{4}°$ and mark them on the brass with the needle

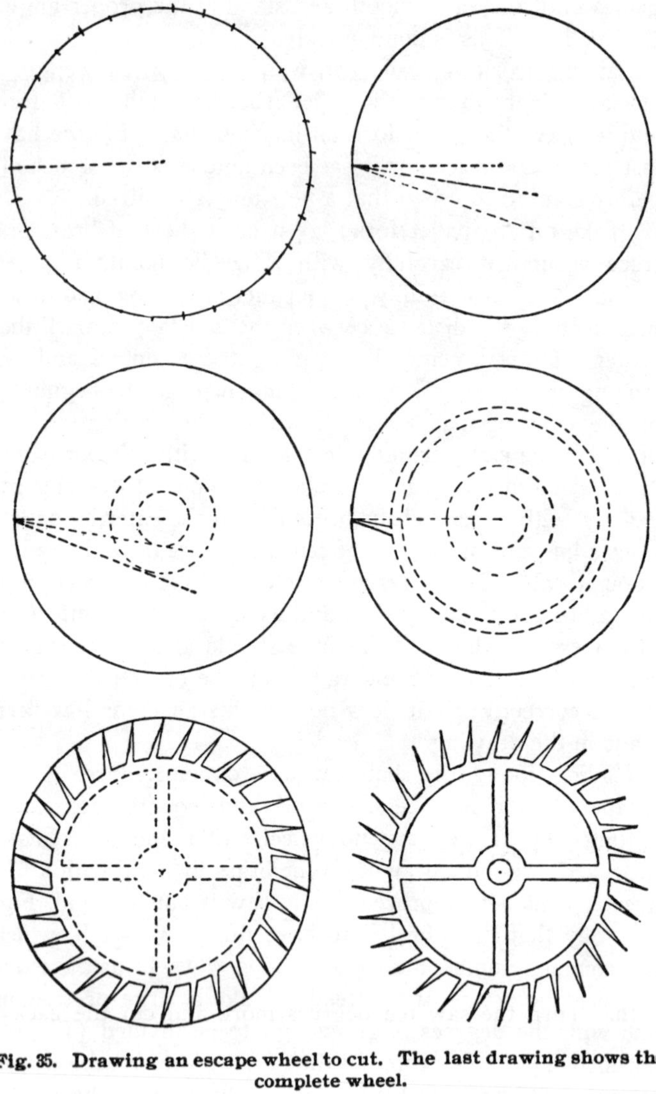

Fig. 35. Drawing an escape wheel to cut. The last drawing shows the complete wheel.

point, at the edge of the protractor. Then take a straight edge and draw a radius from the center to the circumference; change the straight edge to the other mark and mark the point where it crosses the circumference; set your dividers accurately by this mark and space off the teeth on your circumference. If they are set at eleven degrees and fifteen minutes they will come out exactly at the end. Now take your protractor and with its center at the junction of the radius and circumference set off ten degrees and draw a line past the center of the wheel; set off twenty degrees and draw another line the same way. From the center of the escape wheel draw two circles just touching these lines. Outside of these draw two circles defining the inner and outer edges of the rim of the wheel. With the straight edge just touching the inner circle draw in the fronts of the teeth; these will all be set at ten degrees from a radius, so that only the extreme points will touch the locking planes of the pallets and thus reduce the friction during the run. The backs of the teeth are marked out in the same way from the twenty-degree circle. The hub is made to coincide with the ten-degree circle; the spokes are traced in and we are ready to begin sawing out.

If the workman has a wheel cutter the job is much simpler. A piece of brass is mounted on a cement brass with soft solder, faced off, centered and the pitch circle, inner and outer edges of the rim and the hub are traced with the T-rest and graver. The extra metal is then cut away and a suitable index placed on the spindle and locked. The wheel cutter is set up with a fine toothed, smooth cutting saw on the spindle, horizontal, with its upper edge at the line of centers of the lathe. It is then run out to the circumference of the wheel, turned upwards ten degrees and the wheel cut around, Fig. 36. This makes the fronts of the teeth. Turn the saw ten degrees more and cut the backs of the teeth. Then turn the saw so that it will reach from the front of one tooth to the root of the back of the next

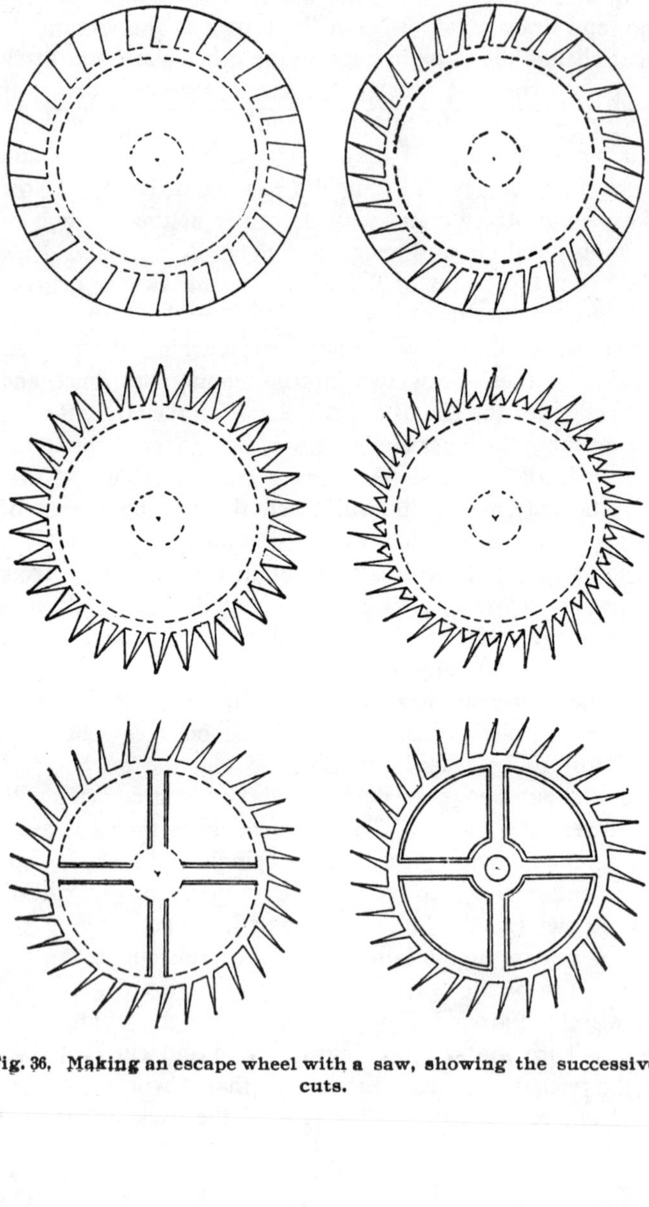

Fig. 36. Making an escape wheel with a saw, showing the successive cuts.

one, without touching either tooth, and cut round again; this cuts out a triangular piece of waste metal between the teeth. Turn the saw again so that it reaches from the bottom of the front of a tooth to the top of the back of the next one and cut around again, thus removing another portion of the waste metal, and leaving only a small triangle between the teeth. Lower the saw its own thickness and cut around the wheel again, repeating the operation until the waste metal is all removed and you have a smooth circular rim between the teeth, Fig. 36.

Set the saw horizontally at the lathe center; raise it one-half the thickness of the spokes; set the index pin of the lathe head firmly at O; feed in the saw the thickness of the wheel and make straight cuts across from the circle of the inner rim to the circle marking the hub, but not cutting either; set the index pin at 30 and repeat; next lower your saw and cut the other side of the spokes the same way.

Next you can mount a lap in place of the saw and smooth the fronts and backs of the teeth and if you have a rather thick disc the outer edge of the rim, between the teeth, may also be smoothed.

If you have a good strong pivot polisher, mount a triangular end mill in the spindle, lock the yoke, and cut the arcs of circles of the hub and rim from edge to edge of the spokes, feeding carefully against the mill with the hand on the lathe pulley.

Put on your jeweling tailstock and open the wheel to fit the pinion, collet, or arbor, if there is no collet.

You now have the wheel all done, except facing the side that was soldered to the cement brass and trimming up the corners of the spokes at the rim and hub, and you have got it round, true and correct in much less time than you could have done in any other way, while an immense amount of work with the file and eye-glass has been avoided. It is true because it was soldered in position at the beginning and has not been removed until finished.

Sometimes what are known from their appearance as club-shaped teeth are used in the wheels of Graham's escapements. Pendulums receive their impulse from escapements made in this manner partly from the lifting planes on the pallets, and partly from the planes on the scape-wheel. The advantage gained by this method is, that wheels made in this way will work with the least possible drop, and consequently, power is saved; but the power saved is thrown away again in the increased friction of the planes of the wheel against those of the pallets, which is considerably more than when plain-pointed teeth are used on the escape wheel.

Clock pallets are usually made of steel, and on the finer classes of work jewels are often set into them to prevent the oil from drying, after the same fashion as jewels are placed in steel pallets in a lever watch; but it is obvious that stone pallets made in this way have to be finished with polishers held in the hand, and that, except in factories, they cannot be made so perfectly regular, especially that pallet that is struck downwards, as the particular action of a fine Graham escapement requires. When great accuracy is required, the pallets are usually made of separate pieces, and the acting circles ground and polished on laps, running in a lathe. This method of constructing pallets also allows a means of adjustment which in some particular instances is very convenient.

There is also a plan of making jeweled pallets adjustable, which is practiced on fine work, such as astronomical and master clocks. The pallet fork consists of two pieces of thin, hard, sheet brass, cut out in the usual form and two mounted on one arbor. Circular grooves are cut in the sides of both plates, at the proper distance, and of the proper size to receive the jewels which are the acting parts of the pallets. When jewels cannot be made of the desired size, pallets of steel are made, and the jewels are then set into the steel large enough for the teeth of the wheel to act

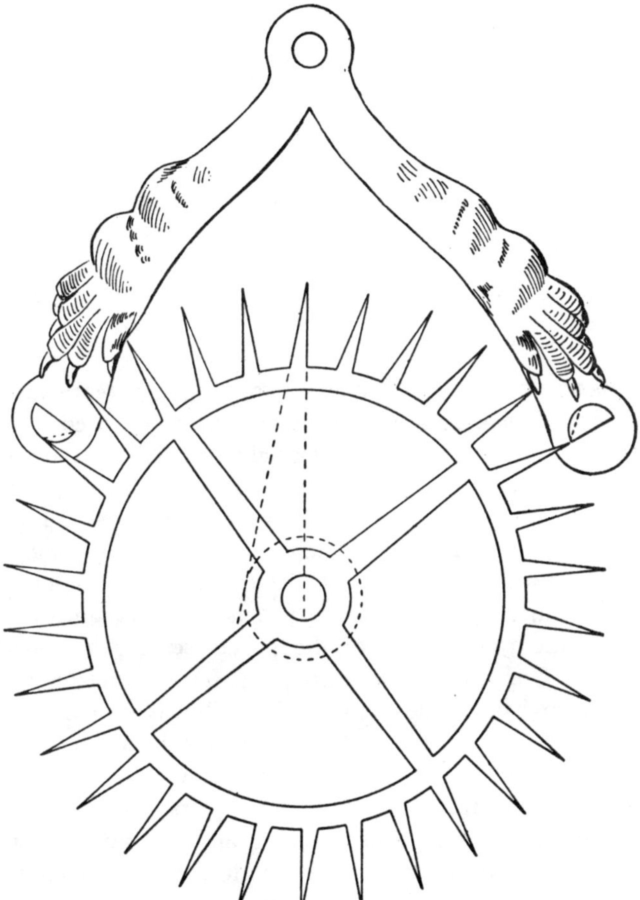

Fig. 37. Brocot's visible escapement, escaping over 120° with pointed teeth. Dotted lines on pallets show where they are cut to avoid stopping.

upon. The two parts of the fork are fastened at a given distance apart, and the jewels, or pieces of steel, go in between them, and, after they have been adjusted to the proper position, are fastened by screws that pull the frames close together and press against the edges of the jewels. Pallets made in this manner have a very elegant appearance. Another method is to have only one frame, and to have it thick enough, where the jewels have to be set in, to allow a groove to be cut in its side as deep as the jewels (or the pieces of steel that hold the jewels) are broad, and which are held in their proper position by screws. This system of jeweling pallets is frequently adopted by the makers of fine mantel clocks.

BROCOT'S VISIBLE ESCAPEMENT.—Fig. 37 represents a system of making and jeweling pallets much used by the French in their small work, especially in visible escapements. The acting parts of the pallets are simply cylinders, generally of colored stones, usually garnets, one-half of each cylinder being cut away. These cylinders extend some distance from the front of the pallet frame, and work into the escape wheel the same as the pallets of a Graham escapement—the round parts of the pallets serving as impulse planes. The neck of the brass pallet frame is cut up in the center, and the width between the pallets is sometimes adjusted by a screw, sometimes by bending the arms.

Clock movements with this escapement, of a careful construction, will frequently come for repairs, accompanied by the complaint of constant stopping and that no attempt at closely regulating can succeed with them, although they appear to have no visible disturbing cause. In such cases the depthing of the escapement is generally wrong. With proper depthing the point of the escape wheel tooth should drop on the center or a little beyond the center of the pallet stone. If it is set in this way the clock will stop when wound, especially if it has a strong spring, as the light

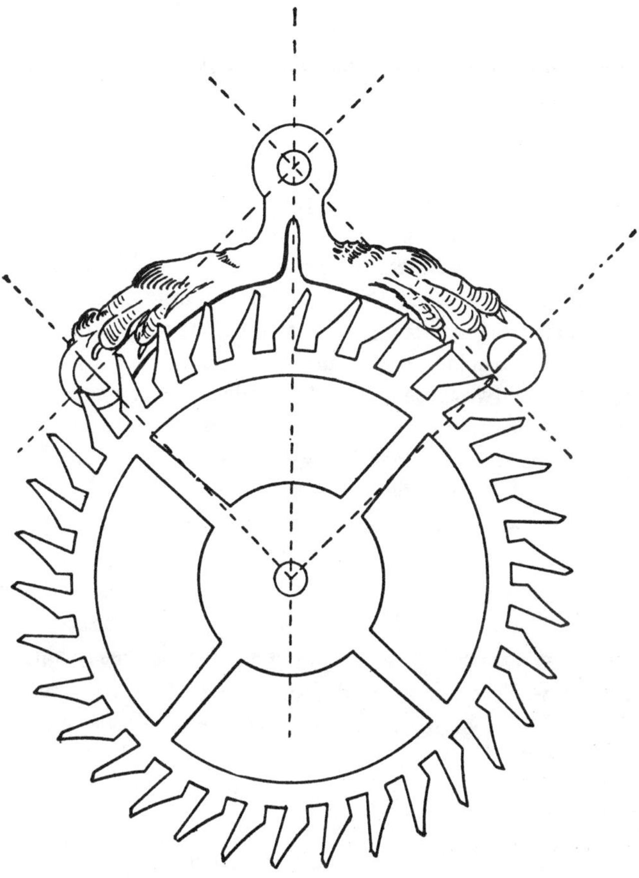

Fig. 38. Brocot's visible escapement escaping over 90° with a small lift on the escape wheel teeth.

pendulum will not then have momentum enough to unlock it against the full power of the spring. If the pallets are set shallow, in order to avoid this difficulty, then the pendulum will take too short a swing and thus the clock will have a gaining rate. Generally the pendulum ball cannot be made enough heavier to correct the defect.

In these movements, in which the length of the pendulum does not exceed 4 inches, the pallet fork embraces, generally about 120°, or the one-third part of the wheel; it will be seen that unless there are stop works on the barrel of the main spring no manner of regulating is possible with these conditions, in view of the considerable influence exercised by the mainspring through the train on the very light pendulum, and by replacing this unduly high anchor by a lower one, I have always been able to produce a very satisfactory rate with movements having pendulums of three and a half to four inches. Fig. 38 shows a 90° escapement with a small lift on the escape wheel teeth.

In spite of its incontestable qualities, the visible escapement possesses one inherent fault. I refer to the formation of its pallets, the semi-circular shape of which renders unequal the action of the train in giving impulse to the pendulum exceeding 50 centimeters (20 inches), since to make it to describe arcs of from one to two degrees only, with pendulums of from 60 centimeters to one meter in length, it became necessary to make the anchor arms extremely long, which considerably impeded the freedom of action, especially when the oil became thick, and this disposition would, therefore, stand in direct contradiction with the principles of modern horology. Both stopping and the irregularity of rate can be obviated by changing the semi-circular form of the pallets for one of an inclinea plane, either by grinding a new plane or turning the stones in such manner as to offer an inclined plane to the action of the wheel, analagous to that of the Graham escapement.

See Fig. 37, the dotted lines on the pallets showing the portion to be ground away.

The importance of this transformation will readily be understood; it suffices to give to these planes a more or less large inclination in order to obtain a greater regularity of lifting, and, at desire, a lifting arc more or less considerable without being compelled to modify the proportions of the fork or to exaggerate the center distance of wheel and pallet arbor.

In adjusting an escapement, perhaps it may be advisable to mention that moving the pallets closer together, or opening them wider, will only adjust the drop on one side, while the other drop can only be affected by altering the distance between the centers of the pallets and scape-wheel. This is accomplished in various ways. The French method consists of an eccentric bush, riveted in the frame just tight enough to be turned by a screw-driver. Another plan, common in America, is simply pieces of brass (cocks) fastened on the sides of the frames. The pivots of the pallet axis are hung in holes in these cocks, and an adjustment of great accuracy may be quickly obtained by loosening the clamping screws. Lock, drop and run should be of the same amount on each pallet. However, we do not approve of adjustments of any kind, except in the very highest class of clocks, where they are always likely to be under the care of skillful people, who understand how to use the adjustments to obtain nicety of action in the various parts.

In making escapements, lightness of all the parts ought to be an object always in view in the mind of the workman, and such materials should be used as will best serve that purpose. The scape-wheel, and the pallets and fork, should have no more metal in them than is necessary for stiffness. The pallet arbor, and also the escape-wheel arbor, should be left pretty thick when the wheel and pallets are placed in the center between the plates, to prevent their springing when giving impulse to the pendulum. We have often been puzzled to find out the necessity or the utility of placing them in the center between the plates, as they are so generally done in English clockwork. The escapement acts much more firmly when it is placed near one of the plates, and it is just as easy to make it in this way as in the other.

It is often assumed that the friction of the teeth on the circular part of the pallets of a dead-beat escapement is small in amount and unimportant in its value. With respect to its amount, we believe it is often not far short of being equal to one-half of the combined retarding forces presented to the pendulum; and with respect to its being unimportant, this assumption is founded on the supposition that it is always a uniform force, when it is easy to show that it is not a uniform force. It is very well known that the force transmitted in clock trains, from each wheel to the next, is very far from being constant. Small defects in the forms of the teeth of the wheels and of the leaves of the pinions, and also in the depths to which they are set into each other, cause irregularities in the amount of power transmitted from each wheel to the next; and the accidental combination of these irregularities in a train of four or five wheels, makes the force transmitted from the first to the last exceedingly variable. The wearing of the parts and the change in the state of the oil, are causes of further irregularities; and, from these causes, it must be admitted that the propelling power of the scape-wheel on the pallets is of a variable amount, and a more important question for consideration than it is usually supposed to be. To avoid the consequences of this irregular pressure of the scape-wheel on the pallets being communicated to the pendulum, is a problem that has puzzled skillful mechanicians for many years; for, although we find the Graham escapement to be pronounced both theoretically and mechanically correct, and by some authorities little short of perfection, we find some of these same authorities—both theoretically and practically—testify their dissatisfaction with it by endeavoring to improve on it. In Europe the experience of generations and the expenditure of small fortunes, in pursuit of this improvement, through the agency of the gravity, and other forms of escapements, proves this fact; while of late years, in the United States, much time and money has been spent on the same subject, and results have been reached which have raised questions that ten years ago were little dreamed of by those clockmakers who are generally engaged on the highest class of work.

While considering this class of escapements, we would say a few words in regard to the sizes of escape wheels generally used. Small wheels can now be cut as accurately as larger ones and there is now no reason or necessity for continuing the use of a wheel of the size Graham and Le Paute used, and which has been the size generally adopted by most European makers who use these escapements. The Germans and Swiss make wheels much smaller for Graham escapements than the English makers do; and the American factories make them smaller still. On the continent of Europe the wheels of Le Paute's escapement are made much larger than they are made in England and in the United States. No wheel, and more especially a scape-wheel, should be larger than will just give sufficient strength for the number of teeth it has to contain, in proportion to the amount of work that it has to perform. The amount of work a scape-wheel has to perform in giving motion to the pendulum is of the lightest description, and not more than one-tenth of what it is popularly supposed to be, which is shown by its variation under slight increase of friction; therefore we do not consider that we take extreme ground in recommending wheels for these escapements to be made nearly half the size their originators made them, and the pallets drawn off in proportion to the reduced size of the wheel. It is plain that by reducing the size of the wheel its inertia will be reduced. When the teeth begin to act on the inclined planes of the pallets, the wheel will be set in motion with greater ease, as it has a shorter leverage, and the amount of the dead friction of the scape-wheel teeth on the inclined planes and circular part of the pallets will also be proportionately reduced by making the wheel smaller. Factory experience and examination of a large number of clocks in repair shops have also shown that smaller and thicker escape wheels will wear much longer than larger and thinner ones, as all the wear is at the points of the teeth and this is the portion to be protected.

CHAPTER IX.

LE PAUTE'S PIN WHEEL ESCAPEMENT.

Probably in no other escapement, except the lever, has there been so many modifications as in the pin wheel; this is so to such an extent that it will be found by the student that nearly every escapement of this kind which he will examine will differ from its fellows if it has been made by a different maker. They will be found to vary in the lengths of the pallet arms from three-fourths to one and a half times the diameter of the escape wheel; some of them will have the longer arm of the pallets outside and some inside; some will have the lift for both pallets laid out on one side of the perpendicular P, Fig. 39, while others will have the lift divided, with the perpendicular in the center. Very old escapements have the pallet center directly over the escape wheel center, while the pallet arms work at an angle of 45°, while others have them with the pallet center planted on a perpendicular, tangent to the pitch line of the escape wheel. Some have the circular rest or locking faces of the pallets rounded slightly to hold the oil in position while others have them flat and still others have them made of hard stone, polished. More than half have the pins in the escape wheel cut away for one-half of their diameters, leaving the bottoms round, as shown in Fig. 39, while others use a wider pin and trim away the bottoms also, as in Fig. 40, leaving the lifting surface on the pins not more than one-fourth the arc of the circle. This is especially true of the larger escapements used in tower clocks, though they are also found in regulators.

In view of the wide variation in practice, therefore, we have endeavored to present in Fig. 39 a conservative statement of the general practice as found in existing clocks. We say existing, because very few of these escapements are made now—none at all in America—and those in use are

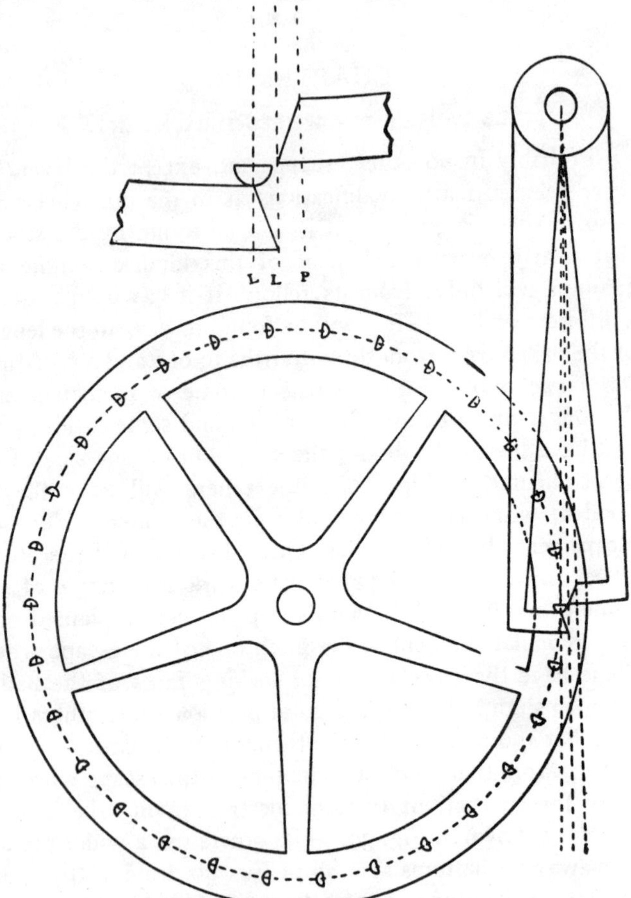

Fig. 39. Pin Wheel Escapement.

generally in imported regulators, which have come from Switzerland or Germany. The Waterbury Clock Co. at one time made this escapement for its regulators and the

Seth Thomas Clock Company made a number of its early tower clocks with it, but both have discontinued it for some years, and it is safe to say that any movement coming into

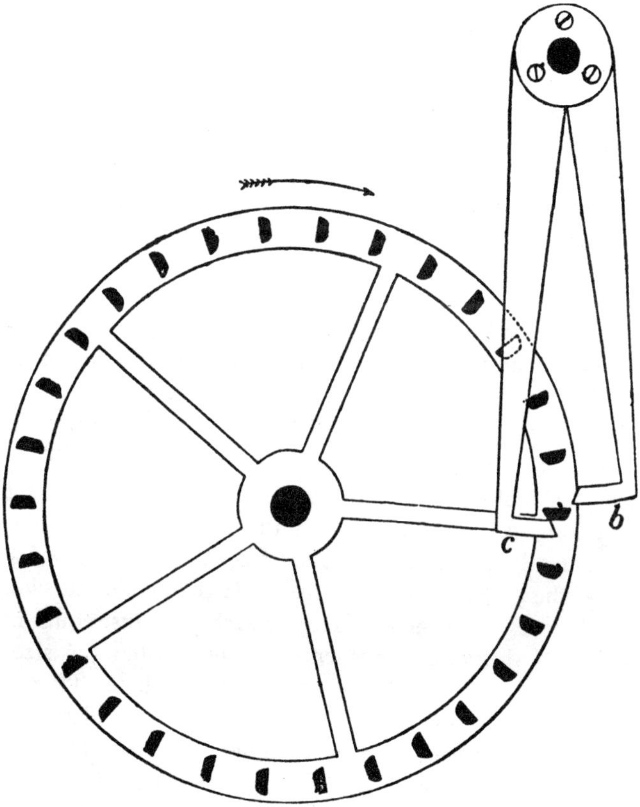

Fig. 40. Pin Wheel With Flattened Teeth.

the watchmaker's hands which has this escapement is imported; or if American, it is out of the market.

Le Paute claimed as an advantage the fact that the impact of the escape wheel teeth is downward on both pallets, whereas in the gravity and recoil escapements one blow is struck upwards and the other downwards. He claimed that by this means a better action was secured after the pivot holes began to wear, as there was less lost motion with both blows in the same direction and any shake would not affect the amount of impulse given to the pendulum. The difference is more theoretical than practical, however, and the escapement possesses one serious fault, which is that the pins forming the escape wheel teeth conduct the oil away from the pallets, so that the clock changes its rate in from eight months to one year after being oiled and cleaned. The most effective means of counteracting this is to round the locking planes of the pallets slightly, so that the oil will be held on them by capillary attraction. Another method is to turn the pins so that they are thicker in diameter at the point of contact with the pallets, but this is seldom tried. The best plan is to keep the pallets as close as they can be to the face of the wheel without touching.

To Draw the Escapement.—In laying out this escapement the first thing to consider is the arc of swing of the pendulum, because one-half of the lift is on the pin and consequently one-half the lift must equal one-half the diameter of the pin, as shown in Fig. 39. If the pendulum swings four degrees, then the diameter of each pin must equal four degrees of the pallet movement. This establishes the size of our pin; it is measured from the pallet staff hole. There are 30 of these pins for a second's pendulum, and unless it is a very large escapement the pins cannot be made less in diameter than one-fourth the distance between the pins, or they will be too weak and will spring; consequently $360 \div 30 = 12°$ and $12° \div 4 = 3°$, so that three degrees of the pitch line of the escape wheel equals the swing of the pallet fork. This establishes the relation as to size between the escape wheel and the opening, or swing of the pallet fork. Draw a perpendicular, P, from the pallet center and on one side of it lay out the lift lines L, L; draw a line at right angles to the perpendicular and where it crosses the inner lift line draw a circle touching the outer lift line. The diameter of this circle equals three degrees of the circumference of the wheel, on its pitch line, and this multiplied by 120 gives 360° or the pitch circumference of the escape wheel. Dividing the sum so found by 3.1415 gives the diameter of the escape wheel and half of this is the radius. After finding the radius draw the pitch circle and set out the other twenty-nine teeth spaced twelve degrees apart, and drawn in half circles as shown in Fig. 39.

Now to get the thickness of the pallet arms. When the pin shown in action in Fig. 39 has just cleared the lower edge of the inner pallet, the succeeding pin should fall safely on the upper corner of the outer pallet; consequently the thickness of these two arms, the pin between them, and the drop (clearance between the pin and the lower edge of the upper pallet) should just equal the distance between two pins, from center to center, or 12° of the escape wheel. With the first or inner lift line as a starting point, draw the lower arcs of the pallets and draw the upper or locking planes from the perpendicular and the outer lift line. Then draw the lifting planes of the pallets by connecting the ends of these arcs. The enlarged view above the escape wheel in Fig. 39 will show how this is done more clearly than the main drawing.

It is best to make the pallet fork of steel, in two pieces, screwed to a collet on the pallet arbor, as the inner arm must be bent, or offset, so that it will clear the pins of the escape wheel, and the pallets should lie in the same plane, as close to the wheel as is possible without touching it. The pallets are hardened.

In tower clocks the escapement is so large that a pin having a diameter of three degrees of the escape wheel gives a half pin of greater strength than is necessary for the work to be done and such pins are cut away on the bottom, as in Fig. 40. In making the wheel it should be drilled in the lathe with the proper index to divide the wheel and the pins riveted in; then the pins are cut with a wheel cutter as if they were teeth of a wheel. Pins should be of hard brass.

Care should be used in handling clocks with this escapement while the pendulum is connected with the pallet fork, as, if the motion of the fork should be reversed while a pin was on one of the lifting planes, it would bend or break the pin.

CHAPTER X.

THE RECOIL OR ANCHOR ESCAPEMENT.

This escapement, always a favorite with clockmakers, has had a long and interesting history and development. Because it started with a suddenly achieved reputation, and because it is adapted to obtain fair results with the cheapest and consequently most unfavorable working conditions, it has won its way into almost universal use in the cheaper classes of clock work; that is to say, it is used in about ninety per cent of the pendulum clocks which are manufactured to-day.

It achieved a sudden reputation at its birth, because it was designed to replace the old verge, which, with its ninety degree pallets close to the arbor, and working into the crown wheel, required a very large swing of the pendulum. This necessitated a light ball, a short rod, required a great force to drive it, and made it impossible to do away with the circular error, while leaving the clock sensitive to variations in power. The recoil escapement was therefore the first considerable advance in accuracy, as its use involved a longer and heavier pendulum, shorter arcs of vibration and less motive power than was practicable with the verge; and as the pendulum was less controlled by the escapement, it was less influenced by variations of power.

In the early escapements the entrance pallet was convex and the exit pallet concave. Escapements of this description may still be met with among the antiquities that occasionally drift into the repair shop. Later on both pallets were made straight, as shown in Fig. 41. It will be seen by studying the direction of the forces that the effect is to wear off the points of the teeth very rapidly, and for this reason the pallets were both made convex (See Fig. 42), so as to bring the rubbing action of the recoil more on the sides of the

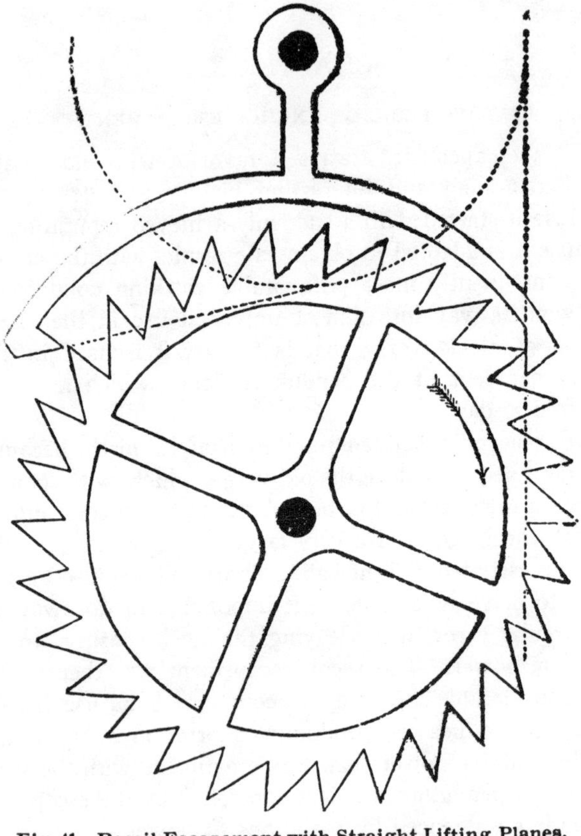

Fig. 41. Recoil Escapement with Straight Lifting Planes.

teeth and do away to a large extent with the butting on the points which destroyed them so rapidly.

The rather empirical methods of laying out the recoil escapement, which have gained general circulation in works on horology, have had much to do with bad depthings of this escapement and the consequent undue wear of the escape wheel teeth and great variation in time keeping of the movements in which such faulty depthings occur, particularly in eight-day movements with short and light pendulums. The escapement will invariably drive the clock faster for an increase of power and slower for a decrease; an unduly great depthing will greatly increase the arc of vibration of the pendulum, as the train exerts pressure on the pendulum for a longer period during the vibration; the consequence is that instead of the pendulum being as highly detached as possible, we have the opposite state of affairs and a combination of a strong spring, light pendulum and excessive depthing will easily make a variation of five minutes a week in an eight-day clock.

The generally accepted method of laying out this escapement is shown in Figs. 41 and 42, as follows: "Draw a circle representing the escape wheel; multiply the radius of the escape wheel by 1.4 and set off this as the center distance between the pallet and escape wheel centers. From the pallet staff center describe a circle with a radius equal to half the distance between escape wheel and pallet centers. Set off on each side of the center line one-half the number of teeth to be embraced by the pallets and from the points of the outside teeth draw lines tangent to the circle described from the pallet center. These lines would then form the faces of the pallets if they were left flat."

We wonder how much information this description and the drawing conveys to the average reader. How long should the pallets be? What is the drop? How much will the escape wheel recoil with such a depthing? What arc will the pallets give the pendulum? Why should the center distance always be the same (seven tenths of the diameter of the wheel) whether the escapement embraces eight, or ten, or six teeth? As a matter of fact it should not be the same. We could ask a few more questions as to other details of this formula, but it will be seen that such a description is practically useless to all but those who are already so skilled that they do not need it.

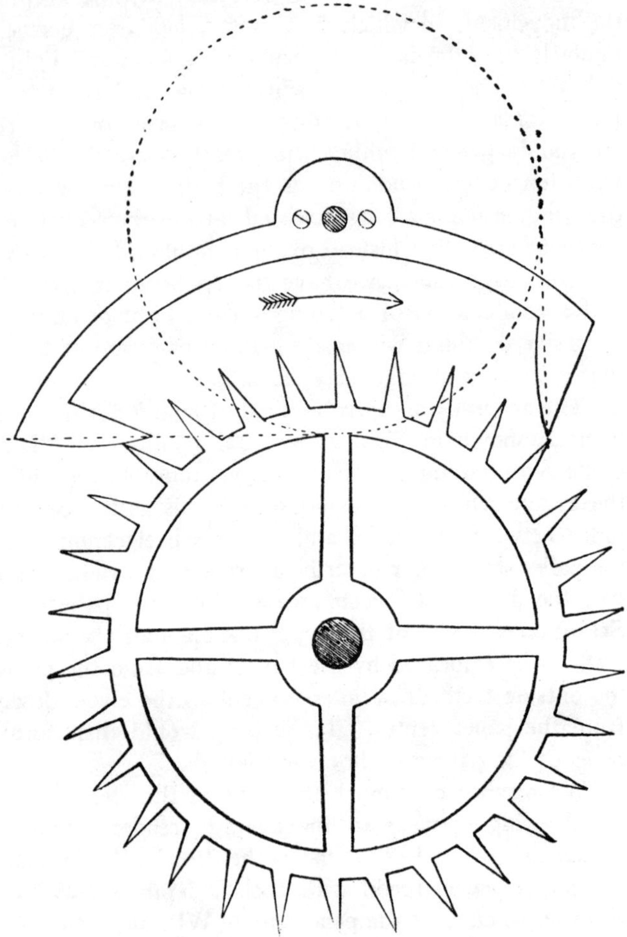

Fig 42. Recoil Escapement with Curved Lifting Planes.

Let us analyze these drawings. A little study of Figs. 41, 42 and 43 will show that there is really only one point of difference between them and Fig. 32, which shows the elements of the Graham, or dead beat. The sole difference is in the fact that there are no separate locking planes in the recoil, the locking and run taking place on an extension of the lifting planes. Otherwise we have the same elements in our problem and it may therefore be laid out and handled

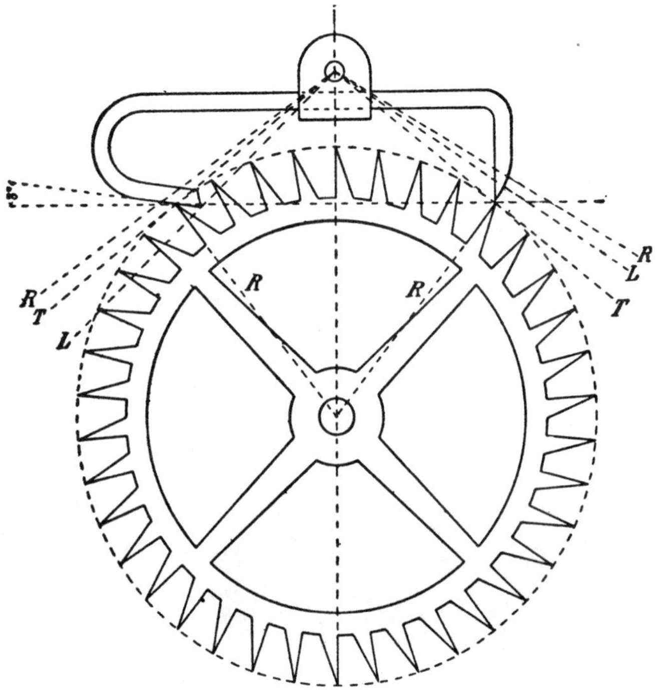

Fig. 43. Drawing the Lock Lift and Recoil of the Usual Form.

in the same manner; indeed, if we were to set off on Fig. 32, the amount of angular motion of the pallet fork which is taken up by the run of the escape wheel teeth on the locking planes, by drawing dotted lines above the tangents, T, we should then have measured all the angles necessary to intelligently set out the recoil escapement. We should have the lock at the tangent, T, the lift and the run (or recoil) being defined by the lines on either side of it, and the length of our running and lifting planes would be found for the entering pallet by drawing a straight line between the points of the two acting teeth of the escape wheel and noting where this line cut the lines of recoil and lift. A similar line traced at right angles to this would in the same way

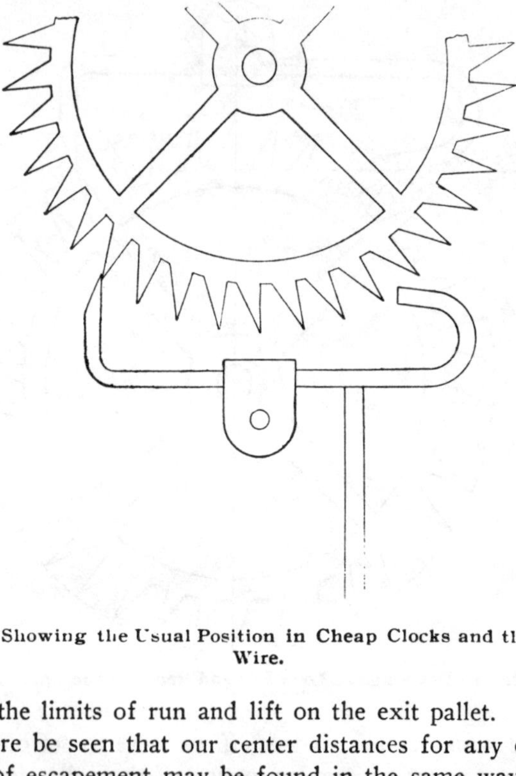

Fig. 45. Showing the Usual Position in Cheap Clocks and the Verge Wire.

define the limits of run and lift on the exit pallet. It will therefore be seen that our center distances for any desired angle of escapement may be found in the same way (Fig. 28), for either escapement, and thus the method of making the pallets for the ordinary American clock, Fig. 43, becomes readily intelligible. The sole object of curving the pallets, as explained previously, was to decrease the butting effect of the run on the points of the teeth. This is accomplished in Fig. 43 by straight planes on the pallets and straight sides to the teeth with 20° teeth on the escape wheel; merely inclining the plane of the entering pallet about six degrees toward the escape wheel center, thus serv-

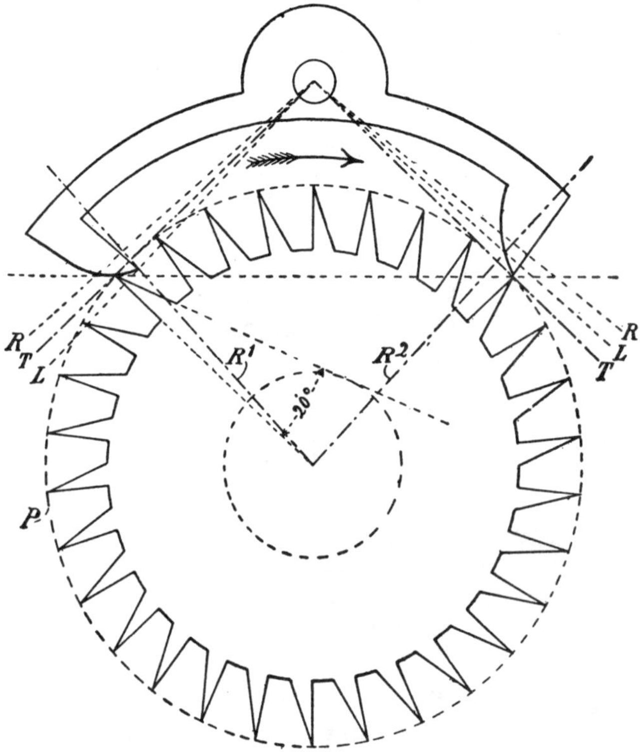

Fig. 44. Recoil with Curved Planes.

ing all purposes, while the gain in the cost of manufacture by using straight instead of curved pallets and wheel teeth is very great.

One factory in the United States is turning out 2,000,000 annually of two movements, or about 1,000,000 of each movement; there are four other larger factories and several with a less product; so it will readily be seen that any decrease in cost, however small it may be on a single movement, will run up enormously on a year's output. Suppose the factory mentioned were enabled to save only one-eighth of a cent on one of its million movements manufactured last year, this would amount to $1,250 per year, a little over $100 per month. Thus it will be seen that close figuring on costs of production is a necessity.

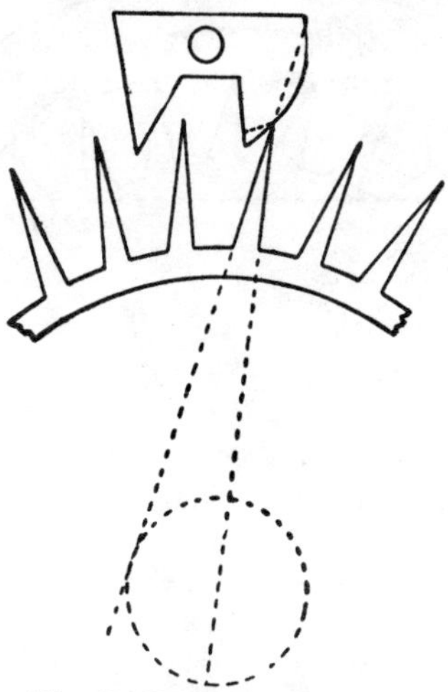

Fig. 46. Drum Escapement.

Fig. 44 shows the method of drawing the escapement according to the common sense deductions given above. As the methods of laying out the angle of escapement, lock, lift, and run, were given in detail in Figs. 28 to 32, they need not be repeated here.

Fig. 46 shows the escapement frequently used in French "drum" clocks and hence called the "Drum" escapement. These are clocks fitted to go in any hole of the diameter of the dial and hence they have very short, light pendulums. An attempt is made to gain control over the pendulum by decreasing the arc of escapement to not more than two and sometimes to only one tooth. This gives an impulse to the pendulum only on one-half of the vibrations, the escape wheel teeth resting and running on the long circular locking pallet during alternate swings of the pendulum. The idea is that the friction of the long lock will tend to reduce the effect of the extra force of the mainspring when the clock is freshly wound. Such clocks often stop when the clock is nearly run down, from deficiency of power, and stop when wound, because the friction of the escape wheel teeth on the locking plane is such as to destroy the momentum of the light pendulum. All that can be done in such cases is to alter the locking planes as shown by the dotted lines, so that the "drum" becomes virtually a recoil escapement of two teeth.

CHAPTER XI.

THE DENNISON OR GRAVITY ESCAPEMENT.

The distinguishing feature of this escapement lies in the fact that it aims to drive the pendulum by applying to it a falling weight at each excursion on each side. As the weight is lifted by the train and applied to the pendulum on its return stroke and there is no other connection, it follows that the pendulum is more highly detached than in any other form of pendulum escapement. This should make it a better time-keeper, as the application of the weight should give a constant impulse and hence errors and variations in the power which drives the train may be neglected.

On tower clocks this is undoubtedly true, as these clocks are interfered with by every wind that blows against the hands, so that a detached pendulum enables a surplus of power to be applied to the train to meet all emergencies With a watchmaker's regulator, however, the case is different. Here every effort is made to favor the clock, vibrations, variations of temperature, variations of power, dirt, dust, wind pressure and irregularities of the mechanism are all carefully excluded and the consequence is that the special advantages of the gravity escapement are not apparent, for the reason that there are practically no variations for the escapement to take care of. Added to this we must consider that the double three-legged form, which is the usual one, is practically an escape wheel of but six teeth, so that another wheel and pinion must be added to the train and this, with the added complications of the fan and the heavier driving weight required, counterbalance its advantages and bring it back to an equality of performance with the simpler mechanism of the well made and properly adjusted dead beat escapement. Theoretically it should work far better than the dead beat, as it is more detached; but theory is always modified by working conditions and if the variations are lacking there is no special advantage in constructing a mechanism to take care of them. This is the reason why so many watchmakers have constructed for themselves a regulator with this escapement, used in the making all the care and skill of which they were capable and then been disappointed to find that it gave no better results with the same pendulum than the dead beat it was to replace. They had eliminated all the conditions under which the detached escapement would have shown superiority.

Although the gravity escapement will not give a superior performance under the most favorable conditions for time-keeping, it is distinctly superior when these conditions are unfavorable and therefore fully merits its high place in the estimation of the horological fraternity. We have instanced its value in tower clock work; it has another advantage in running cheap and poorly made (home made) regulators with rough and poor trains; therefore, it is a favorite escapement with watchmakers who build their own regulators while they are still working at the bench, before entering into business for themselves. As the price of a first-class clock for this purpose is about $300 and the cheapest that is at all reliable is about $75, it will be seen that the temptation to build a clock is very strong and many of them are built annually.

Regulators with the gravity escapement are built by the Seth Thomas Clock Co., the Howard, and one or two others in this country, but they are furnished simply to supply the demand and sales are never pushed for the reasons given previously. Clocks with this escapement are quite common in England and many of them have found their way to America. It is one of the anomalies of trade that our clock-makers are supplying Europe with cheap clocks, while we are importing practically all the high-priced clocks sold in

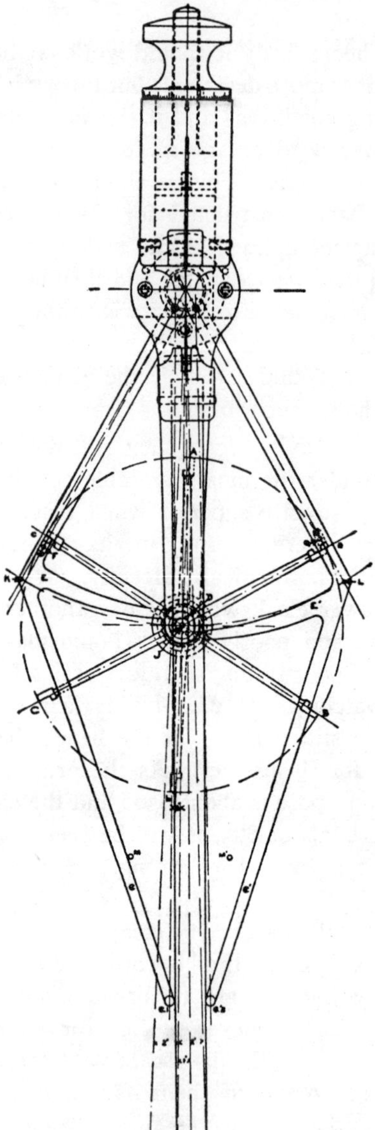

Fig. 47.

the United States and among them are a few having the three-legged and four-legged gravity escapements, therefore the chances are that when a repairer finds such a clock it is likely to be either of English origin or homemade, unless it be a German regulator.

Figs. 47 and 48 show plans and side views of the three-legged escapement. Fig. 48 also shows an enlarged view of the escape wheel, showing how the three-leaved pinion between the two escape wheels, is made where it is worked out of the solid. A, B and C and a, b and c show the escape wheel which is made up of two three-armed wheels, one on each side of a three-leaved pinion marked D^1 and D^2 in the enlarged view of Fig. 48. The pallets in this escapement consist of the two arms of metal suspended from points opposite the point of bending of the pendulum spring and the lifting planes are found on the ends of the center arms in these pallets, which press against the three leaves of the pinion, while the impulse pins e^1 and e^2, Fig. 47 and 48 act directly upon the pendulum in place of the verge wire. The pallets act between the wheels in the same plane as each other. The lifting pins or pinion leaves act on the lifting planes after the line of centers when the long teeth or legs of the escape wheels have been released from the stops, F and G, Figs. 47 and 48, which are placed one on each side of the pallets and act alternately on the wheels. These pallets are pivoted one on each side of the bending point of the suspension spring. To lay out the escapement, draw a circle representing the escape wheel diameter, then draw the line of centers and set off on the diameter of the escape wheel from each side of the line of centers 60° of its circumference, thus marking the positions for the pallet stops 120° apart. Draw radii from the center of the escape wheel to these positions and draw tangents from the ends of these radii toward the center line. The point where these meet will be the bending point of the pendulum spring.

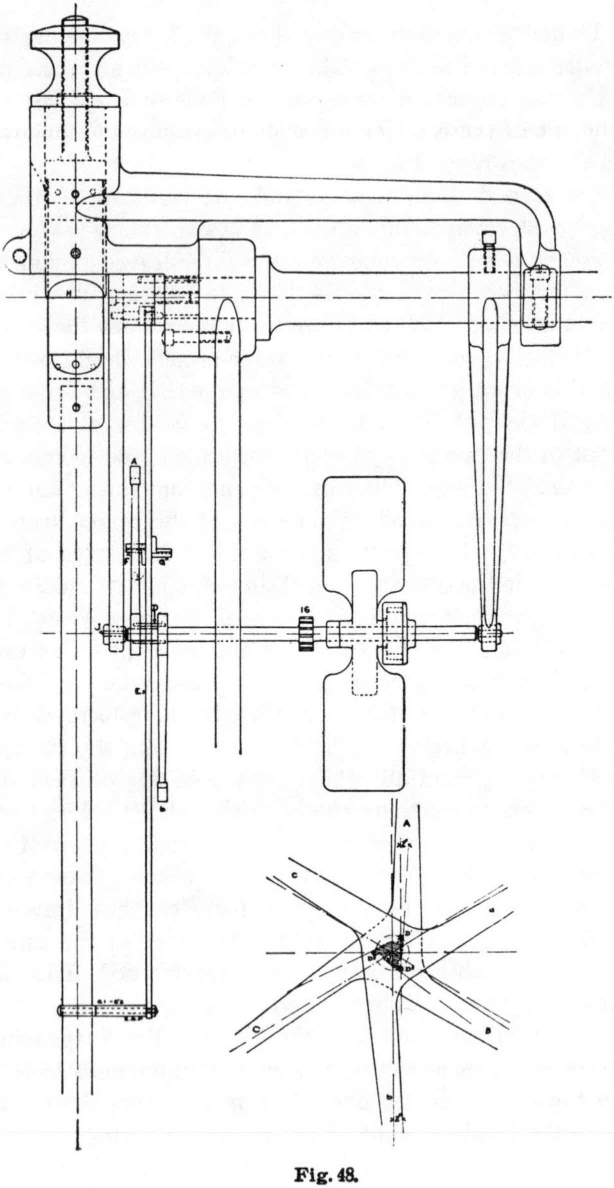

Fig. 48.

This is clearly shown at H, Fig. 47. The points of suspension for the pallets are planted on the line of these tangents and a little below the point H, where the tangents meet on the line of centers. This is done to avoid the mechanical difficulty of having the studs for the two pallets occupy the same place at the same time. The arms of the pallets below the stops may be of any length, but they are generally constructed of the same angle as the upper arms and will be all right if drawn parallel to these upper arms. They are in some instances continued further down, but this is largely a matter of taste and the lower portion of the escapement is generally drawn so as to be symmetrical.

The impulse of the pendulum is given by having pins projecting from the pallet arms and bearing upon the pendulum rod, which pins may be of brass, steel or ivory. In the heavier escapements they are made of ivory in order to avoid any chatter from contact with the pendulum rod of a heavy pendulum. These pallets should be as light as it is possible to make them without having them chatter under the impact of the escape wheel arms on the stops. They have only to counteract the force of the pendulum spring and the resistance of the air and for light pendulums this force is much less than is generally understood. Two ounces of impulse will maintain a 250-pound pendulum, but two pennyweights is more than sufficient for a fifty-pound pendulum. The reader can see that in the case of a pendulum weighing but eight to fourteen pounds, there will be a still greater proportionate drop, as the spring itself is thinner, the rod is thinner, the pendulum ball offers little resistance to the air and the consequence is that it is difficult to get the pallet arms light enough for an ordinary clock.

Watchmakers who make this escapement for themselves, to drive an eight to fourteen pound pendulum, generally make the escape wheel three inches diameter and make the escape wheel and pallet arms all from the steel obtained by buying an ordinary carpenter's saw. The lifting planes should not be more than one-eighth its diameter from the center of the escape wheel, as where this is the case the circular motion of the center pins will be so great that the pallet in action will be thrown out too rapidly and will chatter when striking the pendulum rod. On the other hand it should not be less than one-twelfth of the diameter of the escape wheel, or the pendulum will not be given sufficiently free swing and the motion will be so slow that while such a clock will work under favorable conditions, jarring, shaking in wind storms, etc., will have a tendency to make the pendulum wabble and stop the clock. From what has been said above, it will also be seen that the necessity for slow motion of the pallet arms unfits this escapement for use with short pendulums.

The action of the escapement is as follows: The pendulum traveling to the right, when it has thrown the right pallet arm sufficiently far, will liberate the escape wheel tooth from the stop G and the pinion, acting on the lifting plane, will raise the left pallet arm, allowing the pendulum to continue its course without doing any further work until it has reached its extreme point of excursion, when the weight of the right pallet arm upon the pendulum rod will act upon the pendulum until it has passed the center when the pallet arm will be stopped by the banking pin M^1; exactly the same procedure takes place on the left side of the escapement during the swing of the pendulum to the left. The beat pins M and M^1 should be set so that the impulse pins e^1 and e^2 will just touch the pendulum when the latter is hanging at rest and the escapement will then be in beat. The stops should be cut from sheet steel and the locking faces of the escape wheel arms, stops on the pallets, lifting planes of the pallets and the lifting pins should all be hardened. In some of the very fine escapements the faces of the blocks are jeweled. The arms of the inner part of the escape wheel are usually set at equal angular distances between those of the outer, although this is not absolutely necessary, and the lifting pins are set on radii to the acting faces of the arms of one of the wheels, so as to cross the line of centers at the distance from the center, not exceeding one-eighth of the radius of the wheel, for the reasons explained above.

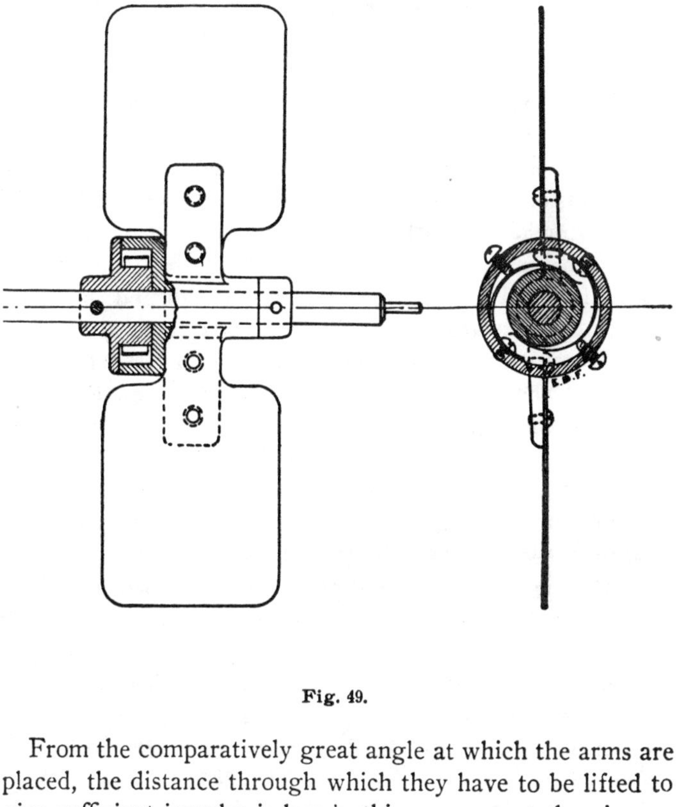

Fig. 49.

From the comparatively great angle at which the arms are placed, the distance through which they have to be lifted to give sufficient impulse is less in this escapement than in one with a larger number of teeth acting in the same plane, as the pallets would then hang more nearly upright. This is a great advantage, as the contact is shorter. The unlocking is also easier for the same reason, and from the greater diameter of the wheel in proportion to other parts of the escapement, the pressure on the stops is considerably less. The two wheels must be squared on the arbor, so there will be no possibility of slipping. The lifting pins D are shouldered between them like a three-tooth lantern pinion. In small escapements the lifting pins are not worked out of the solid arbor, but are made as hardened screws to connect the two portions of the wheel. In tower clocks the pinion is generally made solid on the shaft J, Fig. 48. The wheel, A, B, C, is made to pass over the pinion D and is fitted to a triangular seat, the size of the triangle of the leaves, D, against the collar on the shaft. The other wheel, a, b, c, is fitted to the inside triangle of the pinion, so that the leaves, D, form a shoulder against which it fits. The pallets, E and E^1, also lie in one plane between the wheels, but one stop, F, points forward to receive the A, B, C, teeth and the other, G, points backward to receive the a, b, c teeth alternately. The distance of the pendulum top, H, or cheeks from the center of the escape wheel, J equals the diameter of the escape wheel. The lifting pins should be so placed that the one which is holding up a pallet and the one which is to lift next will be vertical over each other, on the line of centers, the third pin being on the level with the center, and to one side of it, see Fig. 48, enlarged view.

The fly is a very essential part of this escapement, as the angular motion of the escape wheel is such that unless it were checked it would be apt to rebound and unlock; consequently, a large fly is always a feature of this escapement and is mounted upon the scape wheel arbor with spring friction in such a way that the fly can continue motion after the scape wheel has been stopped. This is provided for by a spring pressure, either like the ordinary spring attachment of the fly of striking trains of small clocks, or as shown in Fig. 49 for tower clocks. This fly is effective in proportion to its length and hence a long narrow fly will be better than a shorter and wider one, as the resistance of the air

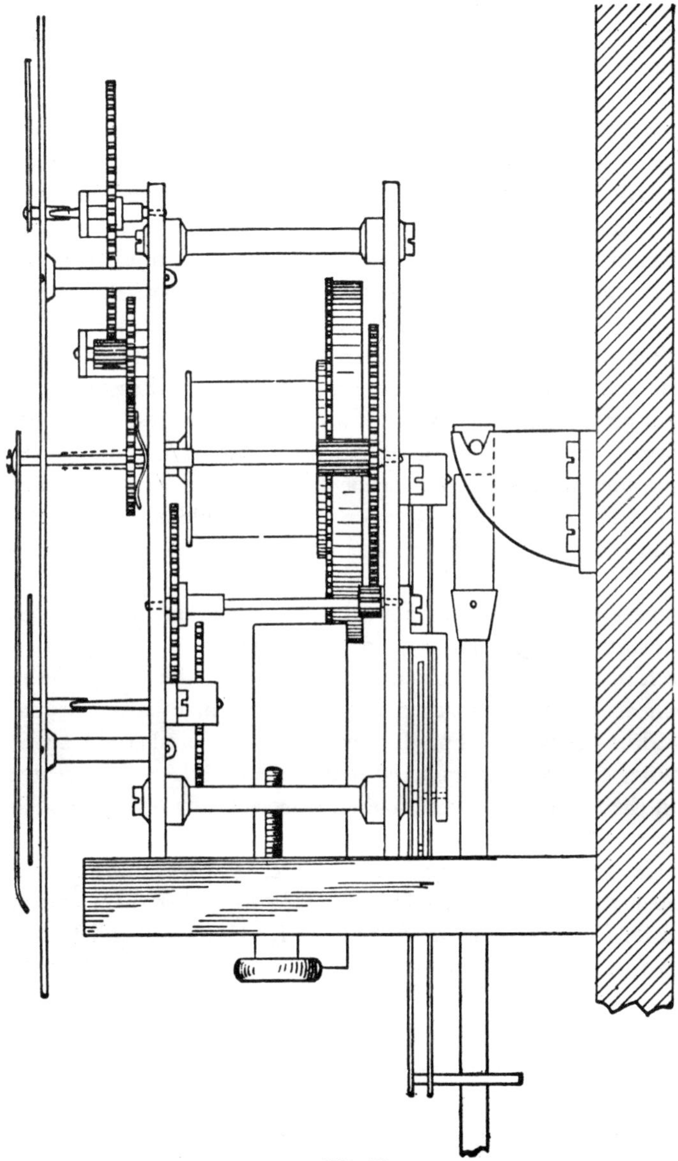

Fig. 50.

striking against the ends of the fly is much greater the further you get from the center.

The pallet stud pins and the impulse pins should on no account be touched with oil or other grease of any kind, but be left dry whatever they are made of, because the slightest adhesion between the impulse pins and the pendulum rod is fatal to the whole action of the escapement. Care must also be taken that one pallet begins to lift simultaneously with the resting of the other, neither before nor after.

The gravity escapement requires a heavier weight or force to operate the train than a dead beat escapement, because it must be strong enough to be sure of lifting the pallets quickly and firmly, and also because the escape wheel having but six teeth necessitates the use of another wheel and pinion between the escape and center and consequently the train is geared back more than it would be for a dead beat escapement, with the seconds hand mounted on the escape wheel arbor. But with this form of escapement the superfluous force does not work the pendulum and it does no harm if the train is good enough not to waste power in getting over rough places left in cutting the teeth of the wheels or any jamming from those which have unequal widths or spaces. For this reason a high numbered train is better than a low numbered one, as these defects are greater on the teeth of a low numbered train and any defect in such cases will show itself.

In the gravity escapement the escape wheel must have a little run at the pallets before it begins to lift them and in order to do this the banking pins, M, M^1, for the pallet arms to rest on, should hold them just clear of the lifting pins or leaves of the escape wheel. The escape wheel should be as light as possible, for every blow heard in the machine means a loss of power and wear of parts. Of course, in an escapement a sudden stop is expected, but the light wheel will reduce it to a minimum if the fan is large enough. Particular attention should therefore be given to the length of

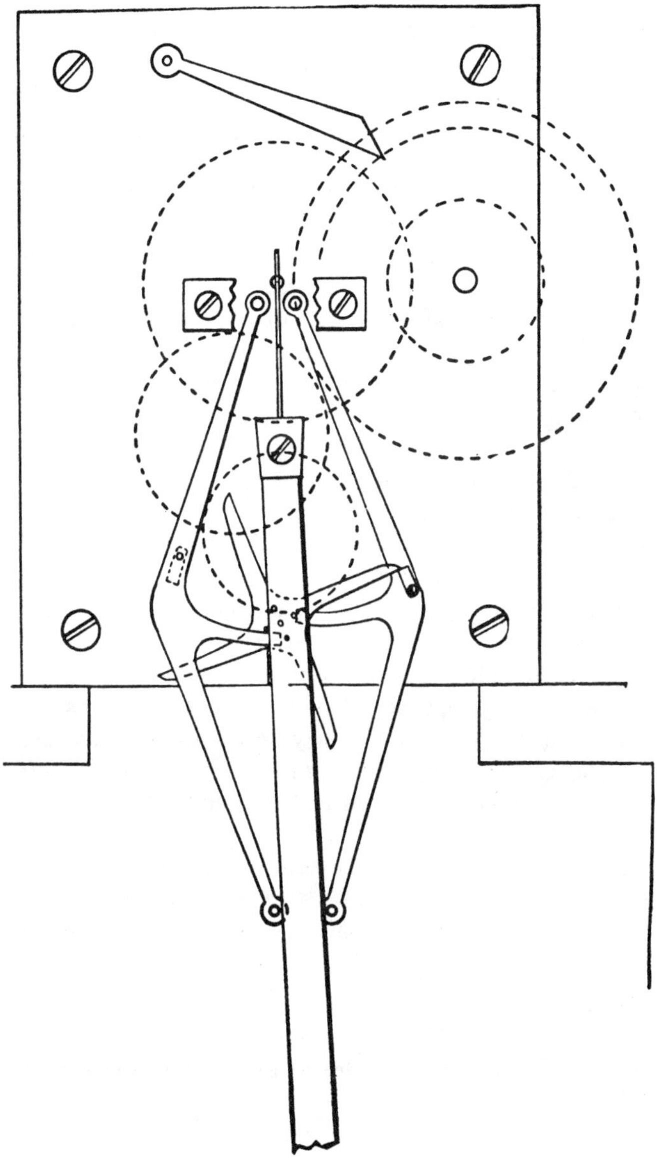

Fig. 51.

this fan and if the stop of the escape wheel seems too abrupt, the fan should be lengthened.

Figs. 50 and 51 show the same escapement with a four-legged wheel instead of the double three-legged. In this case, where there is but one wheel, the pallets must of necessity work on opposite sides of the wheel and hence they are not planted in the same plane with each other, but are placed as close to each side of the wheel as is practicable.

To lay out this escapement, draw the circle of the escape wheel as before, make your line of centers and mark off on the circle 67½° on each side of the line of centers and draw radii to these points, which will indicate the approximate position of the stops. Tangents to these radii, meeting above the wheel on the line of centers will give the theoretical point of the suspension. One set of the lifting pins is planted on radii to the acting faces of the teeth of the escape wheel. The opposite set, on the other side of the wheel, is placed midway between the first set. This secures the lifting at the line of centers. The wheel turns 45° at each beat and its arbor likewise carries a fly.

In case the locking is not secure, the stops may be shifted a little up or down, care being taken to keep them 135° apart. In this way a draw may be given to the locking of the scape wheel arms similar to the draw of the pallets in a detached lever escapement and thus any desired resistance to unlocking may be secured. The stops in either escapement are generally made of steel and it is of the utmost importance that the arms of the escape wheel should leave them without imparting the least suspension of an impulse. Therefore, the stops and the ends of the arms should be cut away (backed off) to rather a sharp angle to insure clearance when the arms are leaving the stops. It is also of equal importance that the legs of the wheels should fall on the stops dead true. The fit of each of the legs should be examined on both stops with a powerful eye glass, so that they should be correct and also see that when the unlocking takes place the wheel is absolutely free to turn.

CHAPTER XII.

THE CYLINDER ESCAPEMENT AS APPLIED TO CLOCKS.

We remarked in a previous chapter that the lifting planes were sometimes on the wheel and sometimes on the anchor. In another chapter we pointed out clearly that the run on the locking surface of the pallets had an important bearing on the freedom of the escapement and hence on the rate of the dead beat escapement. In considering the cylinder escapement, so common in carriage clocks, we shall find that the lift is almost entirely on the curved planes of the escape wheel, and that the locking planes are greatly extended, so that they form the outer and inner surfaces of the cylinder walls. Thus we have here a form of the dead beat escapement, which embraces but one tooth of the escape wheel and is adapted to operate a balance instead of a pendulum. Therefore the points for us to consider are as before, the amount of lift, lock, drop and run, and the shapes of our escape wheel teeth to secure the least friction, as our locking surfaces (the run) being so greatly extended this matter becomes important.

ACTION OF THE ESCAPEMENT.—Fig. 52 is a plan of the cylinder escapement, in which the point of a tooth of the escape wheel is pressing against the outside of the shell of the cylinder. As the cylinder, on which the balance is mounted, is moved around in the direction of the arrow, the wedge-shaped tooth of the escape wheel pushes into the cylinder, thereby giving it impulse. The tooth cannot escape at the other side of the cylinder, for the shell of the cylinder at this point is rather more than half a circle; but its point locks against the inner side of the shell and runs there till the balance completes its vibration and returns, when the tooth which was inside the cylinder escapes, giving an impulse as it does so, and the point of the succeeding tooth is caught on the outside of the shell. The teeth rise on stalks from the body of the escape wheel, and the cylinder is cut away just below the acting part of the exit side, leav-

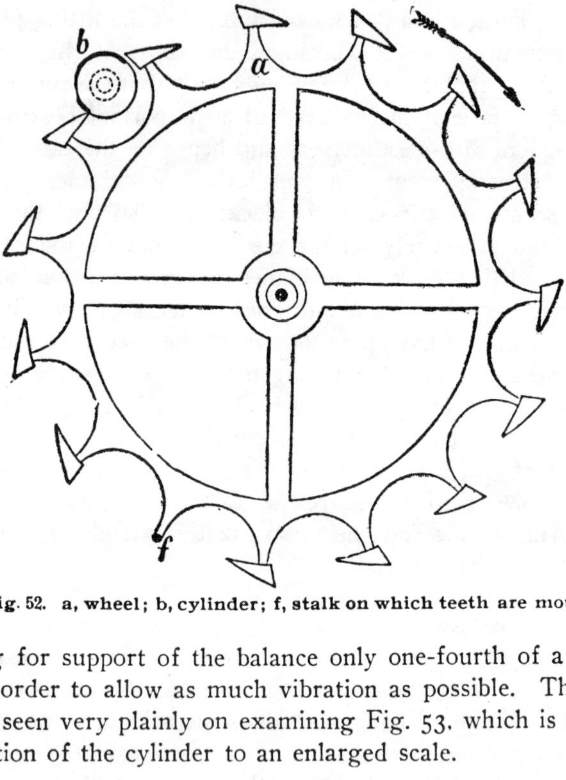

Fig. 52. a, wheel; b, cylinder; f, stalk on which teeth are mounted.

ing for support of the balance only one-fourth of a circle, in order to allow as much vibration as possible. This will be seen very plainly on examining Fig. 53. which is an elevation of the cylinder to an enlarged scale.

Proportion of the Escapement.—The escape wheel has fifteen teeth, formed to give impulse to the cylinder during from 20° to 40° of its vibration each way. Lower angles are as a rule used with large than with small-sized escapements, but to secure the best result either extreme must be avoided. In the escapement with very slight inclines to the wheel teeth, the first part of the tooth does no work, as the tooth drops on to the lip of the cylinder some distance up the plane. On the other hand, a very steep tooth is almost sure to set in action as the oil thickens. The diameter of

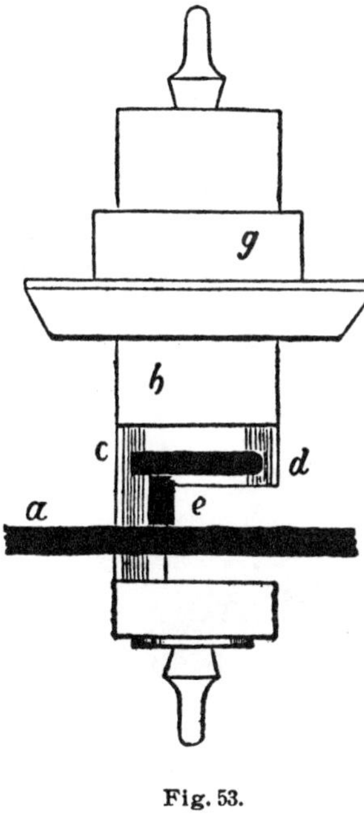

Fig. 53.

the cylinder, its thickness and the length of the wheel teeth are all co-related. The size of the cylinder with relation to the wheel also varies somewhat with the angle of impulse, a very high angle requiring a slightly larger cylinder than a low one. If a cylinder of average thickness is desired for an escapement with medium impulse, its external diameter may be made equal to the extreme diameter of the escape wheel multiplied by 0.115

Then to set out the escapement, if a lift of say 30° be decided on, a circle on which the points of the teeth will fall is drawn within one representing the extreme diameter of the escape wheel, at a distance from it equal to 30° of the circumference of the cylinder. Midway between these

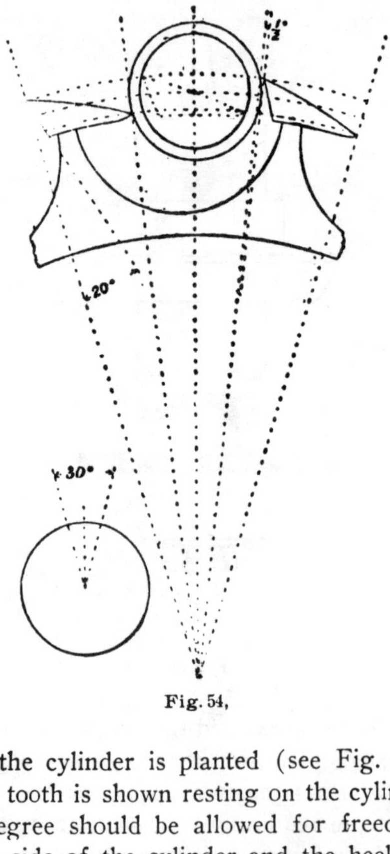

Fig. 54,

two circles the cylinder is planted (see Fig. 54). If the point of one tooth is shown resting on the cylinder, a space of half a degree should be allowed for freedom between the opposite side of the cylinder and the heel of the next tooth. From the heel of one tooth to the heel of the next equal 24° of the circumference of the wheel, 360÷15=24°, and from the point of one tooth to the point of the next also equals 24° so that the teeth may now be drawn. They are extended within the innermost dotted circle to give them a little more strength, and their tips are rounded a little, having the points of the impulse planes on the inner or basing circle. The backs of the teeth diverge from a radial line from 12° to 30°, in order to give the cylinder clearance, a high angled tooth requiring to be cut back more than a low one. A curve whose radius is about two-thirds that of the wheel is suitable for rounding the impulse planes of the teeth. The internal diameter of the cylinder should be such as to allow a little freedom for the tooth. The rule in fitting cylinders is to have equal clearance inside and outside, so as to equalize the drop. The acting part of the shell of the cylinder (where the lips are placed) should be a trifle less than seven-twelfths of a whole circle, with the entering and exit lips which are really the pallets, rounded as shown in the enlarged plan, Fig. 55, the entering lip or pallet rounded both ways and the exit pallet rounded from the inside only. This rounding of the lips of the cylinder adds a little to the impulse beyond what would be given by the angle on the wheel teeth alone. The diameter of the escape wheel is usually half that of the balance, rather under than over.

SIZE OF CYLINDER PIVOT.—To establish the size of the pivot with relation to its hole is apparently an easy thing to do correctly, but to an inexperienced workman it is not so. The side shake in cylinder pivot holes should be greater than that for ordinary train holes; one-sixth is the amount prescribed by Saunier; the size of the pivot relatively to the cylinder about one-eighth the diameter of the body of the cylinder. It is very necessary that this amount of side shake should be correctly recognized; if less than the amount stated, the escapement, though performing well while the oil is fresh, fails to do so when it commences to thicken.

When the balance spring is at rest, the balance should

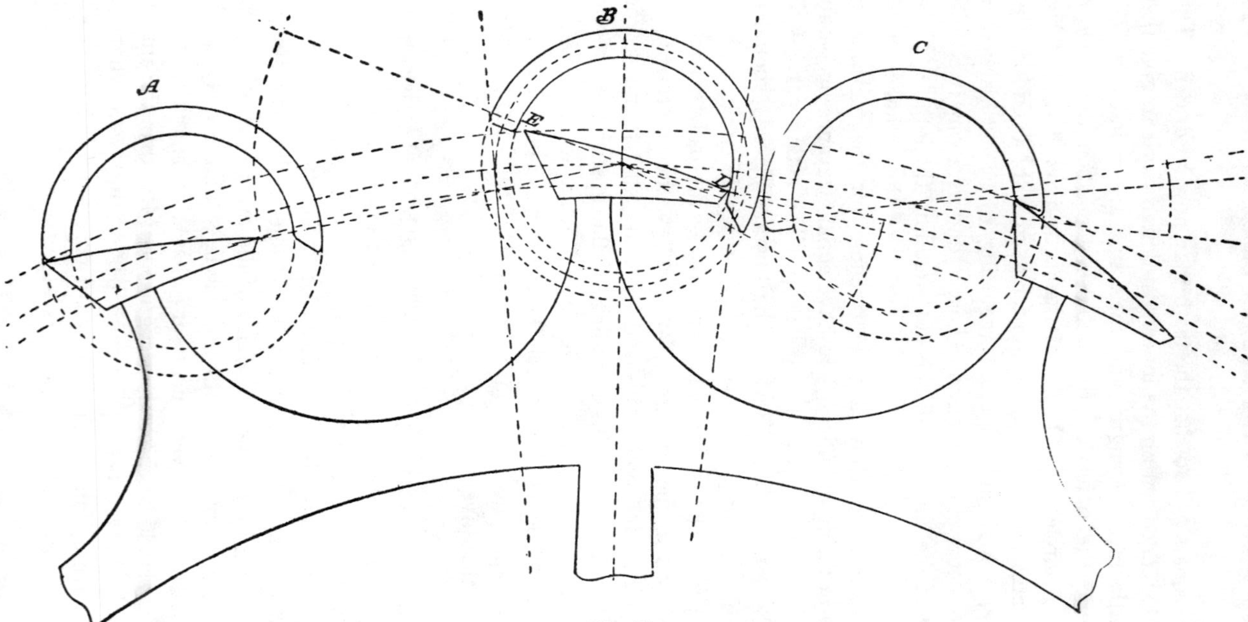

Fig. 55.

have to be moved an equal amount each way before a tooth escapes. By gently pressing against the fourth wheel with a peg this may be tried. There is generally a dot on the balance and three dots on the plate to assist in estimating the amount of lift. When the balance spring is at rest, the dot on the balance should be opposite to the center dot on the plate. The escapement will then be in heat, that is, provided the dots are properly placed, which should be tested. Turn the balance from its point of rest till a tooth just drops, and note the position of the dot on the balance with reference to one of the outer dots on the plate. Turn the balance in the opposite direction till a tooth drops again, and if the dot on the balance is then in the same position with reference to the other outer dot, the escapement will be in beat. The two outer dots should mark the extent of the lifting, and the dot on the balance would then be coincident with them as the teeth dropped when tried in this way; but the dots may be a little too wide or too close, and it will therefore be sufficient if the dot on the balance bears the same *relative* position to them as just explained; but if it is found that the lift is unequal from the point of rest, the balance spring collet must be shifted in the direction of the least lift till the lift is equal. A new mark should then be made on the balance opposite to the central dot on the plate.

When the balance is at rest, the banking pin in the balance should be opposite to the banking stud in the cock, so as to give equal vibration on both sides. This is important for the following reason. The banking pin allows nearly a turn of vibration and the shell of the cylinder is but little over half a turn, so that as the outside of the shell gets round towards the center of the escape wheel, the point of a tooth may escape and jam the cylinder unless the vibration is pretty equally divided. When the banking is properly adjusted, bring the balance round till the banking pin is against the stud; there should then be perceptible shake between the cylinder and the plane of the escape wheel. Try this with the banking pin, first against one and then against the other side of the stud. If there is no shake, the wheel may be freed by taking a little off the edge of the passage of the cylinder where it fouls the wheel, by means of a sapphire file, or a larger banking pin may be substituted at the judgment of the operator. See that the banking pin and stud are perfectly dry and clean before leaving them: a sticky banking often stops a clock when nearly run down. Cylinder timepieces, after going for a few months, sometimes increase their vibration so much as to persistently bank. To meet this fault a weaker mainspring may be used, or a larger balance, or a wheel with a smaller angle of impulse. By far the quickest and best way is to *very slightly* lap the wheel by holding a piece of Arkansas stone against the teeth, afterwards polishing with boxwood and red stuff. So little taken off the wheel in this way as to be hardly perceptible will have great effect.

Sometimes the escape wheel has too much end shake. We must notice in the first place how the teeth are acting in the cylinder slot. Suppose, when the escape wheel is resting upon its bottom shoulder, the cylinder will ride upon the plane of the wheel, which will cause it to kick or give the wheel a trembling motion, then we know that the cylinder is too low for the wheel; therefore, we have not only to lower the escape top cock in order to correct the end shake, but we must also drive the bottom cylinder plug out a little in order to raise the cylinder sufficient to free it from the plane of the wheel. Now, if the end shake of the cylinder is correct previous to this, we shall now either have to raise the cock or drive the top plug in a little. But suppose the end shake of the escape pinion is excessive, and is, when the bottom shoulder is resting on the jewel, a little too low so that the bottom of the escape wheel runs foul of the cylinder shell; in this case we simply drive out the steady pins from the bottom escape wheel cock and file a piece off the cock, leaving it perfectly flat when we have enough off. We then insert the steady pins again, screw it down, and if the end shake is right, the escapement is mostly free and right also.

Now let us consider the frictions; there is the resistance of the pivots, which depends on their radius, on the weight of the balance, the balance spring, the collet, and the weight of the cylinder; these are called locking frictions. Then there are those of the planes, of the teeth of the wheel, of the lips of the cylinder. It is on these that the change and destruction of the cylinder are produced. To prevent this destruction, it is necessary to render the working parts of the cylinder very hard and well polished, as well as the teeth of the escape wheel.

The oil introduced in the cylinder is also a cause as in the dead beat. It may thicken; the dust proceeding from the impact of the escapement forms with the oil an amalgam which wears the cylinder. The firmness and constancy of the cylinder depend on the preservation and fluidity of the oil.

Then there are the accidental frictions; the too close opening of the cylinder, the play of the balance and of the wheel, with the thickening of the oil, changes the arc of vibration a good deal; the teeth of the wheel may not be sufficiently hollowed, so that the cylinder can revolve in the remaining space, for the oil with the dust forms a thickness which also changes the vibration. The drop should not be too great, for it is increased by the thickening of the oil and impedes the vibration.

Examination of Clocks.—In this particular escapement, when used for larger timepieces than watches, it is astonishing the variety of methods which are employed, yet the same results are expected. In examining such clocks we will first notice that the chariot, cock, etc., are so placed, many of them, that the last wheel in the train is a crown wheel, hence it is made to work at 90° with the escape wheel pinion which is set at right angles with the crown wheel pinion, and, as a matter of course, the cylinder is also set the same way. Now, this arrangement needs especial care, for it is quite natural that when the entire friction of the cylinder is only on the bottom part of the bottom pivot, the clock is sure to go faster than when the whole length of both pivots are more in contact with their jewel holes, which is always the case when the cylinder is parallel with all the pinions, instead of standing upon one pivot only. Now, although there must of necessity be a very great difference in timing the clock in the two different positions, yet we find no difference in the strength of mainspring or any part of the train, which is a mistake, for the result is simply this: the clock will gain time for the first few days after winding, and will then gradually go slower and slower until the mainspring is entirely exhausted. It is not very difficult to ascertain why it goes so fast after winding, for then the whole tension of the spring is on, and as there is not sufficient friction on the point of one pivot to counteract this, the banking pin is almost sure to knock, and will continue to knock for the first few days until a part of the spring's pressure is exhausted. Now, in this case the knocking of the banking pin alone would cause the clock to gain time, even if the extra tension of the mainspring did not assist it to do so. Hence, on the whole, the result is anything but satisfactory, for such a clock can never be properly brought to time.

Having said this much about the fault (which is entirely through the want of a little forethought with the manufacturer), I will give as good a remedy as I can suggest to give the reader an idea of how these faults may be put to right, if he is willing to spend the time upon them. In the first place take out the cylinder and make the bottom pivot perfectly flat instead of leaving it with a round end, as they are mostly left, which only allows just one part of the pivot to be in contact with the endstone. By leaving this pivot flat on the bottom, there is more surface in contact; hence. in a sense, more friction.

In some cases the whole pivot left flat would not be sufficient to retard the mainspring's force; then we must resort to other methods to effect a cure.

Well, our next method in order to try and get the clock to be a uniform timekeeper, is to change the mainspring for one well finished and not quite so strong as the original one. Perhaps some will say "why not do this before we go to the trouble of flattening the bottom pivot?" Just this; when a pivot is working only upon the bottom it is best to have a flat surface to work upon, as the balance is then oscillated with more uniformity, even when the mainspring is not exactly uniform in its pressure; therefore we do no harm—but good—by making the bottom pivot flat, and this alone will sometimes be sufficient to cure the fault of the banking knocking if nothing else.

To my mind, when such strong mainsprings are used as we generally see in this class of timepiece, neither of the jewel holes or pivots should be so small as they usually are. Fancy such small pivots as are mostly seen upon the escape wheel pinion being driven by such a strong mainspring. If we allow the clock to run down while the escape wheel is in place, we are very liable to find one or both pivots broken off before it gets run down. I think all such pivots ought to be sufficiently strong to stand the pressure of the mainspring through the train of wheels without coming to grief. But there is another reason why these pivots are liable to get broken off while letting the train run down; that is, the badly pitched depth we often find in the crown wheel and escape wheel pinion. We frequently find too much end shake to the crown wheel which, while resting one shoulder of the arbor against the plate puts the depth too deep, and on the other shoulder the depth is too shallow. Now, when the train is running rapidly this crown wheel is jumping about in the escape wheel pinion, so that the roughness of the running all helps to break off the escape wheel pivots. The best way to correct this depth is to notice how the screws fit in the cylinder plate—for these screws have to act as steady pins as well. If the holes where the screws go through are at all large, we then notice which would be the most convenient side to screw it securely in order to put a collet upon the shoulder of the crown wheel so that the depth will be right by making the end shake right with only fixing a collet to one shoulder. This depth, when correct, will also cause a more uniform pressure upon the escapement, and help to make the clock keep better time. We are supposing that this crown wheel is perfectly true, or it is not much use trying to correct the depth as mentioned above, for even if the end shake be ever so exact and the wheel teeth are out of true, we shall never get the depth to act as it ought, neither can the clock be depended upon for keeping going, regardless of keeping time. When this crown wheel is out of true it is best to rivet it true, not do as I have seen it done, placed in the lathe and topped true, and then the teeth rounded up by hand. This method simply means a faulty depth after all, for in topping the teeth, those teeth which require the most topping will, when they are finished, be shorter from the top to the base than those which do not get topped so much; therefore, some of the teeth are longer than the others, while the shorter ones are thicker; for when the wheel was originally cut the teeth were all cut alike. These remarks will apply to several kinds of wheels; for whenever a wheel is topped to put it true, we may depend we are making a very faulty wheel of it unless we have a proper wheel cutting machine.

The crown wheel must not be too thick because we will find the tooth to act with the inner edge, and what is left outside only endangers touching the pinion leaf which is next to come into action. Make sure the escape pinion is not too large, which sometimes happens. If it is, it must be reduced in size, or better, put in a new one. The crown wheel holes must fit nicely and the end shake be well adjusted. Do not spare any trouble in making this depth as perfect as you are able, as most stoppages happen through the faults in this place. It would be advisable, when sure the depth is correct, to drill two steady pin holes through the escapement plateau into the edge of the plates. When steady pins are inserted this will always ensure the depth being right when put together.

In some of these clocks it is not only the crown wheel, but frequently the escape wheel has too much end shake. The former, as I have said, can be corrected by making a small collet that will just fit over pivot, fasten it on friction tight, place the wheel in the lathe and turn the collet down until it is the same size as the other part of the arbor, then run off the end to the exact place for the end shake to be right. If it is properly done and a steel collet is used, it will not be detected that a collet has been put on. Now, when the escape wheel end shake is wrong we have to proceed differently under different circumstances for we must notice in the first place how the teeth are acting in the cylinder slot.

See that the cylinder and wheel are perfectly upright. Suppose, when the escape wheel is resting upon its bottom shoulder, the cylinder will ride upon the plane of the wheel, which will cause it to kick or give the wheel a trembling motion, then we know that the cylinder is too low for the wheel; therefore, we have not only to lower the escape top cock in order to correct the end shake, but we must also drive the bottom cylinder plug out a little in order to raise the cylinder sufficient to free it from the plane of the wheel. Now, if the end shake of the cylinder is correct previous to this, we shall either have to raise the cock or drive the top plug in a little. But suppose the end shake of the escape pinion is excessive, and is, when the bottom shoulder is resting on the jewel, a little too low so that the bottom of the escape wheel runs foul of the cylinder shell; in this case we simply drive out the steady pins from bottom escape wheel cock and file a piece off the cock, leaving it perfectly flat when we have got enough off. We then insert the steady pins again, screw it down, and, if the end shake is right, the escapement is mostly free and right also. It sometimes happens that the wheel is free of neither the top nor bottom plug, but should this be the case, sufficient clearance may be obtained by deepening the opening with a steel polisher and oilstone dust or with a sapphire file. A cylinder with too high an opening is bad, for the oil is drawn away from the teeth by the escape wheel.

If a cylinder pivot is bent, it may very readily be straightened by placing a bushing of a proper size over it.

These clocks are very good for the novice to exercise his skill in order to thoroughly understand the workings of the horizontal escapement. He is better able to see how the different parts act with each other than he is in the small watch. When the escape is correct he will find that the plane of the escape wheel will work just in the center of the small slot in the cylinder.

If he will notice how the teeth stand in the cylinder when the banking pin is held firmly upon the fixed banking pin, it will give him an idea of how this should be. At one side the lip of the cylinder is just about to touch the inside of the escape tooth, but the banking pin just prevents it from doing so, while on the other side the cylinder goes round just far enough to let the point of the next tooth just get on the edge of the slot, but it cannot get in owing to the intervention of the banking pin. If this is allowed to get in the slot just here, we then have what is called "a locking," which is, in reality, an overturned banking. If the other side is so that the banking pin does not stop it soon enough, the edge of the slot knocks upon the inside of the teeth and causes a trembling of the escape wheel, and the clock left in this form will never keep very good time. We may easily remedy this by taking off the hair spring collet; holding the cylinder firmly in the plyers, and with the left hand turn the balance a little outwards; this will bring the banking pins in contact before the cylinder touches the inside of the wheel teeth, and all is right, providing we are careful in not doing it too much; if so, we shall find the banking knock—a fault which is quite as bad, if not worse, than the one we are trying to remedy. Those particulars are the most important of anything in connection with the cylinder escapement. Yet, as this kind of clock is now being made up at such a low price, these seeming little items are frequently overlooked; hence, when they get into the hands of the inexperienced, there is often more trouble with them

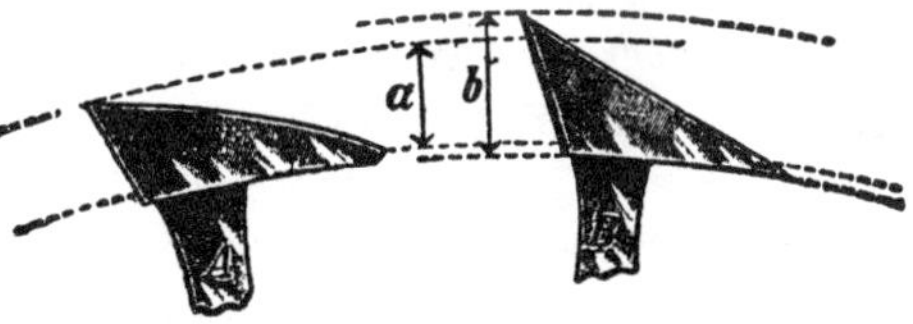

Fig. 56.

than there need be if they knew where to look for some of the faults which I have been endeavoring to bring to light, There are several other things in connection with this particular clock, but we will not comment further just now, but take them up when we are considering the trains, etc.

In the meantime we will resume our study of the cylinder escapement with particular reference to badly worn or otherwise ill fitting escape wheels, as many times, the other points being right, the wheel and cylinder may be such as to give either too great or too small a balance vibration.

A poor motion can also be due to a rough or a badly polished cylinder, but such a cylinder we rarely find. That with a wrong shape of the cylinder lips the motion is not much lessened can be seen in quite ordinary movements, where the quality is certainly not of the best neither are the lips correctly formed, nevertheless they have rather an excessive motion. To cover up these defects in such movements the cylinder wheel teeth are purposely given the shape as shown at B in Fig. 56, and to give sufficient power a strong mainspring is inserted.

With an excessive balance vibration we can usually conclude that it is an intentional deception on the part of the manufacturer, while a poor motion can generally be ascribed to careless methods in making. The continued efforts in making improvements to quicken and cheapen manufacturing processes very frequently result in the introduction of defects which are only found by the experienced and practical watchmaker

As to the causes which induce excessive balance vibrations? As this defect is generally found in the cheaper grades of cylinder escapements, having usually rather small, heavy, and often clumsy balances, those which have balances whose weight is probably less than they ought to be, need not here be further considered, and it only remains for us to look to the cylinder or the escape wheel for the causes which produce these excessive vibrations. It will be found that the cylinder is smaller in diameter than usually employed in such a size of clock; the escape wheel is naturally also smaller, and its teeth generally resemble B, Fig. 56, while A shows the correct shape of a tooth for a wheel of that diameter.

In using small cylinders we can give the escape wheel teeth a somewhat greater angle of inclination than generally used, but that the proper amount of incline is exceeded is proved by the fact that the balance vibrates more than two-thirds of a turn. It can also be readily seen that with a tooth like B a greater impulse must be imparted than one with an easy curve like A, and the impulse is still further increased as the working width of the tooth B (the lift) is greater, indicated by line *b*, while the same line in a correct width of tooth, as shown at *a*, is considerably shorter.

In addition to what has been said of these escapements, we also find them provided with very strong mainsprings to give the necessary power to a tooth like B with its steeply inclined lifting face or impulse angle.

To decrease the great amplitude of the balance vibrations many watchmakers simply replace the strong mainspring with a weaker one. But this proceedure is not advantageous as the power of the escape wheel tooth is insufficient to start the balance going and this is due to two causes. First, the great angle of the escape wheel tooth, and secondly, the inertia of the balance. It is only by violently shaking such a clock that we are enabled to start it going. And the

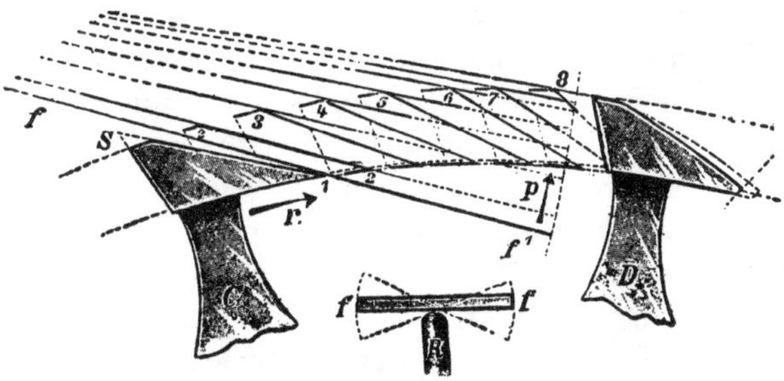

Fig. 57. Fig. 58.

owner soon becomes dissatisfied from its frequent stoppage due to setting of the hands and other causes so that he will be often obliged to shake it until it starts going once more.

For properly correcting these defects the best method to pursue is to replace the cylinder wheel with another one, whose teeth are of the shape as shown at Fig. 55 and without question a good workman will always replace the escape wheel if the clock is of fair quality. But if a low grade one, we would hardly be justified in going to the expense of putting in new wheels, as the low prices for which these clocks are sold preclude such an alteration. As we must improve the wheel some way to get a fair escapement action we can place it in a lathe and while turning, hold an oil stone slip against it, we can remove the point S, Fig. 56. After removing the point the tooth will now have the form as shown at tooth C, Fig. 57. We now take a thin and rather broad watch mainspring, bending a part straight and holding it in the line *f f*, and revolving the wheel in the direction as shown by the arrow, its action being indicated by figures 1 to 8; beginning at the point of the tooth at 1, at 2 it comes in contact with the whole of the lifting face, and from 3 to 8 only on the projecting corner which was left by the oil stone slip in removing the heel of the tooth. In this way all the teeth are acted upon until the corner is entirely removed. Of course oil stone dust and oil is first used upon the spring for grinding, after which the teeth are polished with diamantine. Care must be observed in using the spring so as not to get the end *f* too far into the tooth circle, as it would catch on the heel of the preceding tooth.

After the foregoing operation has been completed any feather edge remaining on the points of the teeth must be removed with a sapphire file and polished; we will now have a tooth as indicated by D, Fig. 57. This shape of tooth can hardly be said to be theoretically correct, nevertheless it is a close approximation of the proper form of tooth, which is shown by the dotted lines, and will then perform its functions much better than in its original condition.

Fig. 58 also shows how the spring must be moved from side to side—indicated by dotted lines—so that the lifting face will have a gentle curve instead of being flat; R represents the tooth.

After the wheel has been finished, as described, and again placed in the clock, it will be found that the balance makes only two-thirds of a turn, and as a result the movement can be easier brought to time and closely regulated.

In the above I have described the cause of excessive balance vibration, the method by which it can be corrected, and in what follows I shall endeavor to make clear the reasons for a diminished balance vibration or poor motion. It has been probably the experience of most watchmakers to repair small cylinders of a low grade, having a poor motion or no motion at all, and it would hardly be profitable to expend much time in repairing them. But considerable time is often wasted in improving the motion by polishing pivots and escape wheel teeth, possibly replacing the cap jewels, or even the hole jewels, increasing the escapement depth or making it shallower, examining the cylinder and finding nothing defective, and as a last effort putting in a stronger mainspring. But all in vain, the balance seems tired and with a slight pressure upon an arm of the center wheel it stops entirely.

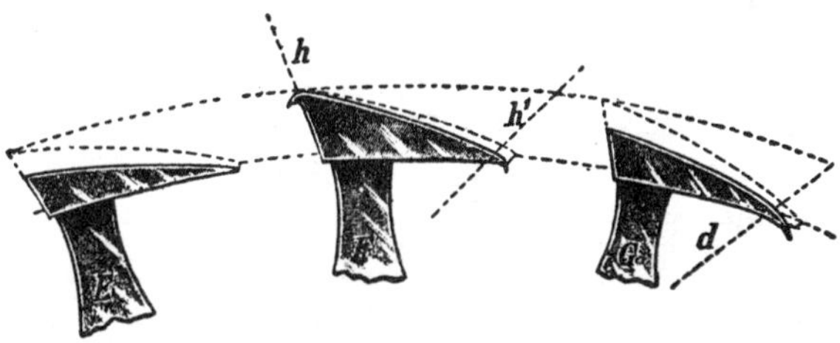

Fig. 59.

In this case, as in a former one, in fact, it is necessary at all times to carefully examine the cylinder wheel. My reason for not considering the cylinder itself so much as the wheel is that the makers of them have made a considerable advance in their methods of manufacture, so we find the cylinders fairly well made and generally of the correct size. Even if the cylinder is incorrectly sized, either too large or small, it does not necessarily follow that the watch would have a bad motion, as I have frequently had old movements where the cylinder was incorrectly proportioned and yet the motion was often a good, satisfactory one. Generally speaking, the cylinder escapement is one which admits of the worst possible constructive proportions and treatment, as we have often examined such clocks when left for repairs, that, notwithstanding their being full of dirt, worn cylinder, broken jewel holes, etc., they have been running until one of the cylinder pivots has been completely worn away.

It only remains to look for the source of the trouble in the escape wheel. If we examine the wheel teeth carefully, we shall find them resembling those in Fig. 59, the dotted lines representing the correct shape of the teeth for a wheel of that diameter.

Why do we find wheels having such defective teeth? This is probably due to their rapid manufacture, as they very likely had the correct shape when first cut, but by careless grinding and polishing they were given improper forms, careless treatment being very evident at tooth F, which we find on examination has a feather edge at the point as well as at the heel of the tooth. If we grind these edges of the tooth with a ruby file, by placing it in the position as indicated by dotted lines h and h^1, and afterwards polishing the tooth point, we will find that the balance makes a better vibration. A wheel, having teeth like E, can still be used, but the balance will have a very poor motion, due to the fact that the impulse angle of the wheel tooth is too small; the impulse faces of the teeth having so small an angle, are nearly incapable of any action. With a tooth like G, if we should remove its bent point at the dotted line *d*, then the tooth would be too short, and as the inclination of the impulse face is incapable to produce a proper action, a new wheel must be used, having teeth as shown at Fig. 55.

The reasons why a tooth, having the shape as shown at F and G (Fig. 59), will cause a bad action of the escapement and also why in such cases with a greater force acting on the wheel, causes a stopping of the clock, I will endeavor to explain with the aid of the illustration Fig. 60. Here we clearly see the curved points of the teeth resting against the outer and inner walls of the cylinder while the escapement is in action.

Teeth H and H^1 represent the defective tooth, while K and K^1 shows a correctly formed tooth for a wheel of the same size, the correct depth and positions where the tooth strikes the inner and outer walls of the cylinder. It will be readily seen that the position of the tooth point upon the cylinder (at *c*) is most favorable in reducing the resistance to the least possible amount. But in the case of the teeth H and H^1 the condition is entirely different. We find that it was necessary to set the escapement very deeply in order that it could perform its functions at all, and, as a conse-

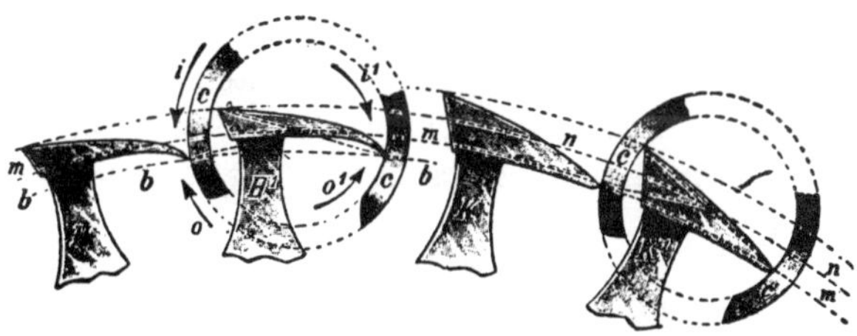

Fig. 60.

quence, we have a false proportion; the effects being considerably increased by the worst possible position of the teeth H and H^1, where they touch the cylinder. While the cylinder *c* is turning in the direction shown by the arrows $i\ i^1$, the tooth does not affect the cylinder to any extent; but during the reverse movement of the cylinder, in the direction of $o\ o^1$, an excessive amount of engaging friction must take place. A close inspection of the drawing will enable us to see that there is a great tendency of the cylinder to drag the tooth along with it during each of these motions. It is evident that in such a case the friction will eventually become so great as to lock the escapement, and if greater pressure is applied by any means to teeth H and H^1, it is easily seen that this effect will take place much more rapidly. Replacing the escape wheel with one of correctly formed teeth and size is the best means at our disposal.

CHAPTER XIII.

THE DETACHED LEVER ESCAPEMENT AS APPLIED TO CLOCKS.

As the clock repairer is almost of necessity a watchmaker, or hopes to become one, and as he must enter deeply into the study of all questions pertaining to the detached lever in its various forms before he can make any progress at all in watchmaking, it would seem unnecessary to repeat in these pages that which has already been so well said and so perfectly drawn, described and illustrated by such authorities as Moritz Grossman, Britten, Playtner and the various teachers in the horological schools, to say nothing of an equally brilliant and more numerous coterie of writers among the French, Germans and Swiss, so that the reader is referred to these writers for the mathematics and drawings which already so fully cover the technical and theoretical properties of the detached lever escapement. A few words as to its adaptation to clocks may, however, not be out of place.

Anyone who sees the clocks of to-day would be inclined to suppose that the first clocks were constructed with pendulums, because this is evidently the most simple and reliable system for clocks, and that the employment of the balance has been suggested by the necessity for portable time pieces. This is, however, not the case, for the first clocks had a verge escapement with a crude balance consisting of two arms, carrying shifting weights for regulation. The pendulum was not used until about three hundred years after the invention of the first clock.

After the invention of the dead beat escapement, with its great gain in accuracy by the reduction of the arc of pendulum oscillation, attempts were made to combine its many virtues with the necessarily large vibrations of a balance and thus get all the advantages of both systems. By placing the lever on the arbor of the anchor, it was possible to multiply the small angle of impulse on the pallets very considerably at the balance, and to make all connection between them cease immediately after the impulse had been given. The dead beat escapement was thus converted into the detached lever escapement and the latter made available for both watches and clocks. Another important feature of this

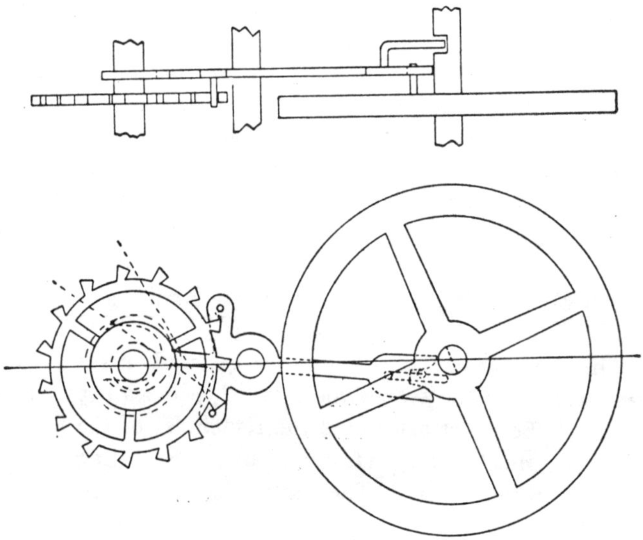

Fig. 61. Pin Escapement for Clocks.

escapement is that when properly proportioned it will not set on the locking or lifting, but will start to go as soon as power is applied to the escape wheel through the train. This cannot be said of the cylinder, duplex, or detent escapements, and it will be seen at once that this has an important influence upon the cost of construction, which must always be considered in the manufacture of cheap clocks in enormous quantities.

The lever escapement with pins for pallets and the lifting planes on the teeth of the escape wheel, which is the one usually put into cheap clocks, is from the theoretical point of view a very perfect form, because its lifting and locking take place at exactly the same center distance and at the same angles, which again allows for greater latitude in cheap construction, while still maintaining a reasonably accurate rate of performance. These are the main reasons why the pin anchor has such universal use in cheap clocks.

As this escapement is generally centered between the plates, banking pins are dispensed with by extending the counterpoise end of the lever far enough so that its crescent shaped sides will perform that office by banking against the scape wheel arbor; see Fig. 61. The fork end of the lever engages with an impulse pin carried in the balance and the balance arbor is cut away to pass the guard point or dart, thus doing away with the roller table. In other constructions the roller table is supplied in the shape of a small brass collet which carries the pin and has a notch for the guard point, thus making a single roller escapement.

The diameter of the lifting pins is generally made equal to 2½ degrees of the scape wheel, which gives a lift of 2 degrees on the pallet arms, and the remainder of the lift, 6½ degrees, must be performed by the lifting planes of the wheel teeth. The front sides of the wheel teeth are generally made with 15 degrees of draw and the lever should bank when the center of the pin is just a little past the locking corner of the tooth. Other details of the pin anchor escapement coincide with the ordinary pallet form, as used in watches, and the reader is referred for them to the works of the various authors mentioned previously.

The trouble with the majority of these clocks is in the escapement and balance pivots, and to these parts are we going to direct particular attention, for often, be it ever so clean, the balance gets up a sort of "caterpillar motion" that is truly distressing, and if no more is done we may expect a "come back" job in a very short time. In taking down the movement the face wheels are left in place, but sometimes it may be necessary to remove the "set wheel" of the alarm in order to proceed as we do. Remove the screws or pins that hold the plates together in the vicinity of the escapement, leaving the others, though if screws they may be loosened slightly; pry up the corner of the plate over the lever to loosen one pivot of same and let it drop away from the scape wheel sufficiently to let the wheel revolve until it is locked by a wire or pegwood previously inserted in the train, after which the plates can be pried apart more conveniently to permit the lever being removed entirely, also the scape wheel and the one next following. As nickel clocks differ in make-up, the operator must, of course, exercise judgment as to the work in hand to accomplish this.

Have ready a straight-sided tin pail, with cover, that will hold at least one-half gallon of gasoline and of diameter large enough to receive the largest brass clock; remove the wire or pegwood and immerse the clock into the fluid and allow it to run down; this will loosen all the dirt and gummy oil and clean the clock very effectually. Let it remain long enough for all the dirt to settle to the bottom of the pail; then remove and wipe as dry as possible with a soft rag; by having no binder on the spring it is permitted to uncoil to its full, and thereby remove all gummy oil between its coils. Now peg out the holes of the wheels removed and of the lever and that portion of our work is complete.

Polish or burnish the pivots of wheels either in a split chuck in the lathe, or by holding in a pin vise, resting the pivot on a filing block (an ivory one is best), and revolving between the fingers, using a smooth back file for burnishing, after the manner of pointing up a pin tongue, only let the file be held flat, so as to maintain a cylindrical pivot as nearly as possible. The scape wheel is now polished, i. e., the teeth, with a revolving bristle wheel on a polishing lathe, charged with kerosene oil and tripoli. This will smooth up the teeth in fine form, especially those wheels that work into a lever with pin pallets. Clean the scape wheel by dipping into gasoline to remove all the oil and tripoli. The other wheel may simply be brushed in the gasoline or dipped and then brushed dry.

We now turn our attention to the lever and closely examine the pallets with a glass; if there are the least signs of wear upon them they must be removed. If the lever with pin pallets it is better to remove the steel pins and insert new ones. See if the holes in the anchor where they are inserted will admit a punch to drive them out from the back; if not, open these holes with a drill until the ends of the pins are reached. Put a hollow stump with a sufficiently large hole in the staking tool, and by placing the pins in the stump they can be driven out successively, being sure that the driving punch is no larger than the pins; drive or insert into their places a couple of needles of the proper size, and then break off at correct lengths; this completes the job in this particular style of lever.

With the other style the job is not quite so easy; with a pair of small round-nose pliers grasp the brass fork close up to the staff and bend it back from the pallets till it lays parallel with the staff; treat the counter poise of the fork in like manner; place a thin zinc lap into the lathe, charged with flour of emery, and with the fingers holding the pallets grind off all wheel teeth marks on both the impulse and locking faces of the pallets. Then polish with a boxwood lap charged with diamantine. It is surprising how speedily this can be done if laps are at hand. The only care necessary is not to round off the corners of the pallets, and as they are so large they can be easily held flat against the laps with the thumb and finger as before stated. Bend back the fork and counterpoise to their original position. The fork must now be attended to; see that no notches are worn in the horns of the fork by the steel impulse pin in the balance; if they appear they must be dressed out and polished, also examine and smooth if necessary the ends of the horns that bank against the balance staff. These may seem small matters, but they are often what cause all the trouble.

We now come to the balance staff and the hardened screws in which the staff vibrates; their irregularities are often the source of much vexation, and there is only one way to go at it and that is with a will and determination to make it right. Examine the points of the staff and see if they are in their normal shapes and are sharp and bright; if so they will probably do their work. But we will suppose we have a bad case in hand and will therefore treat it thoroughly according to our method. We find the staff is large in diameter and the ends are very blunt; the notch in the center has a burr on each side as hard as glass, making an admirable cause for catching the horns of the fork in some of the vibrations or in a certain position; also the round part of the staff back of the notch is rough and looks as if it never had been finished, and, in fact, it has not, for it truly appears as if half, if not all, the nickel clocks are made to be finished by the watchmaker. Remove the hairspring and place the staff between the jaws of your bench vise, with the jaws close up to the staff, but not gripping it, the balance "hub" resting on the jaws with the impulse pin also down between the jaws. Have a block of brass about one-fourth inch square; rest it on top of the staff, or on its pivot end, if it may so be called, holding it with the thumb and finger of the left hand. Strike this block with a hammer and drive out the staff; a hollow punch is apt to be split in doing this, and as the pivot is to be re-pointed no harm will be done to the pivot or to the end of the staff. Draw the temper so it will work easily, insert into a split chuck and turn up new points; have them long and tapering, that is, turn the points to a long slant from the end of the staff to the body of same, or at least twice as much taper as they generally have; polish off the back of the notch or round part of the staff with an oil stone slip. Remove from the chuck, smear all over with powdered boracic acid by first wetting the staff in water, and then heat to a bright red and plunge straight into water; it will now be white and hard; draw the temper from the staff in the vicinity of the notch, leaving the pivot points hard as before; re-insert into the chuck and with diamantine polish the points and also around the staff in the vicinity of the notch. The drawing of the temper from the center of the staff to a spring temper is to make it less liable to breakage while driving on the balance. Fasten the staff tight in the vise and with a rather stout brass tube, large enough to step over the largest staff, drive on the balance to its former position.

If the workman has a pivot polisher with a large lap, the job may be done, without softening the staff or removing the balance, by grinding the pivots. In turning the staff we often find it almost impossible to hold true. We straighten the best we can and then turn up our pivots, and as long as the untruth of the staff will not cause the balance to wabble to such an extent as to give us a headache or cause us to look cross-eyed it will do. We do not wish to be misunderstood or to give the impression that we go on the principle of "good enough"; but as gold dollars cannot be bought for seventy-five cents, neither can a workman devote the time to have everything perfect for fifty cents; and for this very reason do they come in such an unfinished state from the manufacturers.

Next see if the two screws in which the balance vibrates have properly cut countersinks; if rough or irregular, better at once draw the temper, re-drill with a sharp-angled drill and again harden.

Occasionally a bunch of these clocks will come in with both pivots and cones badly rusted. This has generally been caused by acid pickling, or some sort of chemical hardening at the factory; the acid or alkali gets into the pores of the steel and comes out after the clock has been shipped.

They are generally made in such quantities that fifty or a hundred thousand of them have been distributed before finding out that they were not right and then it is a matter of two or three years before the factory hears the last of it. The trouble is attributed to bad oil, or to anything else but the hardening, which is the real cause, and the expense of taking back and refitting the balance arbors and cones, paying freight both ways and standing the abuse of disgruntled jewelers, goes on until life becomes anything but a bed of roses. Every jeweler should warn the factory immediately on finding rust in the cones of a shipment of new clocks and not attempt to fix them himself, as such a fault cannot be discovered at the factory and every day it continues means more thousands of clocks distributed that will give trouble.

Our clock is now ready to be put together. Wind up the spring and slip on the binder; then put in the wheels and lever; then adjust the balance and hairspring to their proper places, slightly wind the mainspring and then see (by bringing either horn against the staff) whether it sticks and holds the balance; if so, shorten the fork slightly by bending; try this until the balance and fork act perfectly free and safe. Slightly oil the balance pivots; an excess will only gather dust and prove detrimental, as the countersinks form an admirable place for holding the dust. Now oil the remaining parts and we are sadly mistaken if our clock does not make a motion that will be gratifying.

The foregoing process may seem tedious and uncalled for and too close mention made of the lesser portions of the work, but we must not "despise the day of small things," and as we are watchmakers, we are expected to do this work, even though troublesome and the pay small; we should also bear in mind that if we only make a nickel clock run and keep fair time, it will be a large advertisement, and possibly repay tenfold. It takes only an hour to do this job complete, while in many cases only the balance staff needs attention.

Sometimes such a clock will be apparently all right mechanically but will continue to lose time; then it is probable that the balance does not make the proper number of vibrations, which causes the clock to lose time. There is one way to tell this, which will soon locate the trouble: count the train to ascertain the number of vibrations the balance should make in one minute. You do this by counting the number of teeth in the center wheel, which we will say is 48; third wheel 48; fourth wheel, 45; escape, 15. Multiply all teeth together, which give us 48x48x45x15 = 1,555,200. Now count the leaves in the third wheel pinion, which is 6; fourth, 6; escape, 6. Multiply these together, 6x6x6 = 216; now divide the leaves into the teeth, 1,555,200 ÷ 216 = 7,200, which is the number of whole vibrations some Ansonia alarm clocks make in one hour. Dividing 7,200 by 60 gives us 120, the number of vibrations per minute. Now the balance must make 120 vibrations in one minute, counting the balance going one way. If the balance only vibrates 118, the clock will lose time and the hairspring must be taken up or made shorter, until it makes the required number of vibrations. If it should vibrate 122 the clock would gain and the hairspring should be let out.

Find out the number of vibrations your balance should make and work accordingly; and if you find that the balance makes the proper number of vibrations in one minute, then the trouble must lie in the center post, which has not enough friction to carry the hands and dial wheels, or the wheel that gears into the hour wheel and regulates the alarm hand is too tight and holds back the hands. You should find some trouble about these wheels or center post. for where a balance makes the proper number of vibrations in one minute, the minute hand cannot help going around if everything else is correct.

Fig. 62 illustrates the escapement of the Western Clock Manufacturing Company for their cheap levers. It has hardened steel pallets placed in a mould and the fork cast around them, thus insuring exact placing of the pallets, and the company claim that they thus secure a detached lever escapement with all the advantages of hardened and polished pallets at a minimum cost.

Mr. F. Dauphin, of Cassel, Germany, on page 387 of Der Deutsche Uhrmacher Zeitung, 1905, has described a serious fault of some of the cheap American alarm clocks in the

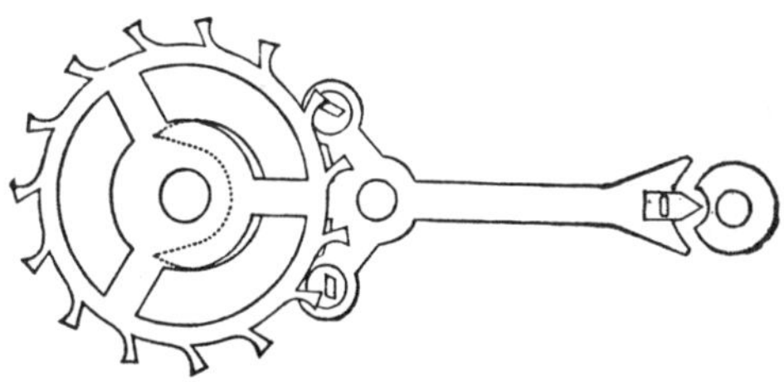

Fig. 62.

depthings of the escapements and how he remedied it by changing the position of the pins. It is to be regretted that Mr. Dauphin did not state the measurements of the parts as nearly as possible in this article and also give the manufacturer's name, simply to enable others not as skilled as he is to do what I would do in such a case; namely, to return it to the jobber and get a new and correct movement in its stead *free of charge*. The American clock manufacturers are very liberal in this respect and never hesitate to take back a movement that was not correct when it left the factory, even when the customer, in the attempt to correct it, has spoiled it; spoiled or not, it goes to the waste pile anyway, when it reaches the factory. I seriously doubt the ability of the average watch repairer to correctly change the position of the pins as suggested; and to change the center of action of the lever is certainly a desperate job. I herewith give a correct drawing of an escape wheel and lever, such as are used in the above cited clocks, made from measurements of the parts of a clock. The drawing is, of course, enlarged. The measurements are: Escape wheel, actual diameter, 18.11 mm.; original diameter, 17 mm.; lever, from pin to pin, outside, 9.3 mm.; distance of centers of wheel and lever, 10.0 mm. I found that all these measurements almost exactly agree with Grossmann's tables, and I do not doubt at all that they were taken from

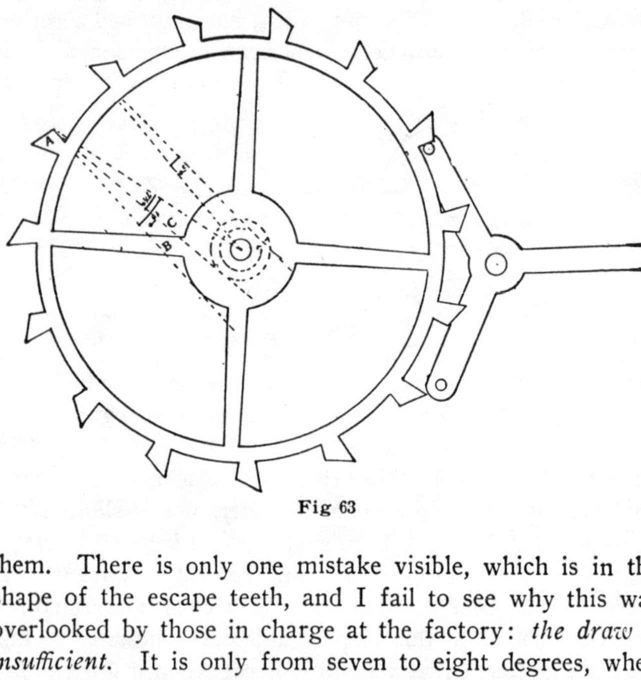

Fig 63

them. There is only one mistake visible, which is in the shape of the escape teeth, and I fail to see why this was overlooked by those in charge at the factory: *the draw is insufficient.* It is only from seven to eight degrees, when it should be fifteen degrees. I show this at tooth A, in the drawing, where you can see both dotted lines, measuring the angle of draw; line C as it is and line B as it should be.

Notwithstanding the deficient draw, this escapement will work safely as long as the pivot holes are not too large, or worn sideways; but if you want to make it safe you should file the locking faces of teeth slightly under; even if you do not make a model job, you have remedied the fault. Make a disk of 18.11 mm. diameter, put it on the arbor of the wheel and lay a straight edge from the point of the tooth to the center of the disk, so as to see how much it needs to be filed away. Even if this undercutting is not very true it will go.

TO MEASURE WHEELS WITH ODD NUMBERS OF TEETH. —This is a job that so frequently comes to the watchmaker who has to replace wheels or pinions that the following simple method should be generally appreciated. It depends upon the fact that the radius of a circle, R, Fig. 64, equals the versed sine E (dotted) plus the cosine B. If we stand such a wheel on the points of the teeth, A C, and measure it we shall get the length of the line T B only, when what we really need is the length of the lines T B E, to give us the real diameter for our wheel, and E we find has been cut away, so that we cannot measure it. Say it is a 15-tooth escape wheel, then by standing the old wheel up on the anvil of a vertical micrometer, resting it on two of its teeth, as shown in Fig. 64, the measuring screw can be brought in contact with the tooth diametrically opposite the space between the two teeth on the anvil, and a measurement taken, which will be less than the full diameter by the versed sine of 12 degrees (half the angle included between two adjoining teeth). By bringing each tooth in succession to the top, such a wheel could be measured in fifteen different directions, which would vary slightly, owing to the fact that some of the teeth may be bent a little, but the mean of these measures should be what the wheel would measure were the teeth in their original shape. If a tooth was badly bent the three measures in which it was involved could be rejected, and the mean of the other twelve measures taken as the correct value and found to be, we will say, 0.732 inch. Consulting a table of natural sines the cosine of 12 degrees is found to be 0.97815, which subtracted from 1 gives

0.02185 as the versed sine. Multiplying this by 0.36 inch (practically one-half of our measured 0.732) to get the approximate radius of the wheel, we get 0.008 inch, the amount to be added to the micrometer measurement in order to get the diameter of the blank.

At first sight it may appear like a vicious principle that we must know the radius of the wheel before we can deter-

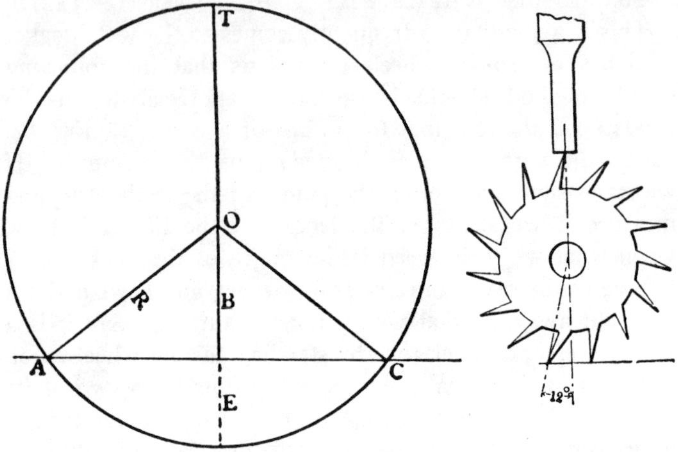

Fig. 64. Getting the full diameter.

mine the value of the correction in question, but we only need to know the radius approximately in order to determine the correction very closely, an error of 1-20 inch in the assumed value of the radius producing an error of only 0.001 inch in the value of the correction.

This method can of course be applied to all wheels and pinions to get the size of the blank; with other wheels than escape wheels, where the pitch line and the full diameter do not coincide, the addendum may be subtracted from the full diameter to get the pitch line.

Cutters for Clock Trains.—In cutting escape wheels or others with wide space between the teeth, it is a matter of some difficulty with many people to enable them to set the cutter properly.

Mr. E. A. Sweet calls attention to the fact that if a cutter be set so that its center touches the circumference of the wheel to be cut, said cutter will be in the proper position for work. For instance, if an escape wheel is to be cut, it is sufficient to set the cutter in such a manner that that portion of the cutter forming the bottom of the cut touch the circumference of the blank at the center of the cutter. It may then be backed off and fed in with the certainty of being properly placed.

CHAPTER XIV.

PLATES, PIVOTS AND TIME TRAINS.

Before going further with the mechanism of our clocks we will now consider the means by which the various members are held in their positions, namely, the plates. Like most other parts of the clock these have undergone various changes. They have been made of wood, iron and brass and have varied in shapes and sizes so much that a great deal may be told concerning the age of a clock by examining the plates.

Most of the wooden clocks had wooden plates. The English and American movements were simply boards of oak, maple or pear with the holes drilled and bushed with brass tubes—full plates. The Schwarzwald movements were generally made with top and bottom boards and stanchions, mortised in between them to carry the trains, which were always straight-line trains. The rear stanchions were glued in position and the front ones fitted friction-tight, so that they could be removed in taking down the clock. This gave a certain convenience in repairing, as, for instance, the center (time) train could be taken down without disturbing the hour or quarter trains, or vice versa. Various attempts have been made since to retain their convenience with brass plates, but it has always added so much to the cost of manufacture that it had to be abandoned.

The older plates were cast, smoothed and then hammered to compact the metal. The modern plate is rolled much harder and stiffer and it may consequently be much thinner than was formerly necessary. The proper thickness of a plate depends entirely upon its use. Where the movement rests upon a seat board in the case and carries the weight of a heavy pendulum attached to one of the plates they must be made stiff enough to furnish a rigid support for the pendulum, and we find them thick, heavy and with large pillars, well supported at the corners, so as to be very stiff and solid. An example of this may be seen in that class of regulators which carry the pendulum on the movement. Where the pendulum is light the plates may therefore be thin, as the only other reason necessary for thickness is that they may provide a proper length of bearing for the pivots, plus the necessary countersinking to retain the oil.

In heavy machinery it is unusual to provide a length of box or journal bearing of more than three times the diameter of the journal. In most cases a length of twice the diameter is more than sufficient; in clock and other light work a "square" bearing is enough; that is one in which the length is equal to the diameter. In clocks the pivots are of various sizes and so an average must be found. This is accomplished by using a plate thick enough to furnish a proper bearing for the larger pivots and countersinking the pivot holes for the smaller pivots until a square bearing is obtained. This countersinking is shaped in such a manner as to retain the oil and as more of it is done on the smaller and faster moving pivots, where there is the greatest need of lubrication, the arrangement works out very nicely, and it will be seen that with all the lighter clocks very thin plates may be employed while still retaining a proper length of bearing in the pivot holes.

The side shake for pivots should be from .002 to .004 of an inch; the latter figure is seldom exceeded except in cuckoos and other clocks having exposed weights and pendulums. Here much greater freedom is necessary as the movement is exposed to dust which enters freely at the holes for pendulum and weight chains, so that such a clock would stop if given the ordinary amount of side shake.

We are afraid that many manufacturers of the ordinary American clock aim to use as thin brass as possible for plates without paying too much attention to the length of bearing. If a hole is countersunk it will retain the oil when a flat surface will not. The idea of countersinking to obtain a shorter bearing will apply better to the fine clocks than to the ordinary. In ordinary clocks the pivots must be longer than the thickness of the plates for the reason that freight is handled so roughly that short pivots will pop out of the plates and cause a lot of damage, provided the springs are wound when the rough handling occurs.

It will be seen by reference to Chapter VII (the mechanical elements of gearing), Figs. 21 to 25, that a wheel and pinion are merely a collection of levers adapted to continuous work, that the teeth may be regarded as separate levers coming into contact with each other in succession; this brings up two points. The first is necessarily the relative proportions of those levers, as upon these will depend the power and speed of the motion produced by their action. The second is the shapes and sizes of the ends of our levers so that they shall perform their work with as little friction and loss of power as possible.

To Get Center Distances.—As the radii and circumferences of circles are proportional, it follows that the lengths of our radii are merely the lengths of our levers (See Fig. 24), and that the two combined (the radius of the wheel, plus that of the pinion) will be the distance at which we must pivot our levers (our staffs or arbors of our wheels) in order to maintain the desired proportions of their revolution. Consequently we can work this rule backwards or forwards.

For instance if we have a wheel and pinion which must work together in the proportion of 7½ to 1; then 7½ + 1 = 8½, and if we divide the space between centers into 8½ spaces we will have one of these spaces for the radius of the *pitch circle* of the pinion and $7\frac{1}{2}$ for the *pitch circle* of the wheel, Fig. 65. This is independent of the number of teeth so long as the *proportions* be observed; thus our pinion may have eight teeth and the wheel sixty, $60 \div 8 = 7.5$, or $75 \div 10 = 7.5$, or $90 \div 12 = 7.5$, or any other combination of teeth which will make the correct proportion between them and the center distances. The reason is that the teeth are added to the wheel to prevent slipping, and if they did not agree with each other and also with the proportionate distance between centers there would be trouble, because the desired proportion could not be maintained.

Now we can also work this rule backwards. Say we have a wheel of 80 teeth and the pinion has 10 leaves but they do not work together well in the clock. Tried in the depthing tool they work smoothly. $80 \div 10 = 8$, consequently our center distance must be as 8 and 1. $8 + 1 = 9$; the wheel must have 8 parts and the pinion 1 part of the radius of the pitch circle of the wheel. Measure carefully the diameter of the pitch circle of the wheel; half of that is the pitch radius, and nine-eighths of the pitch radius is the proper center distance for that wheel and pinion.

Say we have lost a wheel; the pinion has 12 teeth and we know the arbor should go seven and one-half times to one of the missing wheel; we have our center distances established by the pivot holes which are not worn; what size should the wheel be and how many teeth should it have? $12 \times 7.5 = 90$, the number of teeth necessary to contain the teeth of the pinion 7.5 times. $7.5 + 1 = 8.5$, the sum of the center distances; the pitch radius of the pinion can be closely measured; then 7.5 times that is the pitch radius of the missing wheel of 90 teeth. Other illustrations with other proportions could be added indefinitely but we have, we think, said enough to make this point clear.

Conversion of Numbers.—There is one other point which sometimes troubles the student who attempts to follow the expositions of this subject by learned writers and that is the fact that a mathematician will take a totally different set of numbers for his examples, without explaining why. If you don't know why you get confused and fail to follow him. It is done to avoid the use of cumbersome fractions. To use a homely illustration: Say we have one foot, six inches for our wheel radius and 4.5 inches for

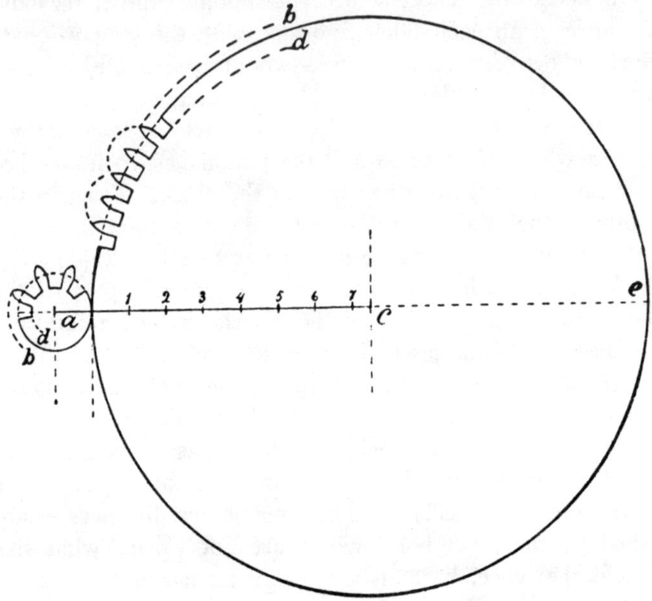

Fig. 65. Spacing off center distances; c, center of wheel; e, pitch circle; d, dedendum; b, addendum; a, center of pinion.

our pinion radius. If we turn the foot into inches we have 18 inches. $18 \div 4.5 = 4$, which is simpler to work with. Now the same thing can be done with fractions. In the above instance we got rid of our larger unit (the foot) by turning it into smaller units (inches) so that we had only one kind of units to work with. The same thing can be done with fractions; for instance, in the previous example we can get rid of our mixed numbers by turning everything into fractions. Eighteen inches equals 36 halves and 4.5 equals 9 halves; then $36 \div 9 = 4$. This is called the conversion of numbers and is done to simplify operations. For instance in watch work we may find it convenient to turn all our figures into thousands of a millimeter, if we are using a millimeter gauge. Say we have the proportions of 7.5 to 1 to maintain, then turning all into halves, $7\frac{1}{2} \times 2 = 15$ and $1 \times 2 = 2$. $15 + 2 = 17$ parts for our center distance, of which the pitch radius of the pinion takes 2 parts and that of the wheel 15.

THE SHAPES OF THE TEETH.—The second part of our problem, as stated above, is the shapes of the ends of our levers or the teeth of our wheels, and here the first consideration which strikes us is that the teeth of the wheels approach each other until they meet; roll or slide upon each other until they pass the line of centers and then are drawn apart. A moment's consideration will show that as the teeth are longer than the distance between centers and are securely held from slipping at their centers, the outer ends *must* either roll or slide after they come in contact and that this action will be much more severe while they are being driven towards each other than when they are being drawn apart after passing the line of centers. This is why the *engaging* friction is more damaging than the *disengaging* friction and it is this *butting* action which uses up the power if our teeth are not properly shaped or the center distances not right. Generally speaking this butting causes serious loss of power and cutting of the teeth when the pivot holes are worn or the pivots cut, so that there is a side shake of half the diameter of the pivots, and bushing or closing the holes, or new and larger pivots are then necessary. This is for common work. For fine work the center distances should be restored long before the wear has reached this point.

If we take two circular pieces of any material of different diameters and arrange them so that each can revolve around its center with their edges in contact, then apply power to the larger of the two, we find that as it revolves its motion is imparted to the other, which revolves in the opposite direction, and, if there is no slipping between the two surfaces, with a velocity as much greater than that of the larger disc as its diameter is exceeded by that of the larger one. We have, then, an illustration of the action of a wheel and pinion as used in timepieces and other mechanisms. It would be impossible, however, to prevent slipping of these smooth surfaces on each other so that power (or motion) would be transmitted by them very irregularly. They simply represent the "pitch" circles or circles of contact of these two mobiles. If now we divide these two discs into teeth so spaced that the teeth of one will pass freely into the spaces of the other and add such an amount to the diameter of the larger that the points of its teeth extend inside the pitch circle of the smaller, a distance equal to about 1⅛ times the width of one of its teeth, and to the smaller so that its teeth extend inside the larger one-half the width of a tooth, the ends of the teeth being rounded so as not to catch on each other and the centers of revolution being kept the same distance apart, on applying power to the larger of the two it will be set in motion and this motion will be imparted to the smaller one. Both will continue to move with the same relative velocity as long as sufficient power is applied. Other pairs of mobiles may be added to these to infinity, each addition requiring the application of increased power to keep it in motion.

These pairs of mobiles as applied to the construction of timepieces are usually very unequal in size and the larger is designated as a "wheel" while the smaller, if having less than 20 teeth, is called a "pinion" and its teeth "leaves." Now while we have established the principle of a train of wheels as used in various mechanisms, our gearing is very defective, for while continuous motion may be transmitted through such a train, we will find that to do so requires the application of an impelling force far in excess of what should be required to overcome the inertia of the mobiles, and the amount of friction unavoidable in a mechanism where some of the parts move in contact with others.

This excess of power is used in overcoming a friction caused by improperly shaped teeth, or when formed thus the teeth of the wheel come in contact with those of the pinion and begin driving at a point in front of what is known as the "line of centers," i. e., a line drawn through the centers of revolution of both mobiles, and as their motion continues the driven tooth slides on the one impelling it toward the center of the wheel. When this line is reached the action is reversed and the point of the driving tooth begins sliding on the pinion leaf in a direction away from the center of the pinion, which action is continued until a point is reached where the straight face of the leaf is on a line tangential to the circumference of the wheel at the point of the tooth. It then slips off the tooth, and the driving is taken up on another leaf by the next succeeding tooth. The sliding action which takes place in front of the line of centers is called "engaging," that after this line has been passed "disengaging" friction.

Now we know that in the construction of timepieces, friction and excessive motive power are two of the most potent factors in producing disturbances in the rate, and that, while some friction is unavoidable in any mechanism, that which we have just described may be almost entirely done away with. Let us examine carefully the action of a wheel and pinion, and we will see that only that part of the wheel tooth is used, which is outside the pitch circle, while the portion of the pinion leaf on which it acts is the straight face lying inside this circle, therefore it is to giving a correct shape to these parts we must devote our attention. If we form our pinion leaves so that the portion of the leaf inside the pitch circle is a straight line pointing to the center, and give that portion of the wheel tooth lying outside the pitch circle (called the addenda, or ogive of the tooth) such a degree of curvature that during its entire action the straight face of the leaf will form a tangent to that point of the curve which it

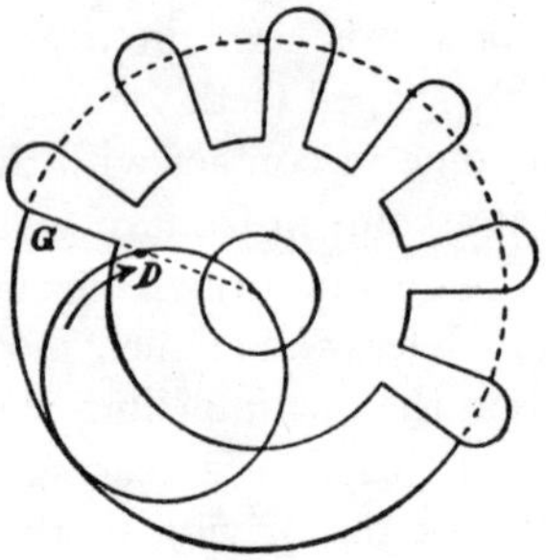

Showing that a hypocycloid of half the pitch circle is a straight line

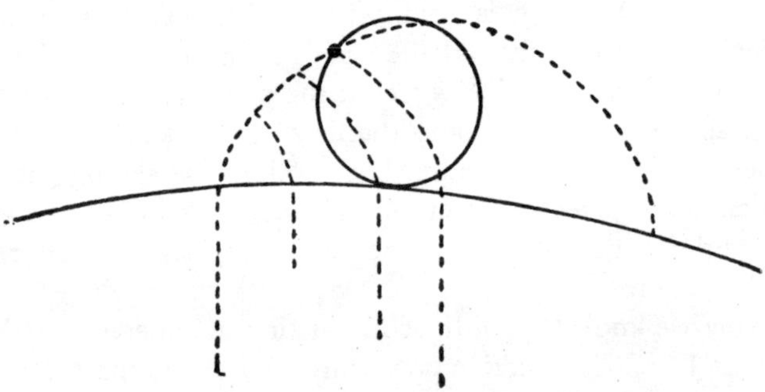

Generating an epicycloid curve for a cut pinion. D, generating circle Dotted line epicycloid curve. Note how the shape varies with the thickness of the tooth.

touches, no sliding action whatever will take place after the line of centers is passed, and if our pinion has ten or more leaves, the "addenda" of the wheel is of proper height, and the leaves of the pinion are not too thick, there will be no contact in front of the line of centers. With such a depth the only friction would be from a slight adhesion of the surfaces in contact, a factor too small to be taken into consideration.

Here, then, we have an ideal depth. How shall we obtain the same results in practice? It is comparatively an easy matter to so shape our cutters that the straight faces of our pinion leaves will be straight lines pointing to the center, but to secure just the proper curve for the addenda of our wheel teeth requires rather a more complicated manipulation. This curve does not form a segment of a circle, for it has no two radii of equal length, and if continued would form, not a circle, but a spiral. To generate this curve, we will cut from cardboard, wood, or sheet metal, a segment of a circle having a *radius* equal to that of our *wheel,* on the pitch circle, and a smaller circle whose *diameter* is equal to the *radius* of the *pinion,* on the pitch circle. To the edge of the small circle we will attach a pencil or metal point so that it will trace a fine mark. Now we lay our segment flat on a piece of drawing paper, or sheet metal and cause the small circle to revolve around its edge without slipping. We find that the point in the edge of the small circle has traced a series of curves around the edge of the segment.

These curves are called "epicycloids," and have the peculiar property that if a line be drawn through the generating point and the point of contact of the two circles, this will always be at right angles to a tangent of the curve at its point of intersection. It is this property to which it owes its value as a shape for the acting surface of a wheel tooth, for it is owing to this that a tooth whose acting surface is bounded by such a curve can impel a pinion leaf through the entire lead with little sliding action between the two surfaces. This, then, is the curve on which we will form the addenda of our wheel teeth.

In Fig. 66, the wheel has a radius of fifteen inches and the pinion a radius of one and one-half, and these two measurements are to be added together to find the distance apart of the two wheels; 16.5 inches is then the distance that the centers of revolution are apart of the wheels. Now, the teeth and leaves jointly act on one another to maintain a sure and equable relative revolution of the pair.

In Fig. 66, the pinion has its leaves radial to the center, inside of the pitch line D, and the ends of the leaves, or those parts outside of the pitch line, are a half circle, and serve no purpose until the depthings are changed by wear, as they never come in contact with the wheel; the wheel teeth only touch the radial part of the pinion and that occurs wholly within the pitch line. So in all pinions above 10 leaves in number the addendum or curve is a thing of no moment, except as it may be too large or too long. In many large pieces of machinery the pinions, or small driven wheels, have no addendum or extension beyond their pitch diameter and they serve every end just as well. In watches there is so much space or shake allowed between the teeth and pinions that the end of a leaf becomes a necessity to guard against the pinion's recoiling out of time and striking its sharp corner against the wheel teeth and so marring or cutting them. In a similar pair of wheels in machinery there are very close fits used and the shake between teeth is very slight and does not allow of recoil, butting, or "running out of time."

Running out of time is the sudden stopping and setting back of a pinion against the opposite tooth from the one just in contact or propelling. This, with pinions of suppressed ends, is a fault and it is averted by maintaining the ends.

The wheel tooth drives the pinion by coming in contact with the straight flank of the leaf at the line of centers, that is a line drawn through the centers of the two wheels; centers of revolution.

The curve or end of the wheel tooth outside of the pitch line is the only part of the tooth that ever touches the pinion and it is the part under friction from pressure and slipping. At the first point of contact the tooth drives the pinion with the greatest force, as it is then using the shortest leverage it has and is pressing on the longest lever of the leaf. As this action proceeds, the tooth is acted on by the pinion leaf farther out on the curve of the wheel tooth, thus lengthening the lever of the wheel and at the same time the tooth thus acts nearer to the center of the pinion by touching the leaf nearer its center of revolution.

By these joint actions it will appear that the wheel first drives with the greatest force and then as its own leverage lengthens and its force consequently decreases, it acts on a shorter leverage of the pinion, as the end of a tooth is nearer to the center of the pinion, or on the shortest pinion leverage, just as the tooth is about ceasing to act.

The action is thus shown from the above to be a variable one, which starts with a maximum of force and ends with a minimum. Practically the variable force in a train is not recognized in the escapement, as the other wheels and pinions making up the train are also in the same relations of maximum and minimum forces at the same time, and thus this theoretical and virtual variability of train force is to a great extent neutralized at the active or escaping end of the movement.

There is another action between the tooth and leaf that is not easy to explain without somewhat elaborate sketches of the acting parts, and as this is not consistent with such an article, we may dismiss it, and merely state that it is the one of maintaining the relative angular velocities of the two wheels at all times during their joint revolutions.

In Fig. 66 will be seen the teeth of the wheel, their heights, widths and spacing, and the epicycloidal curves. Also the same features of the pinion's construction. The curve on the end of the wheel teeth is the only curve in action during the rotation between wheel and pinion. Each flank (both teeth and leaves) is a straight line to the center of each. A tooth is composed of two members—the pillar or body of the tooth inside of the pitch line and the cycloid or curve, wholly outside of this line. The pinion also has two members, the radial flank wholly inside of the pitch line, and its addendum or circle outside of this line.

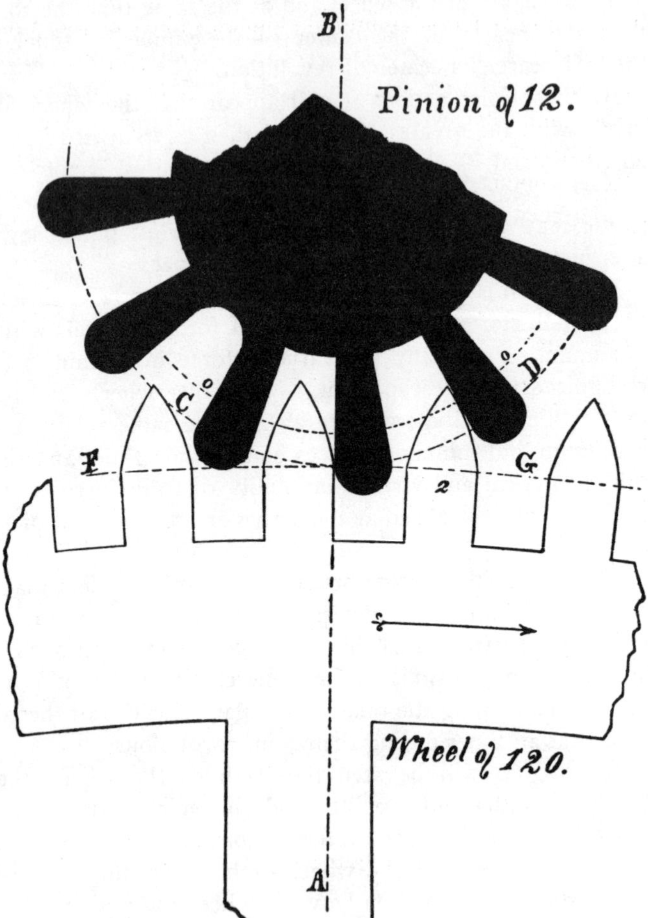

Fig. 66

In Fig. 66 will be seen a tooth on the line of centers A B, just coming in action against the pinion's flank and also one just ceasing action. It will be seen that the tooth just entering is in contact at the joint pitches, or radii, of the two wheels, and that when the tooth has run its course and ceased to act, that it will be represented by tooth 2. Then the exit contact will be at the dotted line o o. From this may be seen just how far the tooth has, in its excursion, shoved along the leaf of the pinion and by the distance the line o o, is from the wheel's pitch line G, at this tooth, No. 2, is shown the extent of contact of the wheel tooth. By these dotted lines, then, it may be seen that the tooth has been under friction for nearly its whole curve's length, while the pinion's flank will have been under friction contact for less than half this distance. In brief, the tooth has moved about 80-100 of its curved surface along the straight flank .35 of the surface of the pinion leaf. From this relative frictional surface may be seen the reason why a pinion is apt to be pitted by the wheel teeth and cut away. In any case it shows the relation between the two friction surfaces. In part a wheel tooth rolls as well as slides along the leaf, but whatever rolling there may be, the pinion is also equally favored by the same action, which leaves the proportions of individual friction still the same.

In Fig. 66 may be seen the spaces of the teeth and pinion. The teeth are apart, equal to their own width and the depths of the spaces are the same measurement of their width—that is, the tooth (inside of the pitch line) is a pillar as wide as it is high and a space between two teeth is of like proportions and extent of surface. The depth of a space between two teeth is only for clearance and may be made much less, as may be seen by the pinion leaf, as the end of the circle does not come half way to the bottom of a space.

The dotted line, o o, shows the point at which the tooth comes out of action and the pointed end outside of this line might be cut off without interfering with any function of the tooth. They generally are rounded off in common clock work.

The pinion is 3 inches diameter and is divided into twelve spaces and twelve leaves; each leaf is two-fifths of the width of a space and tooth. That is one-twelfth of the circumference of the pinion is divided into five equal parts and the leaf occupies two and a space three of these parts. The space must be greater than the width of a leaf, or the end of a leaf would come in contact with a tooth before the line of centers and cause a jamming and butting action. Also the space is needed for dirt clearance. As watch trains actuated by a spring do not have any reserve force there must be allowance made for obstructions between the teeth of a train and so a large latitude is allowed in this respect, more than in any machinery of large caliber. As will be seen by Fig. 66, the spans between the leaves are deep, much more so than is really necessary, and a space at O C shows the bottom of a space, cut on a circle which strengthens a leaf at its root and is the best practice.

Having determined the form of our curve, our next step will be to get the proper proportions. Saunier recommends that in all cases tooth and space should be of equal width, but a more modern practice is to make the space slightly wider, say one-tenth where the curve is epicycloidal. When the teeth are cut with the ordinary Swiss cutters, which, of course, cannot be epicycloidal, it is best to make the spaces one-seventh wider than the tooth. This proportion will be correct except in the case of a ten-leaf pinion, when, if we wish to be sure the driving will begin on the line of centers, the teeth must be as wide as the spaces; but in this case the pinion leaf is made proportionately thinner, so that the requisite freedom is thus obtained.

The height of the addenda of the wheel teeth above the pitch circle is usually given as one and one-eighth times the width of a tooth. While this is approximately correct, it is not entirely so, for the reason that as we use a circle whose diameter is equal to the pitch radius of the pinion for generating the curve, the height of the addenda would be different on the same wheel for each different numbered pinion. So that if a wheel of 60 were cut to drive a pinion of 8, the curve of this tooth would be found too flat if used to drive a pinion of 10. Now, since the pitch diameter of the pinion is to the pitch diameter of the wheel as the number of leaves in the pinion are to the number of teeth in the wheel, in order to secure perfect teeth we must adopt for the height of the addenda a certain proportion of the radius or diameter of the pinion it is to drive, this proportion depending on the number of leaves in the pinion.

A careful study of the experiments on this subject with models of depths constructed on a large scale, shows that the proportions given below come the nearest to perfection.

When the pinion has six leaves the spaces should be twice the width of the leaves and the depth of the space a little more than one-half the total radius of the pinion. The addenda of the pinion should be rounded, and should extend outside the pitch circle a distance equal to about one-half the width of a leaf. The addenda of the wheel teeth should be epicycloidal in form and should extend outside the pitch circle a distance equal to five-twelfths of the pitch radius of the pinion.

With these proportions, the tooth will begin driving when one-half the thickness of a leaf is in front of the line of centers, and there will be engaging friction from this point until the line of centers is reached.

This cannot be avoided with low-numbered pinions without introducing a train of evils more productive of faulty action than the one we are trying to overcome. There will be no disengaging friction.

When a pinion of seven is used, the spaces of the pinion should be twice the width of the leaves, and the depth of a space about three-fifths of the total radius of the pinion. The addenda of the pinion leaves should be rounded, and should extend outside the pitch circle about one-half the width of a leaf. The addenda of the wheel teeth should be epicycloidal, and the height of each tooth above the pitch circle equal to two-fifths of the pitch radius of the pinion.

There is less engaging friction when a pinion of seven is used than with one of six, as the driving does not begin until two-thirds of the leaf is past the line of centers. There is no disengaging friction.

With an eight-leaf pinion the space should be twice as wide as the leaf, and the depth of a space about one-half the total radius of the pinion. The addenda of the pinion leaves should be rounded and about one-half the width of a leaf outside the pitch circle. The addenda of the wheel teeth should be epicycloidal, and the height of each tooth above the pitch circle equal to seven-twentieths of the pitch radius of the pinion.

With a pinion of eight there is still less engaging friction than with one of seven, as three-quarters of the width of a leaf is past the line of centers when the driving begins. As there is no disengaging friction, a pinion of this number makes a very satisfactory depth.

A pinion with nine leaves is sometimes, though seldom, used. It should have the spaces twice the width of the leaves, and the depth of a space one-half the total radius. The addenda should be rounded, and its height above the pitch circle equal to one-half the width of the leaf. The addenda of the wheel teeth should be epicycloidal, and the height of each tooth above the pitch circle equal to three-sevenths of the total radius of the pinion. With this pinion the driving begins very near the line of centers, only about one-fifth of the width of a leaf being in front of the line.

A pinion of ten leaves is the lowest number with which we can entirely eliminate engaging friction, and to do so in this case the proper proportions must be rigidly adhered to. The spaces on the pinion must be a little more than twice as wide as a leaf; a leaf and space will occupy 36° of arc;

of this 11° should be taken for the leaf and 25° for the space. The addenda should be rounded and should extend about half the width of a leaf outside the pitch circle. The depth of a space should be equal to about one-half the total radius. For the wheel, the teeth should be equal in width to the spaces, the addenda epicycloidal in form, and the height of each tooth above the pitch circle, equal to two-fifths the pitch radius of the pinion.

A pinion having eleven leaves would give a better depth, theoretically, than one of ten, as the leaves need not be made quite so thin to ensure its not coming in action in front of the line of centers. It is seldom seen in watch or clock work, but if needed the same proportions should be used as with one of ten, except that the leaves may be made a little thicker in proportion to the spaces.

A pinion having twelve leaves is the lowest number with which we can secure a theoretically perfect action, without sacrificing the strength of the leaves or the requisite freedom in the depths. In this pinion, the leaf should be to the space as two to three, that is, we divide the arc of the circumference needed for a leaf and space into five equal parts, and take two of these parts for the leaf, and three for the space; depth of the space should be about one-half the total radius. The addenda of the wheel teeth should be epicycloidal, and the height of each tooth above the pitch line equal to two-sevenths the pitch radius of the pinion.

As the number of leaves is increased up to twenty, the width of the space should be decreased, until when this number is reached the space should be one-seventh wider than the leaf. As these numbers are used chiefly for winding wheels in watches, where considerable strength is required, the bottoms of the spaces of both mobiles should be rounded.

Circular Pitch. Diametral Pitch.—In large machinery it is usual to take the circumference and divide by the number of teeth; this is called the *circular pitch,* or distance from point to point of the teeth, and is useful for describing teeth to be cut out as patterns for casting.

But for all small wheels it is more convenient to take the diameter and divide by the number of teeth. This is called the *diametral pitch,* and when the diameter of a wheel or pinion which is intended to work into it is desired, such diameter bears the same ratio or proportion as the number required. Both diameters are for their pitch circles. As the teeth of each wheel project from the pitch circle and enter into the other, an addition of corresponding amount is made to each wheel; this is called the *addendum.* As the size of a tooth of the wheel and of a tooth of the pinion are the same, the amount of the addendum is equal for both; consequently the outside diameter of the smaller wheel or pinion will be greater than the arithmetical proportion between the pitch circles. As the diameters are measured presumably in inches or parts of an inch, the number of a wheel of given size is divided by the diameter, which gives the number of teeth to each inch of diameter, and is called the *diametral pitch.* In all newly-designed machinery a whole number is used and the sizes of the wheels calculated accordingly, but when, as in repairing, a wheel of any size has any number of teeth, the diametral number may have an additional fraction, which does not affect the principle but gives a little more trouble in calculation. Take for example a clock main wheel and center pinion: Assuming the wheel to be exactly three inches in diameter at the pitch line, and to have ninety-six teeth, the result will be $96 \div 3 = 32$, or 32 teeth to each inch of diameter, and would be called *32 pitch.* A pinion of 8 to gear with this wheel would have a diameter at the pitch line of 8 of these thirty-seconds of an inch or 8-32 of an inch. But possibly the wheel might not be of such an easily manageable size. It might, say, be 3.25 inches, in which case, 96 being the number of the wheel and 8 of the pinion, the ratio is 8-96 or 1-12, so 1-12 of $3.25 = 0.270$, the pitch diameter of the pinion.

These two examples are given to indicate alternative methods, the most convenient of which may be used. After arriving at the true pitch diameters the matter of the addendum arises, and it is for this that the diametral number is specially useful, as in every case when figuring by this system, whatever the number of a wheel or pinion, two of the pitch numbers are to be added. Thus with the 32 pitch, the outside diameter of the wheel will be 3 in. + 2-32, and if the pinion 8-32 + 2-32 = 10-32. With the other method the same exactness is more difficult of attainment, but for practical purposes it will be near enough if we use 2-30 of an inch for the addendum, when the result will be 3.25 + 2-30 or 3¼ + 2-30 = 3 1-3 in. nearly and the pinion 0.270 + 2-30 = 0.270 + .0666 = 0.3366; or to work by 1-3 of an inch is near enough, giving the outside diameter of the pinion a small amount less than the theoretical, which is always advisable for pinions which are to be driven.

We represent by Figs. 67 to 71 a wheel of sixty teeth gearing with a pinion of six leaves. The wheel, whose pitch diameter is represented by the line mm is the same in each figure. The pinion, which has for its pitch diameter the line kk, is in Fig. 67, of a size proportioned to that of the wheel, and its center is placed at the proper distance; that is to say, the two pitch diameters are tangential.

In Fig. 68 the same pinion, of the proper size, has its center too far off; the depthing is too shallow. In Fig. 69 it is too deep. Figs. 70 and 71 represent gearing in which the pitch circles are in contact, as the theory requires, but the size of the pinions is incorrect. If the wheels and pinion actuated each other by simple contact the velocity of the pinion with reference to that of the wheel would not be absolutely the same; but the ratio of the teeth being the same, the same ratio of motion obtains in practice, and there is necessarily bad working of the teeth with the leaves.

We will observe what passes in each of these cases, and refer to the suitable remedies for obtaining a passable depthing and a comparatively good rate, without the necessity of repairs at a cost out of all proportion with the value of the article repaired.

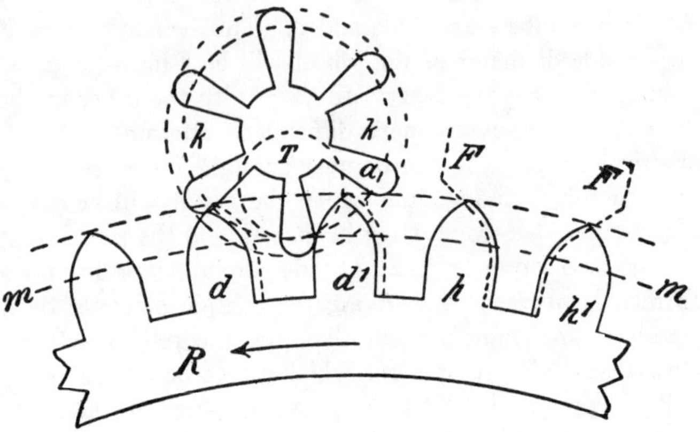

Fig. 67

Fig. 67 represents gearing of which the wheel and pinion are well proportioned and at the proper distance from each other. Its movement is smooth, but it has little drop or none at all. By examining the teeth h, h', of the wheel, it is seen that they are larger than the interval between them. With a cutter FF, introduced between the teeth, they are reduced at d, d', which gives the necessary drop without changing the functions, since the pitch circles mm and kk have not been modified. The drop, the play between the tooth d' and the leaf a, is sufficiently increased for the working of the gearing with safety.

We have the same pair in Fig. 68, but here their pitch circles do not touch; the depthing is too shallow. The drop is too great and butting is produced between the tooth h and the leaf r, which can be readily felt. The remedy is in changing the center distance, by closing the holes, if worn, or moving one nearer the other. But in an ordinary clock this wheel may be replaced with a larger one, whose pitch circle reaches to e. The proportions of the pair are modified, but not sufficiently to produce inconvenience.

It may also answer to stretch the wheel, if it is thick enough to be sufficiently increased in size. A cutter should then be selected for rounding up which will allow the full

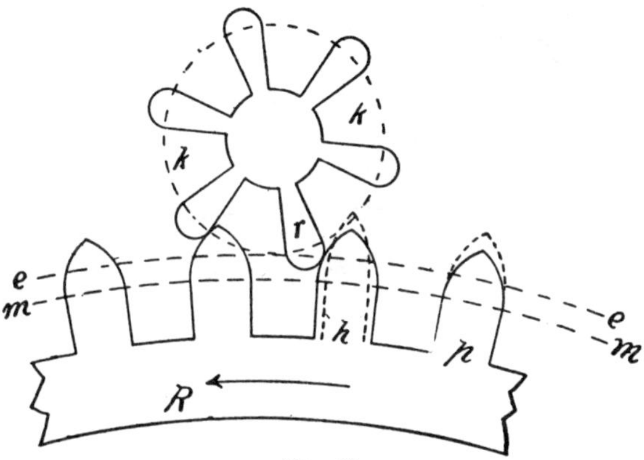

Fig. 68

width to the tooth as at p; but if it is not possible to enlarge the wheel enough, a little of the width of the teeth may be taken off, as is seen at h, which will diminish the butting with the leaf r.

Too great depthing, Fig. 69, can generally be recognized by the lack of drop. When the teeth of the wheel are narrow, the drop may appear to be sufficient. When the train is put in action the depthing that is too great produces scratching or butting and the 'scape wheel trembles. This results from the fact that the points of the teeth of the wheel touch the core of the pinion and cause it to butt against the leaf following the one engaged, as is visible at r in Fig. 69. It should be noticed that in this figure the pitch circles mm and kk overlap each other, instead of being tangential.

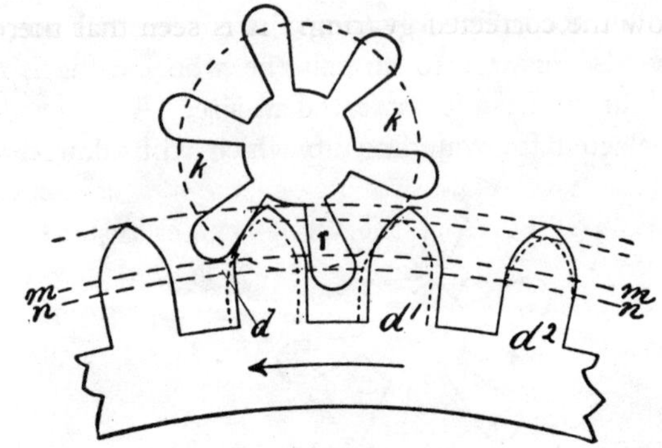

Fig. 69

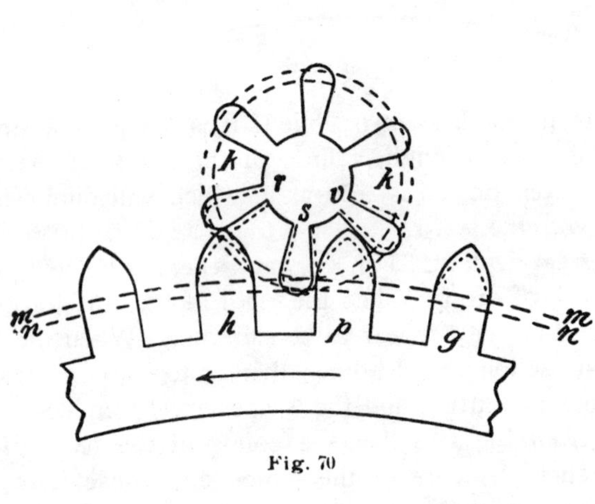

Fig. 70

To correct this gearing, the cutter should act only on the addenda of the teeth of the wheel, so as to diminish them and bring the pitch circle mm to n. The dots in the teeth d, d′, show the corrected gearing. It is seen that there will be, after this change, the necessary drop, and that the end of the tooth d′ will not touch the leaf r.

In the two preceding cases we have considered wheels and pinions of accurate proportion, and the defects of the gearing proceeding from the wrong center distances. We will not speak of the gearing in which the pinion is too small. The only theoretic remedy in this case, as in that of too large a pinion, is to replace the defective piece; but in practice, when time and money are to be saved, advantage must be taken, one way or another, of what is in existence.

The buzzing produced when the train runs in a gearing with too small a pinion proceeds from the fact that each tooth has a slight drop before engaging with the corresponding leaf. If we examine Fig. 70, it will be easy to see how this drop is produced. The wheel revolving in the direction indicated by the arrow, it can be seen that when the tooth h leaves the leaf r, the following tooth, p, does not engage with the corresponding leaf, s; this tooth will therefore have some drop before reaching the leaf. A friction may even be produced at the end or addendum of the tooth p against the following leaf v.

To obtain a fair depthing without replacing the pinion, the wheels can be passed to the rounding up machine, having a cutter which will take off only the points of the teeth, as is indicated in the figure; the result may be observed by the dotted lines. The tooth h being shorter, it will leave the leaf r of the pinion when the latter is in the dotted position; that is to say, a little sooner. At this moment the tooth p is in contact with the leaf s, and there is no risk of friction against the leaf v. Care must be taken to touch only the addendum of the tooth so as not to weaken the teeth. The circumference i will be that of a pinion of accurate size, and if the pinion is replaced, it will be necessary to diminish the wheel so that its pitch circle shall be tangential with i.

With too small a pinion a passable gearing can generally be produced. In any case stoppage can be prevented. This is not so easy when the pinion is too large. In Fig. 71, the

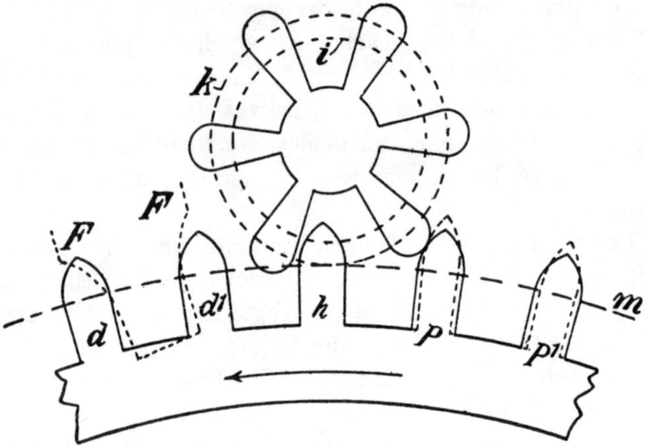

Fig. 71

pinion has as its pitch circle the line k, instead of i, which would be nearer the size with reference to that of the wheel. This is purposely drawn a little small for clearness of illustration. The essential defect of such a gearing can be seen; the butting produced between the tooth p and the leaf s will cause stoppage. How shall this defect be corrected without replacing the pinion?

To remedy the butting as far as possible, some watchmakers slope the teeth of the wheel by decentering the cutter on the rounding-up machine. At FF the cutter is seen working between the teeth d and d′. It is evident that when the wheel becomes smaller it is necessary to stretch it out, and to make use of the cutter afterwards. However, the most rational method is to leave the teeth straight, and to give them the slenderest form possible, after having enlarged the wheel or having replaced it with another. The motive force of the wheel being sufficiently weak, the size of the teeth may be reduced without fear. The essential thing is to suppress the butting. Success will be the easiest when the teeth are thinner.

In conclusion, we recommend verification of all suspected gearings by the depthing tool, which is easier and surer than by the clock itself. One can see better by the tool the working of the teeth with the leaves, and can form a better idea of the defect to be corrected. With the aid of the illustrations that have been given it can be readily noticed whether the depthing is too deep or too shallow, or the pinion too large or too small.

The defects mentioned are of less consequence in a pinion of seven leaves, and they are corrected more readily. With pinions of higher numbers the depthings will be smoother, provided sufficient care has been taken in the choice of the rounding-up cutters.

ROUNDING-UP WHEELS.—It is frequently observed that young watchmakers, and (regretfully be it said) some of the older and more experienced ones, are rather careless when fitting wheels on pinions. In many cases the wheel is simply held in the fingers and the hole opened with a broach, and in doing this no special care is taken to keep the hole truly central and of correct size to fit the pinion snugly, and should it be opened a little too large it is riveted on the pinion whether concentric or not. Many suppose the rounding-up tool will then make it correct without further trouble and without sufficient thought of the irregularities ensuing when using the tool.

To make the subject perfectly clear the subjoined but rather exaggerated sketch is shown, Fig. 72. Of course, it is seldom required to round-up a wheel of twelve teeth, and the eccentricity of the wheel would be hardly as great as shown; nevertheless, assuming such a case to occur the drawing will exactly indicate the imperfections arising from the use of a rounding-up tool.

Presuming from the drawing that the wheel, as shown by dotted lines, had originally been cut with its center at m, but through careless fitting had been placed on the pinion at o, and consequently is very much out of round when tested in the calipers, and to correct this defect it is put in the

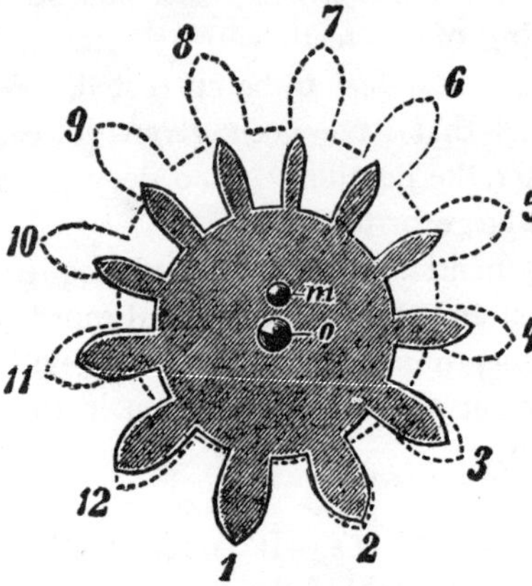

Fig. 72

rounding-up tool. The cutter commences to remove the metal from tooth 7, it being the highest, next the neighboring teeth 6 and 8, then 5 and 9, and so on until tooth 1 comes in contact with the cutter. The wheel is now round. But how about the size of the teeth and the pitch? The result of the action of the cutter is shown by the sectionally lined wheel. Many will ask how such a result is possible, as the cutter has acted equally upon all the teeth. Nevertheless, a little study of the action of the rounding-up cutter will soon make it plain why such faults arise. Naturally the spaces between the teeth through the action of the cutter will be equal, but as the cutter is compelled to remove considerable metal from the point of greatest eccentricity, i. e., at tooth 7 and the adjoining teeth, to make the wheel round, and the pitch circle being smaller the teeth become thinner, as the space between the teeth remains the same. At tooth 1 no metal was removed, consequently it remains in its original condition. The pitch from each side of tooth 1 becomes less and less to tooth 7, and the teeth thinner, and the thickest tooth is always found opposite the thinnest.

In the case of a wheel having a large number of teeth and the eccentricity of which is small, such faults as described cannot be readilv seen, from the fact that there are many teeth and the slight change in each is so gradual that the only way to detect the difference is by comparing opposite teeth. And this eccentricity becomes a serious matter when there are but few teeth, as before explained, especially when reducing an escape wheel. The only proper course to pursue is to cement the wheel on a chuck, by putting it in a step chuck or in any suitable manner so that it can be trued by its periphery and then opening the hole truly. This method is followed by all expert workmen.

A closer examination of the drawing teaches us that an eccentric wheel with pointed teeth—as cycloidal teeth are mostly left in this condition when placed in the rounding-up tool, will not be made round, because when the cutter has just pointed the correct tooth (tooth No. 1 in the drawing) it will necessarily shorten the thinner teeth, Nos. 6, 7, 8, i. e., the pitch circle will be smaller in diameter. We can, therefore, understand why the rounding-up tool does not make the wheel round.

As we have before observed, when rounding-up an eccentrically riveted wheel, the thickest tooth is always opposite the thinnest, but with a wheel which has been stretched the case is somewhat different. Most wheels when stretched become angular, as the arcs between the arms move outward in a greater or less degree, which can be improved to some extent by carefully hammering the wheel near the arms, but some inequalities will still remain. In stretching a wheel with five arms we therefore have five high and as many depressed parts on its periphery. If this wheel is now rounded-up the five high parts will contain thinner teeth than the depressed portions. Notwithstanding that the stretching of wheels, though objectionable, is often unavoidable on account of the low price of repairs, it certainly ought not to be overdone. Before placing the wheel in the rounding-up tool it should be tested in the calipers and the low places carefully stretched so that the wheel is as nearly round as can be made before the cutter acts upon it.

It is hardly necessary to mention that the rounding-up tool will not equalize the teeth of a badly cut wheel, and further should there be a burr on some of the teeth which has not been removed, the action of the guide and cutter in entering a space will not move the wheel the same distance at each tooth, thus producing thick and thin teeth. From what has been said it would be wrong to conclude that the rounding-up tool is a useless one; on the contrary, it is a practical and indispensable tool, but to render good service it must be correctly used.

In the use of the rounding-up tool the following rules are to be observed:

1. In a new wheel enlarge the hole after truing the wheel from the outside and stake it concentrically on its pinion.

2. In a rivetted but untrue wheel, stretch the deeper portions until it runs true, then reduce it in the rounding-up tool. The better method is to remove the wheel from its pinion, bush the hole, open concentrically with the outside and rivet, as previously mentioned in a preceding paragraph. But if the old riveting cannot be turned so that it can be used again it is best to turn it entirely away, making the pinion shaft conical towards the pivot, and after having bushed the wheel, drill a hole the proper size and drive it on the pinion. The wheel will be then just as secure as when rivetted, as in doing the latter the wheel is often distorted. With a very

thin wheel allow the bush to project somewhat, so that it has a secure hold on the pinion shaft and cannot work loose.

3. Should there be a feather edge on the teeth, this should be removed with a scratch brush before rounding it up, but if for some reason this cannot well be done, then place the wheel upon the rest with the feather edge nearest the latter so that the cutter does not come immediately in contact with it. If the feather edge is only on one side of the tooth—which is often the case—place the wheel in the tool so that the guide will turn it from the opposite side of the tooth; the guide will now move the wheel the correct distance for the cutter to act uniformly. Of course, in every case the guide, cutter and wheel, must be in correct position to ensure good work.

4. To obtain a smooth surface on the face of the teeth a high cutter speed is required, and for this reason it is advantageous to drive the cutter spindle by a foot wheel.

MAKING SINGLE PINIONS.—There are two ways of making clock pinions; one is to take a solid piece of steel of the length and diameter needed and turn away the surplus material to leave the arbor and the pinion head of suitable dimensions; the other way is to make the head and the arbor of separate pieces; the head drilled and fixed on the arbor by friction. The latter plan saves a lot of work, and the cutting of the teeth may be easier. One method is as good as the other, as the force on the train is very slight and the pinion head may be driven so tightly on the arbor as to be perfectly safe without any other fastening, provided the arbor is given a very small taper, .001 inch in four inches. The steel for the arbor may be chosen of such a size as to require very little turning, and hardened and tempered to a full or pale blue before commencing turning it, but the piece intended for the pinion head must be thoroughly annealed, or it may be found impossible to cut the teeth without destroying a cutter, which, being valuable, is worth taking care of.

Pinions for ordinary work are not hardened; as they are left soft by the manufacturers it would be nonsense for the repairer to put in one hardened pinion in a clock where all the others were soft. Pinions on fine work are hardened. Turning is done between centers to insure truth.

Before commencing work on the pinion blanks it is advisable to try the cutters on brass rod, turned to the exact size, and if the rod is soft enough it will be found that the cutter will make the spaces before it is hardened, which is a very important advantage, admitting of correction in the form of the cutter if required; only two or three teeth need be cut in the brass to enable one to see if they are suitable, and if found so, or after an alteration of the cutter, the entire number may be cut round and the brass pinion made use of for testing its accuracy as to size and shape by laying the wheel along with it on a flat plate, having studs placed at the proper center distance. By this means the utmost refinement may be made in the diameter of the brass pinion, which will then serve as a gauge for the diameter of the steel pinions, it being recollected, as mentioned in a previous paragraph, that a slight variation in the diameter of a pinion may be made to counterbalance a slight deviation from mathematical accuracy in the form of the wheel-teeth, such as is liable to occur owing to the smallness of the teeth making it impracticable to actually draw the true curves, the only way of getting them being to draw them to an enlarged scale on paper, and copy them on the cutter as truly as possible by the eye.

Supposing the cutter has been properly shaped, hardened and completed and the steel pinion heads all turned to the diameter of the brass gauge, the cutting may be proceeded with without fear of spoiling, or further loss of time which might be spent in cutting the long pinion leaves; and even what is of more importance in work which does not allow of any imperfection, removing the temptation, which might be strong, to let a pinion go, knowing it to be less perfect than it should be.

Assuming the pinion teeth to be satisfactorily cut, the next operation will be hardening and tempering. A good way of doing this is to enclose one at a time in a piece of gas pipe, filling up the space around the pinion with something to keep the air off the work and prevent any of the products of combustion attacking the steel and so injuring the surface. Common soap alone answers the purpose very well, or it may have powdered charcoal mixed with it; also the addition of common salt helps to keep the steel clean and white. The heating should be slow, giving time for the pinion and the outside of the tube to both acquire the same heat. Overheating should be carefully avoided, or there will be scaling of the surfaces, injurious to the steel, and requiring time and labor to polish off. There is no better way of hardening than by dipping the pipe with the pinion enclosed in plain cold water, or if the pinion should drop out of the tube into the water it will do all the same. To be sure the hardening is satisfactory it will be as well not to trust to the clean white color likely to result from this treatment, but try both ends and the center with a file. After all this has been successfully accomplished the pinions will require tempering, the long arbors straightening, and the teeth polishing.

The drilled pinion heads, if hardened at all by the method last mentioned, will, on account of their short lengths, be equally hardened all over, but if the pinion and arbor should be all in one piece care will be needed to ensure equal heating all over, or one part may be burnt and another soft. Also, to guard against bending the long arbors, the packing in the tube will need to be carefully done, so as to produce equal pressure all over; otherwise, while the steel is red hot, and consequently soft enough to bend, even by its own weight, it may get distorted before dropping in the water. A long thin rod like this almost invariably bends if heated on an open fire unless equally supported all along; if hardened so, a little tin tray may be bent up, filled with powdered charcoal, and the pinion bedded evenly in it. Either this way or with a tube the long arbor may get bent before being quenched, but if the arbor, though kept straight up to this point, should happen to be dropped sideways into the water the side cooled first would contract most. To avoid this, the arbor should be dropped endways, as vertically as possible.

TEMPERING THE PINIONS.—For common cheap work the usual and quickest way is what is called "blazing off." That is done either by dipping each piece singly in thick oil and setting the oil on fire, allowing it to burn away, or placing a number of pieces in a suitably sized pan, covering with oil, and burning it. The result is the same either way, the method being simply a matter of convenience regulated by the number of pieces to be tempered at one time. As the result of blazing off is to some extent uncertain, and the pinions apt to be too soft, it will be advisable to adopt the process of bluing, by which the temper desired may be produced with more accuracy. The first thing to do will be to clean the surface of the arbor all along on one side; the pinion head may be left alone. As the pinion head would get overheated before the arbor had reached the blue color, if the piece were simply placed on a bluing pan or a lump of hot iron, it will be necessary to provide a layer of some soft substance to bed the pinion on; iron, steel or brass filings answer well because the heat is soon uniformly distributed through the mass, and by judiciously moving the lamp an equable temper may be got all along, as determined by the color. There is another and very sure way of getting a uniform temper, in using which there is no need to polish the arbors. The heat of lead *at the point of fusion* happens to be just about the same as that required for the tempering of this work; so if a ladle full of lead is available each pinion may be buried in it for a few seconds, holding it down beneath the molten surface with hot pliers. The temper suitable is indicated by a pale blue, a little softer than for springs, and a piece of polished steel set floating on the lead will indicate whether the heat is suitable; if found too great some tin may be added, which will cause the metal to melt at a lower temperature. Overheating the metal must be avoided: it should go no higher than the bare melting point.

STRAIGHTENING BENT ARBORS.—When all care has been taken in the hardening, the long pieces of wire are still apt to become bent more or less, and this is especially the case with solid pinions; so before proceeding further the pieces must be got true, or as nearly so as possible, and it will be found impracticable to do this by simple bending when the steel is tempered. If the piece is placed between centers in the lathe and rotated slowly, the hollow side will be found; this side must be kept uppermost while the steel is held on a smooth anvil, and the pene, or chisel-shaped, end of a small hammer applied crossways with gentle blows, stepping evenly along so that each portion of the steel is struck all along the part which is hollow; this will stretch the hollow side, and, by careful working, trying the truth from time to time, the piece can be got as true as may be wished, and probably keep so during the subsequent turning and finishing, though it is advisable to keep watch on it, and if it shows any tendency to spring out of truth again, repeat the striking process, which should always be done gently and in such a way as to show no hammer marks. Having got the pieces sufficiently true in this way, each arbor may have a collet of suitable size driven on to it for permanency, and as the collets will probably be a little out of truth they may have a finishing cut taken all over them and receive a final polish.

POLISHING.—To polish the steel arbors after turning, a flat metal polisher, iron or steel, is used; this with emery or oilstone dust and oil produces a true surface, with a sharp corner at the shoulder; the polisher will require frequent filing on the flat and the edge to keep it in shape with a sharp corner, and a grain crossing like the cuts on a file to hold the grinding material. The polishing of arbors is not done with the object of making them shine, but to get them smooth and true, so there is no need of using any finer stuff than emery or oilstone dust.

An old way to polish the leaves was to use a simple metal polisher of a suitable thickness, placing the pinion on a cork or piece of wood, or even holding it in the fingers; working away at a tooth at a time until a good enough polish was obtained; but this method, while being satisfactory as to results, was also tedious and very slow. It was in some cases assisted by having guide pinions fitted tight on one or both ends of the arbors to prevent rounding of the teeth, the polisher resting in the guide and the tooth to be polished. On the American lathes an accessory is provided called a "wig wag." This is a rod fastened at one end to a pulley by a crank pin near its circumference; the pulley being rotated by a belt from the counter shaft pulleys causes the rod to move rapidly backwards and forwards. On the other end of the rod a long narrow piece of lead or tin is fixed, the pinion being fitted by its centres into a simple frame held in the slide rest so that it can be rotated tooth by tooth; the lead soon gets cut to the form of the teeth, and the polishing is quickly effected. Another way is to take soft pine or basswood, shape it roughly to about the form of space between two teeth and use it as a file, with emery and oil or oilstone dust. The wood is soon cut to the exact shape of the teeth, and then makes a quick and perfect job. The pinion is held in the jaws of the vise and the wooden polisher used as a file with both hands.

Where there is much polishing to do a simple tool, which a workman can form for himself, produces a result which is all that can be desired. It consists of an arbor to work between the lathe centres, or a screw chuck for wood, with a round block of soft wood, of a good diameter, fixed on it, and turned true and square across; this will get a spiral groove cut in it by the corners of the pinion leaves. The pinion is set between centres in a holder in the slide rest, with the holder set at a slight angle, so that, instead of circular grooves being cut in the wood a screw will be formed, the angle being found by trial. On the wood block being rotated and supplied with fine emery the pinion will be found to rotate, and, being drawn backwards and forwards by the slide rest, can be polished straight, while the circular action of the polisher will cause the sides of the pinion leaves to be made quite smooth and entirely free from ridges.

If it should be desired to face the pinions, like watch pinions, it may be done in the same way, by cutting hollows so as to leave only a fine ring round the bottoms of the teeth, and using a hollow polisher with a flat end held in the fingers while the pinion is rotating. A common cartridge shell with a hole larger than the arbor drilled in the center of the head makes a fine polisher for square facing on the ends of pinions, while a stick of soft wood will readily adapt itself to moulded ends.

The pinion heads being finished and got quite true, the arbors may be turned true and polished. It is not advisable to turn the arbors small; they will be better left thick so as to be stiff and solid, as the weight so near the center is of no importance, the velocity on the small circumference in starting and stopping being also inappreciable. The thickness of the arbors when the pinion heads are drilled is determined by the necessity of having sufficient body inside the bottoms of the teeth; but when solid they may with advantage be left thicker; however, there is no absolute size.

The ends on which the collets for holding the wheels are to be fixed may be turned to the same taper as the broach which will be used for opening the collet holes, while the other ends may be straight.

None of the wheels in a fine clock should be riveted to the pinion heads; even the center wheel, which goes quite up to the pinion head, is generally fixed on a collet. The collets are made from brass cut off a round rod, the outside diameters being just inside the edges of the wheel hubs, and a shoulder turned to fit accurately into the center hole of each wheel. These collets should first have their holes broached to fit their arbors, allowing a little for driving on, as they may be made tight enough in this way without soldering. Be careful to keep the broach oiled to prevent sticking if you want a smooth round hole.

The holes in the wheels being made, each collet may be turned to a little over its final size all over, and then driven on to its place on the pinion, so that a final turning may be made to ensure exact truth from the arbors' own centers. When the collets are thus finished in their places on the arbors, and the wheels fitted to them, if it is a fine clock, such as a regulator, a hole may be drilled through each wheel and its collet to take a screw, the holes in the collet tapped, the holes in the wheels enlarged to allow the screw to pass freely through, and a countersink made to each, so that the screws, when finished, may be flush with the wheels. One hole having been thus made and the wheel fixed with a screw, the other two holes can be made so as to be true, which would not be so well accomplished if all the holes were attempted at once. The spacing of the three screws will be accurate enough if the wheel arms be taken as a guide. If all this has been correctly done, the wheels will go to their places quite true, both in the round and the flat, and may be taken off for polishing, and replaced true with certainty, any number of times.

The polishing of the pivots should be as fine as possible; all should be well burnished, to harden them and make them as smooth as possible if it is a common job; if a fine one with hardened arbors the pivots may be ground and polished as in watch work; if the workman has a pivot polisher and some thin square edged laps this is a short job and should be done before cutting off the centers and rounding the ends of the pivots. During all this work the wheels, as a matter of course, will be removed from the pinions, and may now be again temporarily screwed on, the polishing of them being deferred till the last, as otherwise they would be liable to be scratched.

LANTERN PINIONS.—The lantern pinion is little understood outside of clock factories and hence it is generally underrated, especially by watchmakers and those working generally in the finer branches of mechanics. It will never be displaced in clock work, however, on account of the following specific advantages:

1. It offers the greatest possible freedom from stoppage owing to dirt getting into the pinions, as if a piece large enough to jam and stop a clock with cut pinions, gets into the lantern pinion, it will either fall through at once or be pushed through between the rounds of the pinion by the tooth of the wheel and hence will not interfere with its operation. It is therefore excellently adapted to run under adverse circumstances, such as the majority of common clocks are subjected to.

2. Without giving the reasons it is demonstrable that as smooth a motion may be got by a lantern pinion as by a solid radial pinion of twice the number, and that the force required to overcome the friction of the lantern is therefore much less than with the other. It follows that such pinions can be used with advantage in the construction of all cheap and roughly constructed clocks which are daily turned out in thousands to sell at a low price.

3. We have before pointed out the enormous advantages of small savings per movement in clock factories which are turning out an annual product of millions of clocks, and without going into details, it is sufficient to refer to the fact that where eight or ten millions of clocks are to be made annually the difference in the cost of keeping up the drills and other tools for lantern pinions over the cost of similar work on the cutters for solid pinions is sufficient to have a marked influence upon the cost of the goods. Then the rapidity with which they can be made and the consequent smallness of the plant as compared with that which must be provided for turning out an equal number of cut pinions is also a factor. There are other features, but the above will be sufficient to show that it is unlikely that the lantern pinion will ever be displaced in the majority of common clocks. From seventy-five to ninety per cent of the clocks now made have lantern pinions.

The main difference between lantern and cut pinions mechanically is that as there is no radial flank for the curve of the wheel tooth to press against in the lantern pinion the driving is all done on or after the line of centers, except in the smaller numbers, and hence the engaging or butting friction is entirely eliminated when the pinion is driven, as is always the case in clock work. Where the pinion is the driver, however, this condition is reversed and the driving is all before the line of centers, so that it makes a very bad driver and this is the reason why it is never used as a driving pinion. This, of course, bars it from use in a large class of machinery.

The actual making of lantern pinions will be found to offer no difficulties to those who possess a lathe with dividing arrangements, a slide rest, and a drill holder or pivot polisher to be fixed on it. The pitch circle, being through the centers of the pins, can be got with great accuracy by setting the drill point first to the center of the lathe, reading the division on the graduated head of the slide rest screw, and moving the drill point outwards to the exact amount of the semi-diameter of the pitch circle. This presupposes the slide rest screw being cut to a definite standard, as the inch or the meter, and all measurements of wheels and pinions being worked out to the same standard, the choice of the standard being immaterial. If the slide rest screw is not standardized the pitch circle may be traced with a graver and the drill set to center on the line so traced.

The heads of the pinions may be made either of two separate discs, each drilled separately, and carefully fitted on the arbor so that the pins may be exactly parallel with the arbor; or, of one solid piece bored through the center, turned down deep enough in the middle, and the drill sent right through the pin holes for both sides at one operation. The former way will be necessary when the number of pins is small, but the latter is better when the numbers are large enough to allow of considerable body in the center. In either case it is advisable to drill only part way through one shroud and to close the holes in the other with a thin brass washer pressed on the arbor and turned up to look like part of the shroud after the pins are fitted in the holes. This makes a much neater way of closing the holes than riveting and takes but a moment where only one or two pinions are being made.

There is no essential proportion for the thickness of the pins or rounds. In mathematical investigations these are always taken at first as mere points of no thickness at all; then the diameters are increased to workable proportions, and the width of the wheel-tooth correspondingly reduced until there is a freedom or a little shake. If much power has to be transmitted, the pins, or "staves," as they are called in large work, have to be strong enough to stand the strain, but, as the strain in clockwork is very small, the pins need not be nearly as thick as the breadth of a wheel-tooth. In modern factory practice the custom is to have the diameter of the rounds equal to the thickness of the leaf of a cut pinion of similar size, the measurement being taken at the pitch circle of the cut pinion. As we have already given the proportions observed in good practice on cut pinions they need not be repeated here. Another practice is to have wheel teeth and spaces equal; when this is done the spacing of all pinions above six leaf is to have the rounds occupy three parts and the space five parts.

In some old church clocks, lantern pinions were much used, in many cases with the pins pivoted and working freely in the ends, or, as they called them, "shrouds," but this was a mistake, and they are never made so now. A simple way for clock repair work is to get some of the tempered steel drill rod of exactly the thickness desired, hold one end by a split chuck in the lathe, let the other end run free, and polish with a bit of fine emery paper clipped round it with the fingers, when the wire will be ready for driving through the pinion heads, the holes being made small enough to provide for the rounds being firmly held. The drill may be made of the same wire. The shrouds may be made either of brass or steel; the latter need not be hardened, and, when the rounds are all in place and cut off, the ends may be polished as desired. In the case of a center wheel, where the pinion is close up to the wheel, and space cannot be spared, the collet on which the wheel is mounted may form one end of the pinion head.

THE WHEEL TEETH.—The same principles of calculation belong to these and solid-cut pinions, the only difference being that the round pins require wheel teeth of a different shape from those suited to pinion leaves with radial sides. Both are derived from epicycloidal curves; the curve used for lantern pinions is derived from a circle of the *same size* as the pitch circle of the pinion, while the curve for wheel teeth to drive radial-sided leaves is derived from a circle of *half* that diameter, so that the wheel teeth in the former

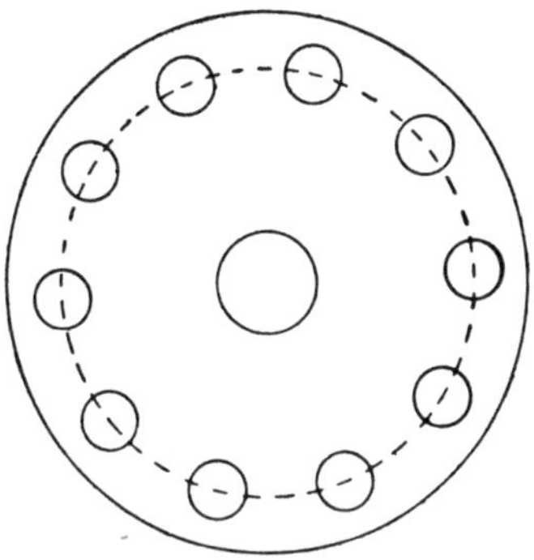

Fig. 73. Lantern pinion showing pitch circle.

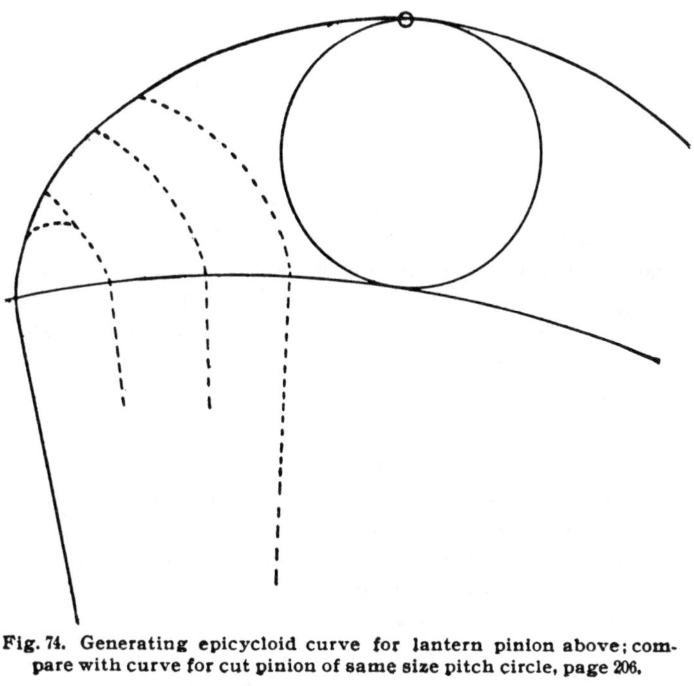

Fig. 74. Generating epicycloid curve for lantern pinion above; compare with curve for cut pinion of same size pitch circle, page 206.

are more pointed than in the latter. There also is a farther difference; as was explained in detail when treating of cut pinions, the curve of the wheel tooth presses upon the radial flank of the leaf inside its pitch circle. Now there is no radial flank in the lantern and the curve is generated from a circle of twice the diameter, so that it is twice as long—long enough to interfere—so it is cut off (rounded) just beyond the useful portion of the working curve of the wheel tooth.

Pillars and arbors are simple parts, yet much costly machinery is used in making them. The wire from which they are made is brought to the factories in large coils, and is straightened and cut into lengths by machines. The principle on which wire is straightened in a machine is exactly the same as a slightly curved piece of wire is made straight in the lathe by holding the side of a turning tool between the revolving wire and the lathe rest, which is an operation most of our readers must have practiced. The rapid revolution of the wire against the turning tool causes its highest side to yield, till finally it presses on the turning tool equally all round, and is consequently straight. However, in straightening wire by machines the wire is not made to revolve, but remains stationary while the straightening apparatus revolves around it. Wire-straightening machines are usually made in the form of a hollow cylinder, having arms projecting from the inside towards the center. The cylinder is open at both ends, and the arms are adjustable to suit the different thicknesses of wire. The wire is passed through the ends of the cylinder, and comes in contact with the arms inside. A rapid rotary motion is then given to the cylinder, which straightens the wire in the most perfect manner, as it is drawn through, without leaving any marks on it when the machine is properly adjusted. The long spiral lines that are sometimes seen on the wire work of clocks is caused by this want of adjustment; and they are produced in the same way as broad circular marks would be made in soft iron wire if the side of the turning tool was held too hard against it when straightening it in the lathe.

After the wire has been straightened it is cut off into the required lengths, and this operation is worthy of notice. If the thick sizes of wire that are used were to be cut by the aid of a file or a chisel, the ends would not be square, and some time and material would be lost in the operation

Fig. 75. A Slide Gauge Lathe.

of squaring them; and as economy of material as well as economy of labor is a feature in American clock manufacture, wire of all sizes is sheared or broken off into lengths, by being fed through round holes in the shears, which act the same as when a steady pin is broken when a cock or bridge gets a sudden blow on the side, or in the same manner as patent cutting plyers work. The wire is not bent in the operation, and both ends of it are smooth and flat. The wire for the pillars is then taken to a machine to have the points made and the shoulders formed for the frames to rest against. This machine is constructed like a machinist's bench lathe, with two headstocks. There is a live spindle running in both heads. In the ends of these spindles, that point towards the center of the lathe, cutters are fastened, and the one is shaped so that it will form the end and shoulder of the pillar that is to be riveted, while the other is shaped so as to form the shoulder and point that is to be pinned. Between these two revolving cutters there is an arrangement, worked by a screw in the end of a handle, for holding the wire from which the pillar is to be made, in a firm and suitable position. The cutters are then made to act simultaneously on the ends of the wire by a lever acting on the spindles, and the points and shoulders are in this way formed in a very rapid manner, all of the same length and diameter. These machines are in some points automatic. The pieces of wire are arranged in quantities in a long narrow feed box that inclines towards the lathe, and the mechanism for holding the wire is so arranged that when its hold is loosened on the newly made pillar, the pillar drops out into a box beneath, and a fresh piece of wire drops in and occupies its place.

In many of the factories, some clocks are manufactured having screws in place of pins to keep the frames together, and the pillars of these clocks are made in a different manner than that we have just described. The wire that is used is not cut into short lengths, but a turret lathe with a hollow spindle is used, through which the wire passes, and is held by a chuck, when a little more than just the length that is necessary to make the pillar projects through the chuck. The revolving turret head of the lathe has cutting tools projecting from it at several points. One tool is adapted to bore the hole for the screw, and when it is bored the next tool taps the hole to receive the screw, while another forms the point and shoulder; and after that end of the pillar is completed another tool attached to the slide of the lathe forms the other shoulder, prepares that end for riveting, and cuts it off at the same time. One thousand of these pillars are in this manner made in a day on each machine. The screws that screw into them are made on automatic screw machines. The latest improvements in this direction being to first turn the blanks and then roll the threads on thread rolling machines.

The pinion arbors, after they have been cut to length, are centered on one end by a milling machine having a conical cutter made for the purpose. The collets for the pinion heads, and the one to fasten the wheel by, are punched out of sheet brass, and a hole is drilled in their centers a little smaller than the wire; and to drive them on, in most in-

Fig. 76. Slide Gauge Tools and Rack.

stances, is all that is necessary to hold them. At one time it was the practice to drive these collets by hand. One was placed on the point of the arbor, and the point was then placed over a piece of steel, with a series of holes in it of such depths that the collets would be in their proper position on the arbor when the point was driven to the bottom of the hole, but this method has now been superseded by automatic machinery, which will be described later. It is impossible to give an intelligible description of these machines without drawings. All we can say at present is that they perform their work in a very rapid and effective manner, and are in use by all the larger clock factories.

The barrels of weight clocks are mostly made from brass castings, and slight projections are raised on the surface of their arbors by swedging, so as to prevent the arbors from getting loose in the barrels after repeated winding of the clock. This swedging and all the other operations in making arbors used to be done on separate machines; but the largest companies now use a powerful and comprehensive machine that works automatically, and straightens any size of wire necessary to be used in a clock, cuts it to the length, centers it, and also swedges the projections on the barrel arbors, or any of the other arbors that may be necessary. A roll of wire is placed on a reel at one end of the machine, first passing through a straightening apparatus, and afterwards to that portion of the machine where the cutting, swedging and centering are executed, and the finished arbors drop into a box placed ready to receive them. The saving effected by the use of this machine is very great, and in some instances amounts to a thousand per cent over the method of straightening, cutting, swedging and centering on different machines, at different operations.

Boring the holes in the arbors of the locking work, to receive the smaller wires, and the pin holes in the points of the pillars, is done by small twist drills, run by small vertical drill presses. The work is held in adjustable frames under the drill, and when more than one hole has to be bored this frame is moved backward or forward between horizontal slides to the desired distance, which is regulated by an adjustable stop, so that every hole in each piece is exactly in the same position. In arbors where holes have to be bored at right angles to each other, the arbor is turned round to the desired position by means of an index. The holes in the locking work arbors are bored just the size to fit the wire that is to go into them, and these small

Fig. 77. Automatic Pinion Making Machine of the Davenport Machine Company.

wires are easily and rapidly fastened in place by holding them in a clamp made for the purpose, and riveting them either with a hammer or with a hammer and punch.

THE SLIDE GAUGE LATHE—The system of turning with the slide gauge lathe, formerly adopted for lantern pinions in the clock factories, would seem to the watchmaker of a peculiarly novel nature. The turning tools are not held in the hand, in the manner generally practiced, neither are they held in the ordinary slide rest, but are used by a combination of both methods, which secures the steadiness of the one plan and the rapidity of the other. Adjustable knees are fastened to the head and tail stocks of the lathe, Figs. 75 and 76, which answer the purpose of a rest; both the perpendicular and horizontal parts of these knees being fastened perfectly parallel with the centers of the lathe. A straight, round piece of iron, of equal thickness, and having a few inches in the center of a square shape, mortised for the reception of cutters, is laid on these knees, and answers the purpose of a handle to hold the cutting tools. Two handles will thus hold eight tools, one set for brass and one for steel. On every side of the square part of this iron bar, or what we will now call the turning tool handle, a number of cutting tools are fastened by set screws, and the method of using them is as follows: The operator holds the tool handle with both hands on to the knees that are fastened to the head and tail stocks of the lathe, with the turning tool that is desired to be used pointing towards the center, and it is allowed to come in contact with the work running in the lathe in the usual manner practiced in turning. Fig. 76 is from a photo furnished by Mr. H. E. Smith of the Smith Novelty Co., Hopewell, N. J., and shows the tools in the rack, which is wound with leather so that the tools may be rapidly thrown in place without injury.

If a plain, straight piece of work is to be turned, the tool is adjusted in the handle so that the work will be of the proper diameter when the round parts of the handle come in contact with the perpendicular part of the knees or rest; and while the handle is thus held and moved gently

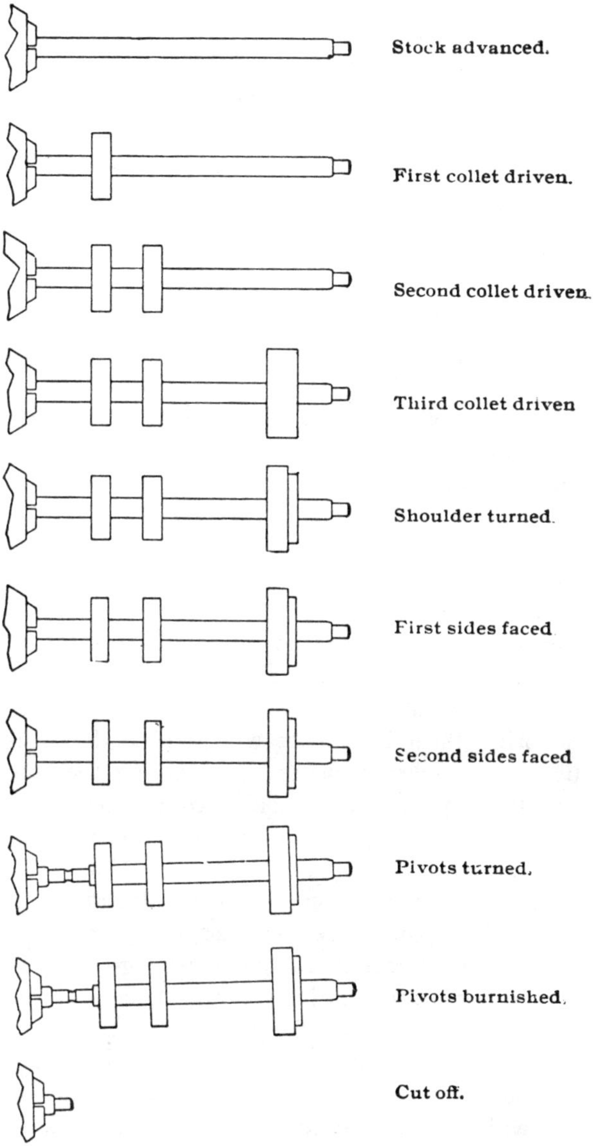

Fig. 78. Showing Successive Steps in Turning on Automatic Pinion Making Machine.

along in the corners of the knees, with the tool sliding on the T-rest, the work is easily turned perfectly parallel, smooth and true. Sometimes a roughing cut is taken by holding the bar loosely and then a finishing cut is made with the same tool by holding it firmly in place. In turning a pinion arbor, for instance, the wire having been previously straightened and cut to length and centered, and the brass collets to make the pinion and to fasten the wheel having been driven on, one end is held in the lathe by a spring chuck fastened to the spindle of the lathe, while the other end works in a center in the other head. One turning tool is shaped and adjusted in the handle for the purpose of turning the brass collets for the pinion to the proper diameter, another turns the sides of the brass work, while others are adapted for the arbors, pivots, and so on, pins being placed in holes in the T-rest to act as stops for the tools. After the brass work has been turned, the positions of the shoulders of the pivots are marked with a steel gauge, and by simply turning round the handle of the turning tool till the proper shaped point presents itself, each operation is accomplished rapidly, and the cutting is so smooth that even for the pivots all that is necessary to finish them is simply to bring them in contact with a small burnisher. The article is not taken from the lathe during the whole process of turning, and when completed the centers are broken off, having been previously marked pretty deep at the proper place with a cutting point. Five hundred to 1,200 arbors per day, per man, is the usual output. All the pinions, arbors, and barrels—in fact every part of an American clock movement that requires turning—were formerly done in this manner, at long rows of lathes in rooms, and by workmen set apart for the purpose. But perhaps it may be well to mention that in the machine shops of these factories, where they make the tools, the ordinary methods of turning with the common hand tool, and by the aid of ordinary and special slide rests, are practiced the same as it

No. 79. Automatic Pinion Drill of the Davenport Machine Company

is among other machinists. In the large factories automatic turret machines are now coming into use and these are shown in Figs. 77, 78 and 79.

The lantern pinions of an American clock have long been a mystery to those unacquainted with the method of their manufacture, and the usual accuracy in the position of the small wires or "rounds," combined with great cheapness, has often been a subject of remark. The holes for the wires in these pinions are drilled in a machine constructed as follows: An iron bed with two heads on it, Fig. 80, one of which is so constructed that by pulling a lever the spindle has a motion lengthwise as well as the usual circular motion, and on the point of this spindle, which is driven at 22,000 revolutions, the drill is fastened that is to bore the holes in the pinions; the other head has an arbor passing through it with an index plate attached, having holes in the plate, and an index finger attached to a strong spring going into the holes, the same as in a wheel-cutting engine; on this head, and on the end of it that faces the drill, there is a frame fastened in which the pinion that is to be bored is placed between centers, and is carried round with the arbor of the index plate, in the same manner as a piece of work is carried round in an ordinary lathe by means of a dog, or carrier; only in the pinion drilling machine the carrier is so constructed that there is no shake in any way between the pinion and the index arbor. This head is carried on a slide having a motion at right angles to the spindle of the other head, by which means the pitch diameter of the proposed pinion is adjusted. The head is moved in the slide by an accurately cut screw, to which a micrometer is attached that enables the workman to make an alteration in the diameter of a pinion as small as the one-thousandth part of an inch. The drill that bores the holes is the ordinary flat-pointed drill, and has a shoulder on its stem that stops the progress of the drill when it has gone through the first part of the pinion head and nearly through the other. All operators make their own drills and the limits of error are for pitch diameter .0005 inch; error of size of drills .0001. The reader can see that these men must know something of drill making.

The action of the machine is simple. The pinion, after it has been turned, pivoted and dogged, is placed in its

Fig. 80. Pinion Drilling Machine.

position in the machine, and by pulling a lever, the drill, which is running at a speed of about 22,000 revolutions a minute, comes in contact with the brass heads of the pinion and bores the one through and the other nearly through. The lever is then let go, and a spring pulls the drill back; the index is turned round a hole, and another hole bored in the pinion, and so on till all the holes are bored. An ordinary expert workman, with a good machine, will bore about fourteen hundred of medium-sized pinions in a day.

The wires or "rounds" are cut from drill rod and are put into the holes by hand by girls who become very expert at it. This is called "filling." We have already stated that the holes are only bored partly through one of the pieces of the brass, and after the wire has been put in, the holes are riveted over, and in this manner the wires are fastened so that they cannot come out. Some factories close the holes by a thin brass washer forced on the arbor, instead of riveting.

Figs. 77, 78 and 79 show the automatic pinion turning machine and its processes in successive operations. These machines are used by most of the large clock manufacturers of the United States and some of the European concerns also. They are entirely automatic, will make 1,500 pinions per day, as an average, and one man can run four machines.

Fig. 79 shows an automatic pinion drilling machine, which takes up the work where it is left by the machine shown in Fig. 77. This machine will drill 4,000 to 5,000 pinions per day according to the size hole and the number of holes. The operator places the pinions in the special chain shown in the front of the machine, from which the transport arms carry them to the spindle, where they are drilled and when completed drop out. One operator can feed three of these machines.

Making Solid Pinions.—The solid steel pinions are not hardened, but are made of Bessemer steel, which could only be case hardened—a thing hardly ever done. The process of making these pinions is as follows: Rods of Bessemer steel are cut into suitable lengths. The pieces obtained are pointed or centered on both ends. The stock not needed for the pinion head is cut away, leaving the arbors slightly tapering, for the purpose of fastening them by this means in a hole on the cutting machine. On the end of the arbor of the index plate are two deep cuts across its center, and at right angles to each other. These cuts are of the same shape that would be made by a knife-edged file. The effect of these cuts is to produce a taper hole in the end of the arbor, with four sharp corners. Into this hole the end of the arbor of the pinion or ratchet that is to be cut is placed, and a spring center presses on the other end, and the sharp corners in the hole hold the work firm enough to prevent it from turning round when the teeth are being cut. The marks that are to be seen on the shoulder of the back pivot of the arbor that carries the minute hand of a Yankee clock is an illustration of this method of holding the pinion when the leaves are being cut, and no injurious effects arise from it. The convenience the plan affords for fastening work in the engine enables twenty-five hundred of these pinions to be cut in a day, one at a time. The pinion head is cut subject to the proper dividing plate by a splitting circular saw, and by a milling tool (running in oil) for forming the shape of the leaves, both of which tools are generally carried on the same arbor, both being shifted into their proper places by an adjusting attachment. Pinion leaves of the better class are generally shaped by two succeeding milling cutters, the second one of which does the finishing, obviating any other smoothing. For very cheap work the arbors receive no further finish. The shaping of the pivots, done by an automatic lathe, finishes the job.

Figure 81 shows an automatic pinion cutting machine which has extensive use in clock factories for cutting pinions up to one-half inch diameter and also the smaller wheels. For wheels the work is handled in stacks suited to the traverse of the machine, the work being treated as if the stacks were long brass pinions.

Wheels are cut in two ways, on automatic wheel cutters as just described and on engines containing parallel spindles for the cutters, carried in a yoke which rises and falls, so that it clears the work while the carriage is returning to the starting point on each trip and engages it on the outward trip. The cutters are about three inches in diameter and rapidly driven; the first is a saw, the second a roughing cutter, and the third a finishing cutter. The carriage is

Fig. 81. Automatic Wheel and Pinion Cutters.

driven by a rack and pinion operated by a crank in the hands of the workman and streams of soda water are used on the cutters and work to carry away the heat, as brass expands rapidly under heat, and if the stack were cut dry the cut would get deeper as the cutting proceeded, owing to the expansion of the brass, and hence the finished wheel would not be round when cold, if many teeth were being cut. The stacks of wheels are about four inches in length and the slide thus travels about twenty inches in

Fig. 82. Wheel Cutting Engine.

order to clear the three arbors and engage with the shifter for the index. The last wheel of the stack has a very large burr formed by the cutters as they leave the brass and this wheel is removed from the stack when the arbor is taken out and placed aside to have the burrs removed by rubbing on emery paper.

This is one of the few instances in which automatic machinery has been unable to displace hand labor, as the work is done so quickly that the time of the attendant would be nearly all taken up in placing and removing the stacks, and so the feeding is done by him as well. About 35,000 wheels per day can be thus cut by one man, with girls to stack the blanks on the arbors, and an automatic feed would not release the man from attendance on the machine, so that the majority of clock wheels are cut to-day as they were forty years ago. Still, some of the factories are adding an automatic feed to the carriage in the belief that the increased evenness of feed will give a more accurately cut wheel, a proposition which the men most vigorously deny. Such a machine, they say, to be truly automatic, must take its stacks of wheels from a magazine and discharge the work when done, so that one attendant could look after a number of machines. This would result in economy, as well as accuracy, but has not been done owing to the great variations in sizes of wheels and numbers of teeth required in clock work.

Figure 82 shows one of these machines, a photograph of which was taken especially for us by the courtesy of the Seth Thomas Clock Company at their factory in Thomaston, Conn.

About every ten years some factory decides to try stamping out the teeth of wheels at the same time they are being blanked; this can, of course, be done by simply using a more expensive punch and die, and at first it looks very attractive; but it is soon found that the cost of keeping up such expensive dies makes the wheels cost more than if regularly cut and for reasons of economy the return is made to the older and better looking cut wheels.

After an acid dip to remove the scale on the sheet brass, followed by a dip in lacquer, to prevent further tarnish, the wheels are riveted on the pinions in a specially constructed jig which keeps them central during the rivetting and when finished the truth of every wheel and its pinions and pivots are all tested before they are put into the clocks. The total waste on all processes in making wheels and pinions is from two to five per cent, so that it will readily be seen that accuracy is demanded by the inspectors. European writers have often found fault with nearly everything else about the Yankee clock, but they all unite in agreeing that the cutting and centering of wheels, pinions and pivots (and the depthing) are perfect, while the clocks of Germany, France, Switzerland and England (particularly France) leave much to be desired in this respect; and much of the reputation of the Yankee clock in Europe comes from the fact that it will run under conditions which would stop those of European make.

We give herewith a table of clock trains as usually manufactured, from which lost wheels and pinions may be easily identified by counting the teeth of wheels and pinions which remain in the movement and referring to the table. It will also assist in getting the lengths of missing pendulums by counting the trains and referring to the corresponding length of pendulums. Thus, with 84 teeth in the center wheel, 70 in the third, 30 in the escape and 7-leaf pinions, the clock is 120 beat and requires a pendulum 9.78 inches from the bottom of suspension to the center of the bob.

To Calculate Clock Trains.—Britten gives the following rule: Divide the number of pendulum vibrations per hour by twice the number of escape wheel teeth; the quotient will be the number of turns of escape wheel per hour. Multiply this quotient by the number of escape pinion teeth, and divide the product by the number of third wheel. This quotient will be the number of times the teeth of third wheel pinion must be contained in center wheel.

Take a pendulum vibrating 5,400 times an hour, escape wheel of 30, pinions of 8, and third wheel of 72. Then 5,400÷60=90. And 90×8÷72=10. That is, the center

Clock Trains and Lengths of Pendulums.

Wheels	Pinions	Escape Wheel	Vibrations of Pendulum—Min.	Length of Pendulum in Inches
120 90 75	10 10 9	Double 3legged	*30	156.56
120 90 90	10 9 9	Do.	*40	88.07
128 120	16	30	60	39.14
112 105	14	30	60	39.14
96 90	12	30	60	39.14
80 75	10	30	60	39.14
64 60	8	30	60	39.14
68 64	8	30	68	30.49
70 64	8	30	70	28.75
72 64	8	30	72	27.17
75 60	8	32	75	25.05
72 65	8	32	78	23.15
75 64	8	32	80	22.01
84 64	8	30	84	19.97
86 64	8	30	86	19.06
88 64	8	30	88	18.19
84 78	7	20	89.1	17.72
80 72	8	30	90	17.39
84 78	7	21	93.6	16.08
94 64	8	30	94	15.94
84 78	8	28	95.5	15.45
108 100	12 & 10	32	96	15.28
84 84	9 & 8	30	98	14.66
84 78	7	22	98	14.66
84 78	8	29	98.9	14.41
80 80	8	30	100	14.09
85 72	8	32	102	13.54
84 78	8	30	102.4	13.44
84 78	7	23	102.5	13.4
105 100	10	30	105	12.78
84 78	8	31	105.8	12.59
84 78	7	24	107	12.3
96 72	8	30	108	12.08
84 78	8	32	109.2	11.82
88 80	8	30	110	11.64
84 77	7	25	110	11.64
84 78	7	25	111.4	11.35
84 80	8	32	112	11.22
84 78	8	33	112.6	11.11

Wheels	Pinions	Escape Wheel	Vibrations of Pendulum—Min.	Length of Pendulum in Inches
96 76	8	30	114	10.82
115 100	10	30	115	10.65
84 78	7	26	115.9	10.49
96 80	8	30	120	9.78
84 70	7	30	120	9.78
84 78	7	27	120.3	9.73
90 84	8	31	122	9.46
84 78	7	28	124.8	9.02
100 80	8	30	125	9.01
90 84	8	32	126	8.87
100 96	10	40	128	8.59
84 78	7	29	129.3	8.42
100 78	8	32	130	8.34
84 77	7	30	132	8.08
84 78	7	30	133.7	7.9
90 90	8	32	135	7.73
84 78	7	31	138.2	7.38
84 80	8	40	140	7.18
120 71	8	32	142	6.99
84 78	7	32	142.6	6.93
100 87	8	32	145	6.69
84 78	7	33	147.1	6.5
100 96	8	30	150	6.26
84 78	7	34	151.6	6.1
96 95	8	32	152	6.09
84 77	7	35	154	5.94
104 96	8	30	156	5.78
84 78	7	35	156	5.78
120 96	9 & 8	30	160	5.5
84 78	7	36	160.5	5.47
84 78	7	37	164.9	5.15
132 100	9 & 8	27	165	5.17
84 78	7	38	169.4	4.88
128 102	8	25	170	4.87
84 78	7	39	173.8	4.65
36 36 35	6	25	175	4.6
84 77	7	40	176	4.55
84 78	7	40	178.3	4.43
45 36 36	6	20	180	4.35
47 36 36	6	20	188	3.99

*These are good examples of turret clock trains; the great wheel (120 teeth) makes in both instances a rotation in three hours. From this wheel the hands are to be driven. This may be done by means of a pinion of 40 gearing with the great wheel, or a pair of bevel wheels bearing the same proportion to each other (three to one) may be used, the larger one being fixed to the great wheel arbor. The arrangement would in each case depend upon the number and position of the dials. The double three-legged gravity escape wheel moves through 60° at each beat, and therefore to apply the rule given for calculating clock trains it must be treated as an escape wheel of three teeth.

wheel must have ten times as many teeth as the third wheel pinion, or ten times 8=80.

The center pinion and great wheel need not be considered in connection with the rest of the train, but only in relation to the fall of the weight, or turns of mainspring, as the case may be. Divide the fall of the weight (or twice the fall, if double cord and pulley are used) by the circumference of the barrel (taken at the center of the cord); the quotient will be the number of turns the barrel must make. Take this number as a divisor, and the number of turns made by the center wheel during the period from winding to winding as the dividend; the quotient will be the number of times the center pinion must be contained in the great wheel. Or if the numbers of the great wheel and center pinion and the fall of the weight are fixed, to find the circumference of the barrel, divide the number of turns of the center wheel by the proportion between the center pinion and the great wheel; take the quotient obtained as a divisor, and the fall of the weight as a dividend (or twice the fall if the pulley is used), and the quotient will be the circumference of the barrel. To take an ordinary regulator or 8-day clock as an example—192 (number of turns of center pinion in 8 days)÷12 (proportion between center pinion and barrel wheel)=16 (number of turns of barrel). Then if the fall of the cord= 40 inches, 40×2÷16=5, which would be circumference of barrel at the center of the cord.

If the numbers of the wheels are given, the vibrations per hour of the pendulum may be obtained by dividing the product of the wheel teeth multiplied together by the product of the pinions multiplied together, and dividing the quotient by twice the number of escape wheel teeth.

The numbers generally used by clock makers for clocks with less than half-second pendulum are center wheel 84, gearing with a pinion of 7; third wheel 78, gearing with a pinion of 7.

The product obtained by multiplying together the center and third wheels=84×78=6,552. The two pinions multiplied together=7×7=49. Then 6,552÷49=133.7. So that for every turn of the center wheel the escape pinion turns 133.7 times. Or 133.7÷60=2.229, which is the number of turns in a minute of the escape pinion.

The length of the pendulum, and therefore the number of escape wheel teeth, in clocks of this class is generally decided with reference to the room to be had in the clock case, with this restriction, the escape wheel should not have less than 20 nor more than 40 teeth, or the performance will not be satisfactory. The length of the pendulum for all escape wheels within this limit is given in the preceding table. The length there stated is of course the theoretical length, and the ready rule adopted by clockmakers is to measure from the center arbor to the bottom of the inside of the case, in order to ascertain the greatest length of pendulum which can be used. For instance, if from the center arbor to the bottom of the case is 10 inches, they would decide to use a 10-inch pendulum, and cut the escape wheel accordingly with the number of teeth required as shown in the table. But they would make the pendulum rod of such a length as just to clear the bottom of the case when the pendulum was fixed in the clock.

In the clocks just referred to the barrel or first wheel has 96 teeth, and gears with a pinion of eight.

Month clocks have an intermediate wheel and pinion between the great and center wheels. This extra wheel and pinion must have a proportion to each other of 4 to 1 to enable the 8-day clock to go 32 days from winding to winding. The weight will have to be four times as heavy, plus the extra friction, or if the same weight is used there must be a proportionately longer fall.

Six-months clock have two extra wheels and pinions between the great and center wheels, one pair having a proportion of 4½ to 1 and the other of 6 to 1. But there is an enormous amount of extra friction generated in these clocks, and they are not to be recommended.

The pivot holes and all the other holes in the frames, are punched at one operation after the frames have been blanked and flattened. They are placed in the press, and a large die having punches in it of the proper size and in the right position for the holes, comes down on the frame and makes the holes with great rapidity and accuracy. These holes are finished afterwards by a broach. In some kinds of clocks, where some of the pivot holes are very small, the small holes are simply marked with a sharp point in the die, and afterwards drilled by small vertical drills. These machines are very convenient for boring a number of holes rapidly. The drill is rotated with great speed, and a jig or plate on which the work rests is moved upwards towards the drill by a movement of the operator's foot. All the boring, countersinking, etc., in American clocks, is done through the agency of these drills. Bending the small wires for the locking work, the pendulum ball, etc., is rapidly effected by forming. As no objectionable marks have been made on the surface of either the thick or smaller wires during any process of construction, all that is necessary to finish the iron work is simply to clean it well, which is done in a very effective manner by placing a quantity of work in a revolving tumbling box, which is simply a barrel containing a quantity of saw-dust.

Milling the winding squares on barrel arbors is an ingenious operation. The machine for milling squares and similar work is made on the principle of a wheel-cutting engine. The work is held in a frame, attached to which is a small index plate, like that of a cutting engine. In the machine two large mills or cutters, with teeth in them like a file, are running, and the part to be squared is moved in between the revolving cutters, which operation immediately forms two sides of the square. The work is then drawn back, and the index turned round, and in a like manner the other two sides of the square are formed. The cutting sides of the mills are a little bevelled, so that they will produce a slight taper on the squares.

Winding keys have shown great improvements. Some manufacturers originally used cast iron ones, but the squares were never good in them, and brass ones were adopted. At first the squares were made by first drilling a hole and driving a square punch in with a hammer; and to make the squares in eighteen hundred keys by this method was considered a good day's work. Restless Yankee ingenuity, however, has contrived a device by which twenty or twenty-five thousand squares can be made in a day, while at the same time they are better and straighter squares than those by the old method; but we are not at liberty to describe the process at present, but only to state that it is done by what machinists call drilling a square hole.

Pendulum rods are made from soft iron wire, and the springs on the ends rolled out by rollers. Two operations are necessary. The first roughs the spring out on rollers of eccentric shape, and the spring is afterwards finished on plain smooth rollers. The pendulum balls in the best clocks are made of lead, on account of its weight, and cast in an iron mold in the same manner as lead bullets, at the rate of about eighteen hundred a day. A movable mandrel is placed in the mold to produce the hole that is in the center of the ball. The balls are afterwards covered with a shell of brass, polished with a blood-stone burnisher. The various cocks used in these clocks are all struck up from sheet brass, and the pins in the wheels in the striking part are all swedged into their shape from plain wire. The hands are die struck out of sheet steel, and afterwards polished on emery belts, and blued in a furnace.

All the little pieces of these clocks are riveted together by hand, and the different parts of the movement, when complete, are put together by workmen continually employed in that department. Although the greatest vigilance is used in constructing the different parts to see that they are perfect, when they come to be put together they are subjected to another examination, and after the movements are put in the case the clocks are put to the test by actual trial before they are packed ready for the market. As a general rule, all the different operations are done by workmen employed only at one particular branch; and in the largest factories from thirty to fifty thousand clocks of all classes may be seen in the various stages of construction.

Such is a description of the main points in which the manufacture of American clock movements differs from those manufactured by other systems. All admit that these clocks perform the duties for which they are designed in an admirable manner, while they require but little care to manage, and when out of order but little skill is necessary to repair them. Of late years there has been a growing demand for ornamental mantel-piece clocks in metallic cases of superior quality, and large numbers of these cases of both bronze and gold finish are being manufactured, which, for beauty of design and fine execution, in many instances rival those of French production. The shapes of the ordinary American movements were, however, unsuitable for some patterns of the highest class of cases, and the full plate, round movements of the same size as the French. but with improvements in them that in some respects render them more simple than the French, are now manufactured. Exactly the same system is employed in the manufacture of the different parts of these clocks that is practiced in making the ordinary American movements.

CHAPTER XV.

SPRINGS, WEIGHTS AND POWER.

We see by the preceding calculations that there is one definite point in the time train of a clock; the center arbor, which carries the minute hand, must revolve once in one hour; from this point we may vary the train both ways, toward the escape wheel to suit the length of pendulum which we desire to use, and toward the barrel to suit the length of time we want the clock to run. The center arbor is therefore generally used as the point at which to begin calculations, and it is also for this reason that the number of teeth in the center wheel is the starting point in train calculations toward the escape wheel, while the center pinion is the starting point in calculations of the length of time the weight or spring is to drive the clock. Most writers on horology ignore this point, because it seems self-evident, but its omission has been the cause of much mystification to so many students that it is better to state it in plain terms, so that even temporary confusion may be avoided.

Sometimes there is a second fixed point in a time train; this occurs only when there is a seconds hand to be provided for; when this is the case the seconds hand must revolve once every minute. If it is a seconds pendulum the hand is generally carried on the escape wheel and the relation of revolutions between the hour and seconds wheels must then be as one is to sixty. This might be accomplished with a single wheel having sixty times as many teeth as the pinion on the seconds arbor; but the wheel would take up so much room, on account of its large circumference, that the movement would become unwieldly because there would be no room left for the other wheels; so it is cheaper to make more wheels and pinions and thereby get a smaller clock. Now the best practical method of dividing this motion is by giving the wheels and pinions a relative velocity of seven and a half and eight, because $7.5 \times 8 = 60$.

Thus if the center wheel has 80 teeth, gearing into a pinion of 10, the pinion will be driven eight times for each revolution of the center wheel, while the third wheel, with 75 teeth, will drive its pinion of 10 leaves 7.5 times, so that this arbor will go 7.5 times eight, or 60 times as fast as the center wheel.

If the clock has no seconds hand this second fixed point is not present in the calculations and other considerations may then govern. These are generally the securing of an even motion, with teeth of wheels and pinions properly meshing into each other, without incurring undue expense in manufacture by making too many teeth in the pinions and consequently in the wheels. For these reasons pinions of less than seven or more than ten leaves are rarely used in the common clocks, although regulators and fine clocks, where the depthing is important, frequently have 12, 14 or 16 leaves in the pinions, as is also the case with tower clocks, where the increased size of the movement is not as important as a smoothly running train. Clocks without pendulums, carriage clocks, locomotive levers and nickel alarms, also have different trains, many of which have the six leaf pinion, with its attendant evils, in their trains.

WEIGHTS.—Weights have the great advantage of driving a train with uniform power, which a spring does not accomplish: They are therefore always used where exactness of time is of more importance than compactness or portability of the clock. In making calculations for a weight movement, the first consideration is that as the coils of the cord must be side by side upon the barrel and each takes up a definite amount of space, a thicker movement (with longer arbors) will be necessary, as the barrel must give a sufficient number of turns of the cord to run the clock the desired time and the length of the barrel, with the wheel and maintaining power all mounted upon the one arbor, will determine the thickness of the movement. If the clock is to have striking trains their barrels will generally be of more turns and consequently longer than the time barrel and in that case the distance between the plates is governed by the length of the longest barrel and its mechanism.

The center wheel, upon the arbor of which sits the canon pinion with the minute hand, must, since the hand has to accomplish its revolution in one hour, also revolve once in an hour. When, therefore, the pinion of the center arbor has 8 leaves and the barrel wheel 144, then the 8 pinion leaves, which makes one revolution per hour, would require the advancing of 8 teeth of the barrel wheel, which is equal to the eighteenth part of its circumference. But when the eighteenth part in its advancing consumes 1 hour, then the entire barrel wheel will consume 18 hours to accomplish one revolution. If, now, 10 coils of the weight cord were laid around the barrel, the clock would then run $10 \times 18 = 180$ hours, or 7½ days, before it is run down.

Referring to what was said in a previous chapter on wheels being merely compound levers, it will be seen that as we gain motion we lose power in the same ratio. We shall also see that by working the rule backwards we may arrive at the amount of force exerted on the pendulum by the pallets. If we multiply the circumference of the escape wheel in inches by the number of its revolutions in one hour we will get the number of inches of motion the escape wheel has in one hour. Now if we multiply the weight by the distance the barrel wheel travels in one hour and divide by the first number we shall have the force exerted on the escape wheel. It will be simpler to turn the weight into grains before starting, as the division is less cumbersome.

Another way is to find how many times the escape wheel revolves to one turn of the barrel and divide the weight by that number, which will give the proportion of weight at the escape wheel, or rather would do so if there were no power lost by friction. It is usual to estimate that three-quarters of the power is used up in frictions of teeth and pivots, so that the amount actually used for propulsion of the pendulum is very small, being merely sufficient to overcome the bending moment of the suspension spring and the resistance of the air.

It is for this reason that clocks with finely cut trains and jeweled pivots, thus having little train friction, will run with very small weights. The writer knows of a Howard regulator with jeweled pivots and pallets running a 14-pound pendulum with a five-ounce driving weight. Of course this is an extreme instance and was the result of an experiment by an expert watchmaker who wanted to see what he could do in this direction.

Usually the method adopted to determine the amount of weight that is necessary for a movement is to hang a small tin pail on the weight cord and fill it with shot sufficient to barely make the clock keep time. When this point has been determined, then weigh the pail of shot and make your driving weight from eight to sixteen ounces heavier. In doing this be sure the clock is in beat and that it is the lack of power which stops the clock; the latter point can be readily determined by adding or taking out shot from the pail until the amount of weight is determined. The extra weight is then added as a reserve power, to counteract the increase of friction produced by the thickening of the oil.

Many clock barrels have spiral grooves turned in them to assist in keeping the coils from riding on each other, as where such riding occurs the riding coils are farther from the center of the barrel than the others, which gives them a longer leverage and greater power while they are unwinding, so that the power thus becomes irregular and affects the rate of the clock, slowing it if the escapement is dead beat and making it go faster if it is a recoil escapement.

Clock cords should be attached to the barrel at the end which is the farthest from the pendulum, so that as they unwind the weight is carried away from the pendulum. This is done to avoid sympathetic vibrations of the weight as it passes the pendulum, which interfere with the timekeeping when they occur. If the weight cannot be brought far enough away to avoid vibrations a sheet of glass may be drilled at its four corners and fixed with screws to posts placed in the back of the case at the point where vibration occurs, so that the glass is between the pendulum rod and the weight, but does not interfere with either. This looks well and cures the trouble.

We have, heretofore, been speaking of weights which hang directly from the barrel, as was the case with the older clocks with long cases, so that the weight had plenty of room to fall. Where the cases are too short to allow of this method, recourse is had to hanging the weight on a pulley and fastening one end of the cord to the seat board. This involves doubling the amount of weight and also taking care that the end of the cord is fastened far enough from the slot through which it unwinds so that the cords will not twist, as they are likely to do if they are near together and the cord has been twisted too much while putting it on the barrel. Twisting weight cords are a frequent source of trouble when new cords have been put on a clock. The pulley is another source of trouble, especially if wire cords (picture cords) or cables are used. Wire cable should not be bent in a circle smaller than forty times its diameter if flexibility is to be maintained, hence pulleys which were all right for gut or silk frequently prove too small when wire is substituted and kinks, twisted and broken cables frequently result from this cause. This is especially the case with the heavy weight of striking trains of hall and chiming clocks, where double pulleys are used, and also leads to trouble by jamming and cutting the cables and dropping of the weights in tower clocks where a new cable of larger size is used to replace an old one which has become unsafe from rust, or cut by the sheaves.

Weight cords on the striking side of a clock should always be left long enough so that they will not run down and stop before the time train has stopped. This is particularly the case with the old English hall clocks, as many of them will drop or push their gathering racks free of the gathering pinion under such conditions and then when the clock is wound it will go on striking continuously until the dial is taken off and the rack replaced in mesh with the gathering pinion. As clocks are usually wound at night, the watchmaker can see the disturbance that would be caused in a house in the "wee sma' hours" by such a clock going on a rampage and striking continuously.

Oiling Cables.—Clock cables, if of wire and small in size, should be oiled by dipping in vaseline thinned with benzine of good quality. Both benzine and vaseline must be free from acid, as if the latter is present it will attack the cable. This thinning will permit the vaseline to permeate the entire cable and when the benzine evaporates it will leave a thin film of vaseline over every wire, thus preventing rust. Tower clock cables should be oiled with a good mineral oil, well soaked into them to prevent rusting. Gut clock cords, when dry and hard, are best treated with clock oil, but olive oil or sperm oil will also be found good to soften and preserve them. New cords should always be oiled until they are soft and flexible. If the weight is under ten pounds silk cords are preferable to gut or wire as they are very soft and flexible.

In putting on a new cable or weight cord the course of the weight and cord should be closely watched at all points, to see that they remain free and do not chafe or bind anywhere and also that the coils run evenly and freely, side by side; sometimes, especially with wire, a new cable gets kinked by riding the first time of winding and is then very difficult to cure of this serious fault. Another point to watch is to see that the position of the cord when wound up will not cause an end thrust upon the barrel, which will interfere with the time keeping if it is overwound, so that the weight is jammed against the seatboard; this frequently happens with careless winding, if there is no stop work.

To determine the lengths of clock cords or weights, we may have to approach the question from either end. If the clock be brought in without the cords, we first count the number of turns we can get on the barrel. This may be done by measuring the length of the barrel and dividing it by the thickness of the cord, if the barrel is smooth, or by counting the grooves if it be a grooved barrel. Next we caliper the diameter and add the thickness of one cord, which gives us the diameter of the barrel to the center of the cords, which is the real or working diameter. Multiply the distance so found by 3.14156, which gives the circumference of the barrel, or the length of cord for one turn of the barrel. Multiply the length of one turn by the number of turns and we have the length of cord on the barrel, when it is fully wound. If the cord is to be attached to the weight, measure the distance from the center of barrel to the bottom of the seat board and leave enough for tieing. If the weight is on a pulley it will generally require about twelve inches to reach from the barrel through the slot of the seat board, through the pulley to the point of fastening.

To get the fall of the weight, stand it on the bottom of the case and measure the distance from the top of the point of attachment to the bottom of the seat board. This will generally allow the weight to fall within two inches of the bottom and thus keep the cable tight when the clock runs down; thus avoiding kinks and over-riding when we wind again after allowing the clock to run down. If the weight has a pulley and double cord, measure from the top of the pulley to the seatboard, with the weight on the bottom, and then double this measurement for the length of the cord.

This measure is multiplied by as many times as there are pulleys in the case of additional sheaves. Striking trains are frequently run with two coils or layers of cord on the barrel, time trains never have but one.

Now, having the greatest available length of cord determined according either of the above conditions, we can determine the number of turns for which we have room on our barrel and divide the length of cord by the number of turns. This will give us the length of one turn of the cord on our barrel and thus having found the circumference it is easy to find the diameter which we must give our barrel in suiting a movement to given dimensions of the case. This is frequently done where the factory may want a movement to fit a particular style and size of case which has proved popular, or when a watchmaker desires to make a movement for which he has, or will buy, a case already made.

As to tower clock cables, getting the length of cable on the barrel is, of course, the same as given above, but the rest of it is an individual problem in every case, as cables are led so differently and the length of fall varies so that only the professional tower clock men are fitted to make the measurements for new work and they require no instruction from me. It might be well to add, however, that in the tower clocks by far the greater part of the cable is always outside the clock and only the inner end coils and uncoils about the barrel. It is for this reason that the outer ends of the cables are so generally neglected by watchmakers in charge of tower clocks and allowed to cut and rust until they drop their weights. Caretakers of tower clocks should remember that the inner ends of cables are always the best ends; the parts that need watching are those in the sheaves or leading to the sheaves. Tower clocks should have the cables marked where to stop to prevent overwinding.

In chain drives for the weights of cuckoo and other clocks with exposed weights, we have generally a steel sprocket wheel with convex guiding surfaces each side of the sprocket and projecting flanges each side of the guides; one of these flanges is generally the ratchet wheel. The ratchet wheel, guide, sprocket, guide and flange, form a built-up wheel which is loose on the arbor and is pinned close to the great wheel, which is driven by a click on the wheel working into the ratchet of the drive. It must be loose on the arbor, because the clock is wound by pulling the sprocket and ratchet backward by means of the chain until the weight is raised clear up to the seat board. There are no squares on the arbors, which have ordinary pivots at both ends, and the great wheel is fast on the arbor. The diameter of the convex portion of the wheel each side of the sprocket is the diameter of the barrel, and the chain should fit so that alternate links will fit nicely in the teeth of the sprocket; where this is not the case they will miss a link occasionally and the weight will then fall until the chain catches again, when it will stop with a jerk; bent or jammed links in the chain will do the same thing. Sometimes a light chain on a heavy weight will stretch or spread the links enough to make their action faulty. If examination shows a tendency to open the links, they should be soldered; if they are stretching, a heavier chain of correct lengths of links should be substituted. Twisted chains are another characteristic fault and are usually the result of bent or jammed links. A close examination of such a chain will generally reveal several links in succession which are not quite flat and careful straightening of these links will generally cure the tendency to twist.

Mainsprings for Clocks.—There are many points of difference between mainsprings for clocks and those for watches. They differ in size, strength, number of coils and in their effect on the rates of the clock.

Watch springs are practically all for 30-hour lever escapements, with a few cylinder, duplex and chronometer escapements. If a fusee watch happens into a shop nowadays it is so rare as to be a curiosity worth stopping work to look at.

The clocks range all the way from 30 hours to 400 days in length of time between windings and include lever, cylinder, duplex, dead beat, half dead beat, recoil and other escapements. Furthermore some of these, even of the same form of escapements, will vary so in weight and the consequent influence of the spring that what will pass in one case will give a wildly erratic rate in another instance. Many of the small French clocks have such small and light pendulums that very nice management of the stop works is necessary to prevent the clock from gaining wildly when wound or stopping altogether when half run down.

Nothing will cause a clock with a cylinder escapement to vary in time more than a set or gummy mainspring, for it will gain time when first wound and lose when half run down, or when there is but little power on the train. In such a case examine the mainspring and see that it is neither gummy nor set. If it is set, put in a new spring and you can probably bring it to time.

With a clock it depends entirely on the kind of escapement that it contains, whether it runs faster or slower, with a stronger spring; if you put a stronger mainspring in a clock that contains a recoil escapement the clock will gain time, because the extra power, transmitted to the pallets will cause the pendulum to take a shorter arc, therefore gain time, where the reverse occurs in the dead-beat escapement. A stronger spring will cause the dead-beat pendulum to take a longer arc and therefore lose time.

If a pendulum is short and light these effects will be much greater than with a long and heavy pendulum.

At all clock factories they test the mainsprings for power and to see that they unwind evenly; those that do are marked No. 1, and those that do not are called "seconds." The seconds are used only for the striking side of the clocks, while the perfect ones are used for the running, or time side.

Sometimes, however, a seconds' spring will be put on the time side and will cause the clock to vary in a most erratic way. This changing of springs is very often done by careless or ignorant workmen in cleaning and then they cannot locate the trouble.

All mainsprings for both clocks and watches should be smooth and well polished. Proper attention to this one item will save many dollars' worth of time in examining movements to try to detect the cause of variations.

A rough mainspring (that is, an emery finished mainspring) will lose one-third of its power from coil friction, and in certain instances even one-half. The deceptive feature about this to the watchmaker is that the clock will take a good motion with a rough spring fully found, but will fall off when partly unwound, and the consequence is that he finds a good motion when the spring is put in and wound, and he afterward neglects to examine the spring when he examines the rate as faulty. The best springs are cheap enough, so that only the best quality should be used, as it is easy for a watchmaker to lose three or four dollars' worth of time looking for faults in the escapement, train and everywhere else, except the barrel, when he has inserted a rough, thick, poorly made spring. The most that he can save on the cheaper qualities of springs is about five cents per spring and we will ask any watchmaker how long it would take to lose five cents in examination of a movement to see what is defective.

Here is something which you can try yourself at the bench. Take a rough watch mainspring; coil it small enough to be grasped in the hand and then press on the spring evenly and steadily. You will find it difficult to make the coils slide on one another as the inner coils get smaller; they will stick together and give way by jerks. Now open your hand slowly and you will feel the spring uncoiling in an abrupt, jerky way, sometimes exerting very little pressure on the hand, at other times a great deal. A dirty, gummy spring will do the same thing. Now take a clean, well polished spring and try it the same way; notice how much more even and steady is the pressure required to move the coils upon each other, either in compressing or expanding. Now oil the well polished spring and try it again. You will find you now have something that is instantly responding, evenly and smoothly, to every variation of pressure. You can also compress the spring two or three turns farther with the same force. This is what goes on in the barrel of every clock or watch; you have merely been using your hand as a barrel and *feeling* the action of the springs.

Now a well finished mainspring that is gummy is as irregular in its action as the worst of the springs described above, yet very few watchmakers will take out the springs of a clock if they are in a barrel. One of them once said to me, "Why, who ever takes out springs? I'll bet I clean a hundred clocks before I take out the springs of one of them!" Yet this same man had then a clock which had come back to him and which was the cause of the conversation.

There must be in this country over 25,000 fine French clocks in expensive marble or onyx cases, which were given as wedding presents to their owners, and which have never run properly and in many instances cannot be made to run by the watchmakers to whom they were taken when they stopped. Let me give the history of one of them. It was an eight-day French marble clock which cost $25 (wholesale) in St. Louis and was given as a wedding present. Three months later it stopped and was taken to a watchmaker well known to be skillful and who had a fine run of expensive watches constantly coming to him. He cleaned the clock, took it home and it ran three hours! It came back to him three times; during these periods he went over the movement repeatedly; every wheel was tested in a depthing tool and found to be round; all the teeth were examined separately under a glass and found to be perfect; the pinions were subjected to the same careful scrutiny; the depthings were tried with each wheel and pinion separately; the pivots were tested and found to be right; the movement was put in its case and examined there; it would run all right on the watchmaker's bench, but not in the home of its owner. It would stop every time it was moved in dusting the mantel. He became disgusted and took the clock to another watchmaker, a railroad time inspector; same results. In this way the clock moved about for three years; whenever the owner heard of a man who was accounted more than ordinarily skillful he took him the clock and watched him "fall down" on it. Finally it came into the hands of an ex-president of the American Horological Society. He made it run three weeks. When he found the clock had stopped again he refused pay for it. Three months later he called and got the clock, kept it for three weeks, brought it back without explanation and lo, the clock ran! It would even run considerably out of beat! When asked what he had done to the clock, he merely laughed and said "Wait."

A year later the clock was still going satisfactorily and he explained. "That was the first time I ever got anything I couldn't fix and it made me ashamed. I kept thinking it over. Finally one night in bed I got to considering why a clock wouldn't run when there was nothing the matter with it. The only reason I could see was lack of power. Next morning I got the clock and put in new mainsprings, the best I could find. The clock was cured! None of these other men who had the clock took out the springs. They came to me all gummed up, while the rest of the clock was clean, bright and in perfect order. I cleaned the springs and returned the clock; it ran three weeks. When I took it back I put in stronger springs, because I found them a little soft on testing them. If any of your friends have French clocks that won't go, send them to me."

Three-quarters of the trouble with French clocks is in the spring box; mainspring too weak, gummy or set; stop works not properly adjusted, or left off by some numskull who thought he could make the clock keep time without it when the maker couldn't; mainspring rough, so that it uncoils by jerks; spring too strong, so that the small and light pendulum cannot control it. These will account for far more cases than the "flat wheel" story that so often comes to the front to account for a failure on the part of the workman. Of course he must say something to his boss to account for his failure and the "wheels out of round" and "the faulty depthing" have been standard excuses for French clocks for a century. Of course they do occur, but not nearly as often as they are credited with, and even then such a clock may be made to perform creditably if the springs are right.

Another source of trouble is buckled springs, caused by some workman taking them out or putting them in the barrel without a mainspring winder. There are many men who will tell you that they never use a winder; they can put any spring in without it. Perhaps they can, but there comes a day when they get a soft spring that is too wide for this treatment and they stretch one side of it, or bend, or kink it, and then comes coil friction with its attendant evils. These may not show with a heavy pendulum, but they are certain to do so if it happens to be an eight-day movement with light pendulum or balance, and this is particularly true of a cylinder.

All springs should be cleaned by soaking in benzine or gasoline and rubbing with a rag until all the gum is off them before they are oiled. Heavy springs may be wiped by wrapping one or two turns of a rag around them and pushing it around the coils. The spring should be well cleaned and dried before oiling. A quick way of cleaning is to wind the springs clear up; stick a peg in the escape wheel; remove the pallet fork; plunge the whole movement into a pail of gasoline large enough to cover it; let it stand until the gasoline has soaked into the barrels; remove the peg and let the trains run down. The coils of the spring will scrub each other in unwinding; the pivots will clean the pivot holes and the teeth of wheels and pinions will clean each other. Then take the clock apart for repairs. Springs which are not in barrels should be wound up and spring clamps put on them before taking down the clock. About six sizes of these clamps (from 2½ inches to ¾ inch) are sufficient for ordinary work.

Rancid oil is also the cause of many "come-backs." Workmen will buy a large bottle of good oil and leave it standing uncorked, or in the sun, or too near a stove in winter time, until it spoils. Used in this condition it will dry or gum in a month or two and the clock comes back, if the owner is particular; if not, he simply tells his friends that you can't fix a clock and they had better go elsewhere with their watches.

For clock mainsprings, clock oil, such as you buy from material dealers, is recommended, provided it is intended for French mainsprings. If the lubricant is needed for coarse American springs, mix some vaseline with refined benzine and put it on liberally. The benzine will dissolve the vaseline and will help to convey the lubricant all over the spring, leaving no part untouched. The liquid will then evaporate, leaving a thin coating of vaseline on the spring.

It is best to let springs down with a key made for the purpose. It is a key with a large, round, wooden handle, which fills the hand of the watchmaker when he grasps it. Placing the key on the arbor square, with the movement held securely in a vise, wind the spring until you can release the click of the ratchet with a screwdriver, wire or other tool; hold the click free of the ratchet and let the handle of the key turn slowly round in the hand until the spring is down. Be careful not to release the pressure on the key too much, or it will get away from you if the spring is strong, and will damage the movement. This is why the handle is made so large, so that you can hold a strong spring.

It is of great importance, if we wish to avoid variable coil friction, that the spring should wind, from the very starting, concentrically; i. e., that the coils should commence to wind in regular spirals, equidistant from each other, around the arbor. In very many cases we find, when we commence to wind a spring, that the innermost coil bulges out on one side, causing, from the very beginning, a greater friction of the coils on that side, the outer ones pressing hard against it as you continue to wind, while on the outer side of the arbor they are separated from each other by quite a little space between them, and that this bulge in the first coil is overcome and becomes concentric to the arbor only after the spring is more than half way wound up. This necessarily produces greater and more variable coil friction. When a spring is put into the barrel the innermost coil should come to the center around the arbor by a gradual sweep, starting from at least one turn around away from the other coils. Instead of that, we more often find it laying close to the outer coils to the very end, and ending abruptly in the curl in the soft end that is to be next the arbor. When this is the case in a spring of uniform thickness throughout, it is mainly due to the manner of first winding it from its straight into a spiral form. To obviate it, I generally wind the first coils, say two or three, on a center in the winder, a trifle smaller than the regular one, which is to be of the same diameter of the arbor center in the barrel. You will find that the substitution of the regular center, afterwards, will not undo the extra bending thus produced on the inner coils, and that the spring will abut by a more gradual sweep at the center, and wind more concentrically.

The form of spring formerly used with a fusee in English carriage clocks and marine chronometers is a spring tapering slightly in thickness from the inner end for a distance of two full coils, the thickness incre~sing as we move away from the end, then continuing of ı ıiform thickness until within about a coil and a half from the other end, when it again increases in thickness by a gradual taper. The increase in the thickness towards the outer end will cause it to cling more firmly to the wall of the barrel. The best substitute for this taper on the outside is a brace added to some of the springs immediately back of the hole. With this brace, and the core of the winding arbor cut spirally, excellent results are obtained with a spring of uniform thickness throughout its entire length. Something, too, can be done to improve the action of a spring that has no brace, by hooking it properly to the barrel. The hole in the spring on the outside should never be made close to the end; on the contrary, there should be from a half to three-quarters of an inch left beyond the hole. This end portion will act as a brace.

When the spring is down, the innermost coil of it should form a gradual spiral curve towards the center, so as to meet the arbor without forcing it to one side or the other. This curve can be improved upon, if not correct, with suitably shaped pliers; or it can be approximated by winding the innermost coils first on an arbor a little smaller in diameter than the barrel arbor itself.

Another and very important factor in the development of the force of the spring is the proper length and thickness of it. For any diameter of barrel there is but one length and one thickness of spring that will give the maximum number of turns to wind. This is conditioned by the fact that the volume which the spring occupies when it is down must not be greater nor less than the volume of the empty space around the arbor into which it is to be wound, so that the outermost coil of the spring when fully wound will occupy the same place which the innermost occupies when it is down. In a barrel, the diameter of whose arbor is one-third that of the barrel, the condition is fulfilled when the measure across the coils of the spring as it lays against the wall of the barrel, is 0.39 of the empty space, or, taking the diameter of the barrel as a comparison, 0.123 of the latter; in other words, nearly one-eighth of the diameter of the barrel. This is the width that will give the greatest number of turns to wind, whatever may be the length or thickness of any spring. If now we desire a spring to wind a given number of turns, there is but one thickness and one length of it that will permit it to do so. The thickness remaining the same, if we make the spring longer or shorter, we reduce the number of turns it will wind; more rapidly by making it shorter, less so by making it longer. It is therefore not only useless, but detrimental, to put into a barrel a greater number of coils, or turns, than are necessary, not only because it will reduce the number of turns the barrel will wind, but it will produce greater coil friction by filling up the space with more coils than are necessary.

A mainspring in the act of uncoiling in its barrel always gives a number of turns equal to the difference between the number of coils in the up and the down positions. Thus, if 17 be the number of coils when the spring is run down, and 25 the number when against the arbor, the number of turns in uncoiling will be 8, or the difference between 17 and 25.

The cause of breakage is usually, that the inner coils are put to the greatest strain, and then the slightest flaw in the steel, a speck of rust, grooves cut in the edges of the spring by allowing a screwdriver to slip over them, or an unequal effect of change of temperature, causes the fracture, and leaves the spring free to uncoil itself with verv great rapidity.

Now this sudden uncoiling means that the whole energy of the spring is expended on the barrel in a very small fraction of a second. In reality the spring strikes the inner side of the rim of the barrel, a violent blow in the direction the spring is turning, that is, backwards; this is due to the mainspring's inertia and its very high mean velocity. The velocity is nothing at the outer end, where the spring is fixed, but rises to the maximum at the point of fracture, and the kinetic energy at various points of the spring could no doubt be calculated mathematically or otherwise.

For instance, take a going barrel spring of eight and a half turns, breaking close up to the center while fully wound. A point in the spring at the fracture makes eight turns in the opposite direction to which it was wound, a point at the middle four turns, and a point at the outer end nothing, an effect similar to the whole mass of the spring making four turns backwards. At its greatest velocity it is suddenly stopped by the barrel wheel teeth engaging its pinion; this stoppage or collision is what breaks center pinions, third pivots, wheel teeth, etc., unless their elasticity, or some interposed contrivance, can safely absorb the stored-up energy of the mainspring, the spring being, as every one knows, the heaviest moving part in an ordinary clock, except where the barrel is exceptionally massive.

Stop Works.—Stop works are devices that are but little understood by the majority of workmen in the trade. They are added to a movement for either one or both of two distinct purposes: First, as a safety device, to prevent injury to the escape wheel from over winding, or to prevent undue force coming on the pendulum by jamming the weight against the top of the seat board and causing a variation in time in a fine clock; or, second, to use as a compromise by utilizing only the middle portion of a long and powerful spring, which varies too much in the amount of its power in the up and down positions to get a good rate on the clock if all the force of the spring were utilized in driving the movement.

With weight clocks, the stop work is a safety device and should always be set so that it will stop the winding when the barrel is filled by the cord; consequently the way to set them is to wind until the barrel is barely full and set the stops with the fingers locked so as to prevent any further action of the arbor in the direction of the winding and the cord should then be long enough to permit the weight to be free. Then unwind until within half a coil of the knot in the cord where it is attached to the barrel and see that the weight is also free at the bottom of the case, when the stops again come into action. This will allow the full capacity of the barrel to be used.

When stop work is found on a spring barrel, it may be taken for granted that the barrel contains more spring than is being wound and unwound in the operation of the clock and it then becomes important to know how many coils are thus held under tension, so that we may put it back correctly after cleaning. Wind up the spring and then let it slowly down with the key until the stop work is locked, counting the number of turns, and writing it down. Then hold the spring with the letting down key and take a screw driver and remove the stop from the plate; then count the number of turns until the spring is down and also write that down. Then take out the spring and clean it. You may find such a spring will give seventeen turns in the barrel without the stop work on, while it will give but ten with the stop work; also that the arbor turned four revolutions after you removed the stop. Then the spring ran the clock from the fourth to the fourteenth turns and there were four coils unused around the arbor, ten to run the clock and three unused at the outer end around the barrel. This would indicate a short and light pendulum or balance, which is very apt to be erratic under variations of power, and if the rate was complained of by the customer you can look for trouble unless the best adjustment of the spring is secured. Put the spring back by winding the four turns and putting on the stop work in the locked position; then wind. If the clock gains when up and loses when down, shift the stop works half a turn backwards or forwards and note the result, making changes of the stop until you have found the point at which there is the least variation of power in the up and down positions. If the variation is still too great a thinner spring must be substituted.

There are several kinds of stop work, the most common being what is known as the Geneva stop, a Maltese cross and a finger such as is commonly seen on watches. For watches they have five notches, but for clocks they are made with a greater number of notches, according to the number of turns desired for the arbor. The finger piece is mounted on a square on the barrel arbor and the star wheel on the stud on the plate. In setting them see that the finger is in line with the center of the star wheel when the stop is locked, or they will not work smoothly.

There is another kind of stopwork which is used in some American clocks, and as there is no friction with it, and no fear of sticking, nor any doubt of the certainty of its action, it is perhaps the most suitable for regulators and other fine clocks which have many turns of the barrel in winding. This stop is simple and sure. It consists of a pair of wheels of any numbers with the ratio of odd numbers as 7 and 6, 9 and 10, 15 and 16, 30 and 32, 45 and 48, etc.; the smaller wheel is squared on the barrel arbor and the larger mounted on a stud on the plate. These wheels are better if made with a larger number of teeth. On each wheel a finger is planted, projecting a little beyond the outsides of the wheel teeth, so that when the fingers meet they will butt securely. The meeting of these fingers cannot take place at every revolution because of the difference in the numbers of the teeth of the wheels; they will pass without touching every time till the cycle of turns is completed, as one wheel goes round say sixteen times while the other goes fifteen, and when this occurs the fingers will engage and so stop further winding. When the clock has run down sixteen turns of the barrel the fingers will again meet on the opposite side, and so the barrel will be allowed to turn backwards and forwards for sixteen revolutions, being stopped by the fingers at each extreme. When in action the fingers may butt either at a right or an obtuse angle, only not too obtuse, as this would put a strain on, tending to force the wheels apart. If preferred the fingers may be made of steel, but this is not necessary.

MAINTAINING POWERS.—Astronomical clocks, watchmaker's regulators and tower clocks are, or at least should be, fitted with maintaining power. A good tower clock should not vary in its rate more than five to ten seconds a week. Many of them, when favorably situated and carefully tended, do not vary over five to ten seconds per month. It requires from five to thirty minutes to wind the time trains of these clocks and the reader can easily see where

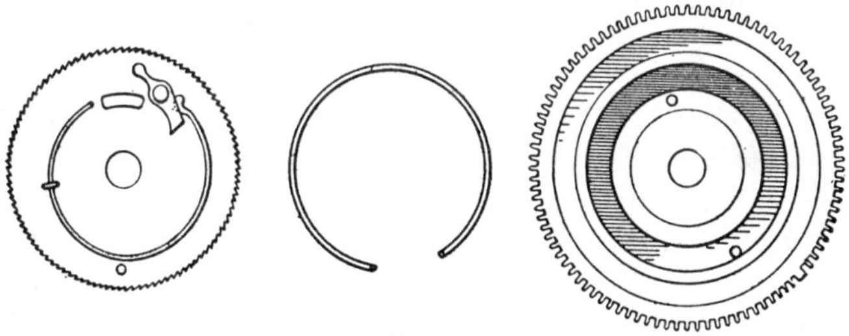

Fig. 83

the rate would go if the power were removed from the pendulum for that length of time; hence a maintaining power that will keep nearly the same pressure on the escape wheel as the weight does, is a necessity. Astronomical clocks and fine regulators have so little train friction, especially if jeweled, that when the barrel is turned backwards in winding the friction between the barrel head and the great wheel is sufficient to stop the train, or even run it backwards, injuring the escape wheel and, of course, destroying the rate of the clock; therefore they are provided with a device that will prevent such an occurrence. Ordinary clocks do not have the maintaining power because only the barrel arbor is reversed in winding, and that reversal is never for more than half a turn at a time, as the power is thrown back on the train every time the winder lets go of the key to turn his hand over for another grip.

Figs. 83, 84 and 85 show the various forms of maintaining powers, which differ only in their mechanical details. In all of them the maintaining power consists of two ratchet wheels, two clicks and either one or two springs; the springs vary in shape according to whether the great wheel is provided with spokes or left with a web. If the great wheel has spokes the springs are attached on the outside of the large ratchet wheel so that they will press on opposite spokes of the great wheel and are either straight, curved or coiled, according to the taste of the maker of the clock and the amount of room. If made with a web a circular recess is cut in the great wheel, see Fig. 83, wide and deep enough for a single coil of spring wire which has its ends bent at right angles to the plane of the spring and one end slipped in a hole of the ratchet and the other in a similar hole in the recess of the great wheel. A circular slot is cut at some portion of the recess in the great wheel where it will not interfere with the spring and a screw in the ratchet works back and forth in this slot, limiting the action of the spring. Stops are also provided for the spokes of the great wheel in the case of straight, curved or coiled springs, Figs. 84 and 85. These stops are set so as to give

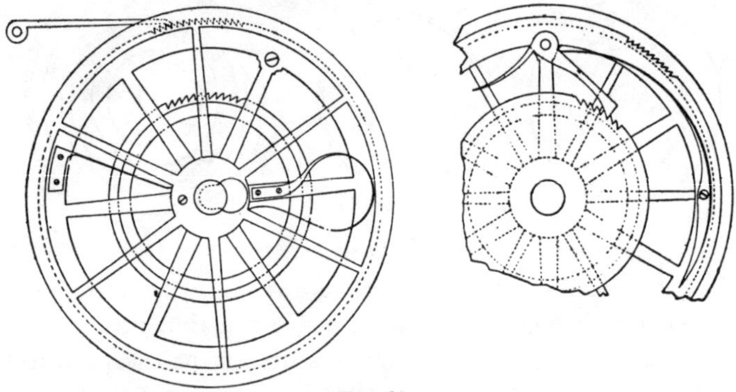

Fig. 84

an angular movement of two or three teeth of the great wheel in the case of tower clocks and from six to eight teeth in a regulator. The springs should exert a pressure on the great wheel of just a little less than the pull of the weight on the barrel; they will then be compressed all the time the weight is in action, and the stops will then transmit the power from the large ratchet to the great wheel, which drives the train. Both the great wheel and the large ratchet wheel are loose on the arbor, being pinned close to the barrel, but free to revolve. A smaller ratchet, having its

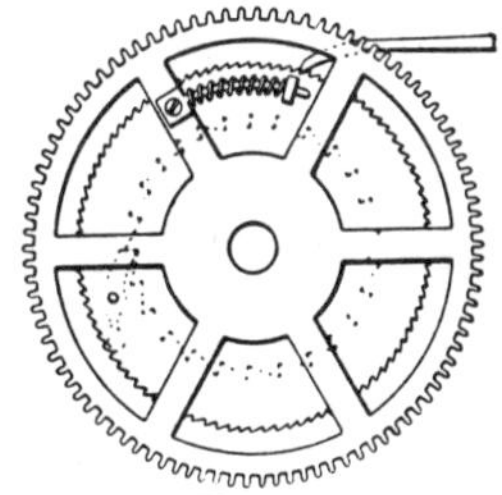

Fig. 85

teeth cut in the reverse direction from those of the larger one, is fast to the end of the barrel. A click, called the winding click, on the larger ratchet acts in the teeth of the smaller one during the winding, holding the two ratchets together at all other times. A longer click, called the detent click, is pivoted to the clock plate, and drags idly over the teeth of the larger ratchet while the clock is being driven by the weight and the maintaining springs are compressed. When the power is taken off by the reversal of the barrel in winding, the friction between the sides of the two ratchets and great wheel would cause them to also turn backward, if it were not for this detent click, with its end fast to the plate, which drops into the teeth of the large ratchet and prevents it from turning backward. We now have the large ratchet held motionless by the detent click on the clock plate and the compressed springs which are carried between the large ratchet and the great wheel will then begin to expand, driving the loose great wheel until their force has been expended, or until winding is completed, when they will again be compressed by the pull of the weight. In some tower clocks curved pins are fixed to opposite spokes of the great wheel and coiled springs are wound around the pins, Fig. 85; eyes in the large ratchet engage the outer ends of the pins and compress the springs.

The clicks for maintaining powers should not be short, and the planting should be done so that lines drawn from the barrel center to the click points and from the click centers to the points, will form an obtuse angle, like B, Fig. 86.

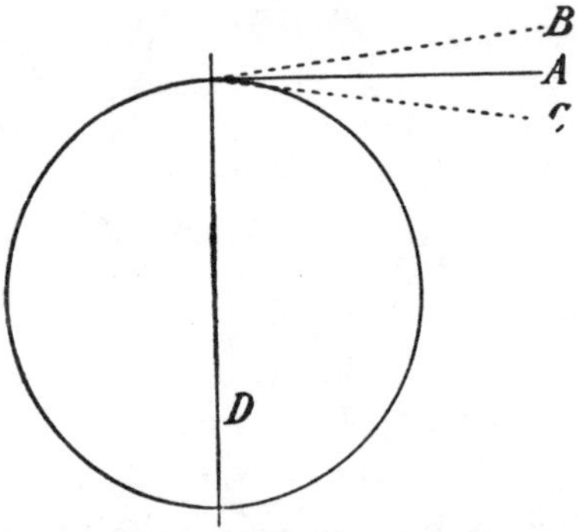

Fig. 86

giving a tendency for the ratchet tooth to draw the click towards the barrel center. The clicks should be nicely formed, hardened and tempered and polished all over with emery. Long, thin springs will be needed to keep the winding clicks up to the ratchet teeth. The ratchet wheel must run freely on the barrel arbor, being carried round by the clicks while the clock is going, and standing still while the weight is being wound up. It is retained at this time by a long detent click mounted on an arbor having its pivots fitted to holes in the clock frame. The same remark as to planting applies to this click as well as the others, and to all clicks having similar objects; but as this click has its own weight to cause it to fall no spring is required. To prevent it lying heavily on the wheel, causing wear, friction and a diminution of driving power, it is as well to have it made light. There is no absolute utility in fixing the click to its collet with screws, but if done, it can be taken off to be polished, and the appearance will be more workmanlike. This click should have its point hardened and tempered, as there is considerable wear on it.

If the great wheel has spokes the best form for the two springs for keeping the train going whilst being wound is that of the letter U, as shown to the left of Fig. 84, one end enlarged for the screw and steady pin and the blade tapering all along towards the end which is free. The springs may be made straight and bent to the form while

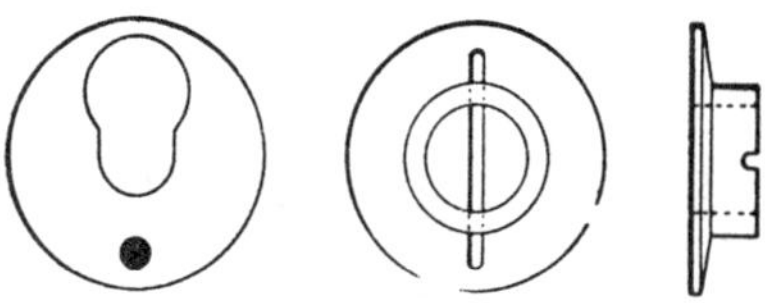

Fig. 87

soft, then hardened and tempered to a full blue. They are best when as large as the space between two arms of the main wheel will allow. When screwed on the large ratchet the backs of both should bear exactly against the respective arms of the mainwheel, and a pair of pins is put in the ratchet, so that any opposite pair of the mainwheel arms may rest upon them when the springs are set up by the clock weight. The strength of the springs can be adjusted by trial, reducing them till the weight of the clock sets them up easily to the banking pins.

There are two methods of keeping the loose wheels against the end of the barrel, while allowing them to turn freely during winding; one is a sliding plate with a keyhole slot, Fig. 87, to slip in a groove on the arbor, as is generally adopted in such house clocks as have fuzees, as well as on the barrels of old-fashioned weight clocks; the other is a collet exactly the same as on watch fuzees. They are both sufficiently effective, but perhaps the latter is the best of the two, because the collet may be fitted on the arbor with a pipe, and being turned true on the broad inside face, gives a larger and steadier surface for the mainwheel to work against, whereas the former only has a small bearing on the shoulder of the small groove in the arbor, which fitting is liable to wear and allow the main and the other loose wheel to wobble sideways, displacing the contact with the detent click and causing the mainwheel to touch the collet of the center wheel if very near together; so, on the whole, a collet, as on a watch fuzee, seems the better arrangement, where there is plenty of room for it on the arbor.

There is an older form of maintaining power which is sometimes met with in tower clocks and which is sometimes imitated on a small scale by jewelers who are using a cheap regulator and wish to add a maintaining power where there is no room between the barrel and plates for the ratchets and great wheel.

The maintaining power, Fig. 88, consists of a shaft, A, a straight lever, B, a segment of a pinion, C, a curved, double lever, D, a weight, E. The shaft, A, slides endwise to engage the teeth of the pinion segment with the teeth of the great wheel, No. 2, the straight lever has a handle at both ends to assist in throwing the pinion out or in and a shield at the outer end to cover the end of the winding shaft, No. 3, when the key is not on it.

The curved lever is double, and the pinion segment turns loosely between the halves and on the shaft, A; it is held up in its place by a light spring, F; the weight, E, is also held between the two halves of the double lever.

The action is as follows: The end of the lever, B, covers the end of the winding shaft so that it is necessary to raise it before putting the key on the winding shaft; it is raised till it strikes a stop, and then pushed in till the pinion seg-

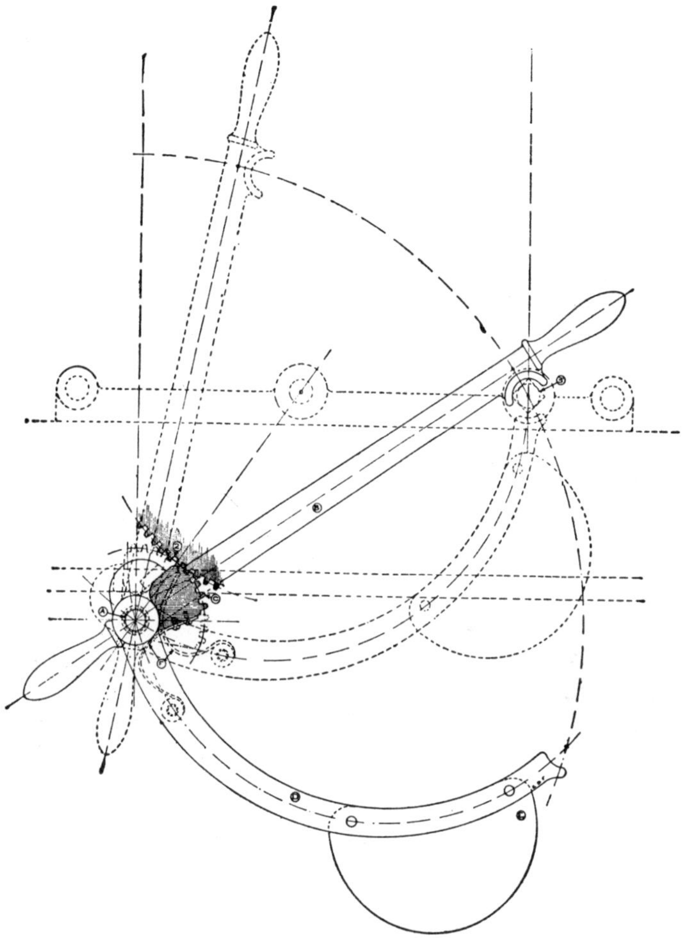

Fig. 88. Maintaining Power.

ment engages with the going wheel of the train, when the weight, E, acting through the levers, furnishes power to drive the clock-train while the going weight is being wound up. Of course the weight on the maintaining power must be so proportioned to the leverage that it will be equal to the power of the going barrel and its weight, a simple proposition in mechanics.

The number of teeth on the pinion segment, C, is sufficient to maintain power for fifteen minutes, at the end of which time the lever, B, will come down and again cover the end of the winding shaft; or, it may be pumped out of gear and dropped down. In case it is forgotten, the spring, F, will allow the segment to pass out of gear of itself and will simply allow it to give a click as it slips over each tooth in the going wheel; if this were not provided for, it would stop the clock.

CHAPTER XVI.

MOTION WORK AND STRIKING TRAINS.

Motion work is the name given to the wheels and pinions used to make the hour hand go once around the dial while the minute hand goes twelve times. Here a few preliminary observations will do much toward clearing up the operations of the trains. The reader will recollect that we started at a fixed point in the time train, the center arbor which must revolve once per hour, and increased this motion by making the larger wheels drive the smaller (pinions) until we reached sixty or more revolutions of the escape wheel to one of the center arbor. This gearing to increase speed is called "gearing up" and in it the pinions are always driven by the wheels. In the case of the hour hand we have to obtain a slowing effect and we do so by making the smaller wheels (pinions) drive the larger ones. This is called "gearing back" and it is the only place in the clock where this method of gearing occurs.

We drew attention to a common usage in the gearing up of the time trains—that of making the relations of the wheels and pinions 8 to one and 7.5 to one; $7.5 \times 8 = 60$. So we find a like usage in our motion work, viz., 3 to one and 4 to one; $3 \times 4 = 12$. Say the cannon pinion has twelve teeth; then the minute wheel generally has 36, or three to one, and if the minute wheel pinion has 10, the hour wheel will have 40, or four to one. Of course, any numbers of wheels and pinions may be used to obtain the same result, so long as the teeth of the wheels multiplied together give a product which is twelve times that of the pinions multiplied together; but three and four to one have been settled upon, just as the usage in the train became fixed, and for the same reasons; that is, these proportions take up the least room and may be made with the least material. Also, the pinion with the greatest number of teeth, being the larger, is usually selected as the cannon pinion, as it gives more room to be bored out to receive the cannon, or pipe. If placed outside the clock plate, the minute wheel and pinion revolve on a stud in the clock plate; but if placed between the frames, they are mounted on arbors like the other wheels. The method of mounting is merely a matter of convenience in the arrangement of the train and is varied according to the amount of room in the movement, or convenience in assembling the movement at the factory, little attention being paid to other considerations.

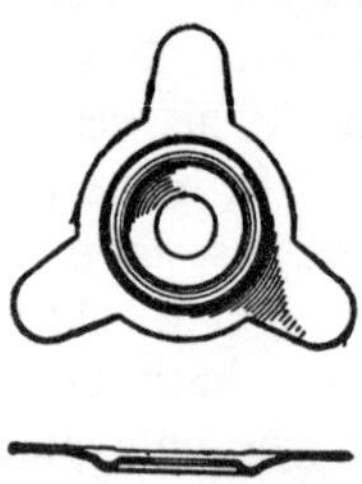

Fig. 89.

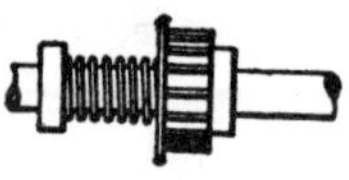

Fig. 90.

The cannon pinion is loose on the center arbor and behind it is a spring, called the center spring, or "friction," Figs. 89 and 90, which is a disc that is squared on the arbor at its center and presses at three points on its outer edge against the side of the cannon pinion; or it may be two or three coils of brass wire. This center spring thus produces friction enough on the cannon to drive it and the hour hand, while permitting the hands to be turned backward or forward without interfering with the train. In French mantel clocks the center spring is dispensed with and a portion of the pipe is thinned and pressed in so as to produce a friction between the pipe and the center arbor which is sufficient to drive the hands; this is similar to the friction of the cannon pinion in a watch.

In some old English house clocks with snail strike, the cannon pinion and minute wheel have the same number of teeth for convenience in letting off the striking work by means of the minute wheel, which thus turns once in an hour. Where this is the case the hour wheel and its pinion

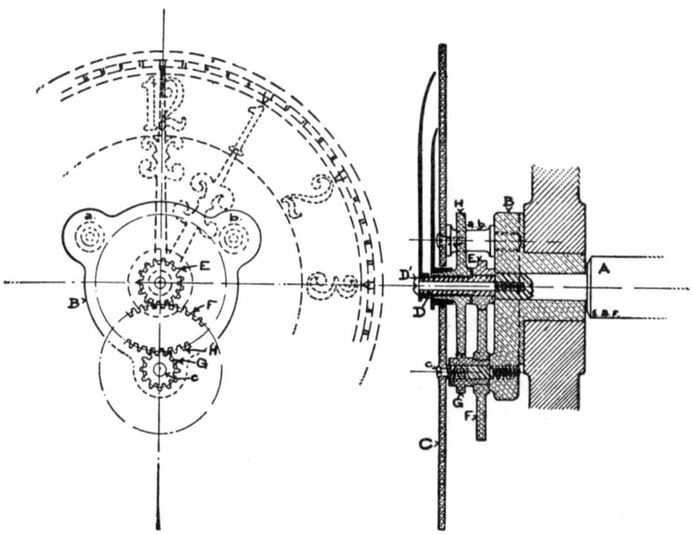

Fig. 91.

bear a proportion to each other of twelve to one; usually there is a pinion of six leaves engaging a wheel of 72 teeth, or seven and eighty-four are sometimes found.

In tower clocks, where the striking is not discharged by the motion work, the cannon pinion is tight on its arbor and the motion work is similar to that of watches. See Fig. 91.

The cannon pinion drives the minute wheel, which, together with its pinion, revolves loosely on a stud in the clock plate, or on an arbor between the frames. The meshing of the minute wheel and cannon pinion should be as deep as is consistent with perfect freedom, as should also that of the hour wheel and minute pinion in order to prevent the hour hand from having too much shake, as the minute wheel and pinion are loose on the stud and the hour wheel is loose on the cannon, so that a shallow depthing here will give considerable back lash, which is especially noticeable when winding.

The hour wheel has a short pipe and runs loosely on the cannon pinion in ordinary clocks. In quarter strike cuckoos a different train is employed and the wheels for the hands are both on a long stud in the plate and both have pipes; the minute wheel has 32 teeth and carries four pins on its under side to let off the quarters. The hour wheel has 64 teeth and works close to the minute wheel, its pipe surrounding the minute wheel pipe, and held in position by a screw and nut on the minute pipe. A wheel of 48 and a pinion of 8 teeth are mounted on the sprocket arbor with a center spring for a friction, the wheel of 48 meshing with the minute wheel of 32 and the 8-leaf pinion with the hour wheel of 64. It will be recollected that the sprocket wheel takes the place of the barrel in this clock and there is no center arbor as it is commonly understood. The sprocket arbor in this case turns once in an hour and a half, hence it requires 48 teeth to drive the minute wheel of 32 once in an hour, as it turns one-third of a revolution (or 16 teeth) every half hour. The sprocket arbor, turning once in an hour and a half, makes eight revolutions in twelve hours and its pinion of eight leaves working in the hour wheel of 64 teeth turns the hour hand once in twelve hours.

In ordinary rack and snail striking work the snail is generally mounted on the pipe of the hour wheel, so that it will always agree with the position of the hour hand and the striking will thus be in harmony with the position of the hands.

STRIKING TRAINS.—It is only natural, after finding certain fixed relations in the calculations of time trains and motion work, that we should look for a similar point in striking trains, well assured that we shall find it here also. It is evident that the clock must strike the sum of the numbers 1, 2, 3, 4, 5, 6, 7, 8, 9, 10, 11, 12, or 78 blows of the hammer, in striking from noon to midnight; this will be repeated from midnight to noon, making 156 blows in 24 hours, and if it is a 30-hour clock, six hours more must be added; blows for these will be 21 more, making a total of 177 blows of the hammer for a 30-hour strike train. The hammer is raised by pins set in the edge of a wheel, called the pin wheel, and as one pin must pass the hammer tail for every blow, it is evident that the number of pins in this wheel will govern the number of revolutions it must make for 177 blows, so that here is the base or starting point in our striking train. If there are 13 pins in the pin wheel, it must revolve 13.5 times for 177 blows; if there are 8 pins, then the wheel must revolve 22.125 times in giving 177 blows; consequently the pinions and wheels back to the spring or barrel must be arranged to give the proper number of revolutions of the pin wheel with a reasonable number of turns of the spring or weight cord, and it is generally desirable to give the same, or nearly the same, number of turns to both time and striking barrels.

If it is an eight-day clock the calculation is a little different. There are 156 blows every 24 hours; then as the majority of "eight-day" clocks are really calculated to keep time for seven and a half days, although they will run eight, we have: $156 \times 7.5 = 1{,}070$ blows in 7.5 days. With 13 pins we have $1{,}070 \div 13 = 80$ and 4-13ths revolutions in the 7.5 days. If now we put an 8-leaf pinion on the pin wheel arbor and 84 teeth in the great wheel or barrel, we will get 10.5 turns of the pin wheel for every turn of the spring or barrel; consequently eight turns of the spring will be enough to run the clock for the required time, as such clocks are wound every seventh day.

Figuring forward from the pin wheel, we find that we shall have to lock our striking train after a stated number of blows of the hammer each hour; these periods increase by regular steps of one blow every hour, so that we must have our locking mechanism in position to act after the passage of each pin, whether it is then used or not; so the pinion that meshes with the pin wheel, and carries the locking plate or pin on its arbor must make one revolution every time it passes a pin. If this is a 6-leaf pinion, the pins on the pin wheel must therefore be 6 teeth apart; or an 8-leaf pinion must have the pins 8 teeth apart; and vice versa. For greater convenience in registering, the pins are set in a radial line with the spaces of the teeth in the pin wheel, as this allows us to measure from the center of the pinion leaf.

It will thus be seen that the calculation of an hour striking train is a simple matter; but if half hours are also to be struck from the train, it will change these calculations. For a 30-hour train 24 must be added to the 156 blows for 24 hours, 180 blows being required to strike hours and half hours for 24 hours. These blows may be provided for by more turns of the spring, or different numbers of the wheels and pinions, which would then also vary the spacing of the pins.

Half hours may also be struck directly from the center arbor, by putting an extra hammer tail on the hammer arbor, further back, where it will not interfere with the hammer tail for the pin wheel, and putting a cam on the center arbor to operate this second hammer tail. This simplifies the train, as it enables the use of a shorter spring or smaller wheels while providing a cheap and certain means of striking the half hours. Half-hour trains are frequently provided with a separate bell of different tone for the half hours, as with only one bell the clock strikes one

Fig. 92. Eight Day Hour and Half Hour Strike.

blow at 12:30, 1 and 1:30, making the time a matter of doubt to one who listens without looking, as frequently happens in the night.

Fig. 92 shows an eight-day, Seth Thomas movement, which strikes the hours on a count wheel train and the half hours from the center arbor. All the wheels, pinions, arbors, pins, levers and hooks are correctly shown in proper position, but the front plate has been left off for greater clearness. The reader will therefore be required to remember that the escape wheel, pallets, crutch, pendulum and the stud for the pendulum suspension are really fixed to the front plate, while in the drawing they have no visible means of support, because the plate is left off.

The time train occupies the right-hand side of the movement and the striking train the left-hand. Running up the right hand from the spring to the escape wheel, we find an extra wheel and pinion which is provided to secure the eight days' run. We also see that what would ordinarily be the center arbor is up in the right corner and does not carry the hands; further, the train is bent over at a right angle, in order to save space and get the escape wheel in the center at the top of the movement. The striking train is also crowded down out of a straight line, the locking cam being to the right of the pin wheel and the warning wheel and fly as close to the center as possible. This leaves some space between the pin wheel and the intermediate wheel of the time train and here we find our center arbor, driven from the intermediate wheel by an extra pinion on the minute wheel arbor, the minute wheel meshing with the cannon pinion on the center arbor. This rearranging of trains to save space is frequently done and often shows considerable ingenuity and skill; it also will many times serve to identify the maker of a movement when its origin is a matter of doubt and we need some material, so that the planting of trains is not only a matter of interest, but should be studied, as familiarity with the methods of various factories is frequently of service to the watchmaker.

Fig. 93 is the upper portion of the same striking train, drawn to a larger scale for the sake of clearness. It also shows the center arbor, both hammer tails and the stop on the hammer arbor, which strikes against the bottom of the front plate to prevent the hammer spring from throwing the hammer out of reach of the pins. The pin wheel, R, and count wheel, E, are mounted close together and are about the same size, so that they are shown broken away for a part of their circumferences for greater clearness in explaining the action of the locking hook, C, and the locking cam, D.

Fig. 94 shows the same parts in the striking position, being shown as just about to strike the last blow of 12. Similar parts have similar letters in both figures.

The count wheel, E, is loose on a stud in the plate, concentric with the arbor of the pin wheel, R. The pivot of R runs through this stud. The sole office of the count wheel is to regulate the distance to which the locking hook C, is allowed to fall. The count hook, A, and the locking hook, C, are mounted on the same arbor, B, so that they move in unison. If A is allowed to fall into a deep slot of the count wheel, C will fall far enough to engage the locking face of the cam D and stop the train, as in Fig. 93. If, on the contrary, A drops on the rim of the wheel, C will be held out of the locking position as D comes around (see Fig. 94), and the train will keep on running. It will be seen that after passing the locking notch, D, Fig. 94, will in its turn raise the hook C, which will ride on the edge of D, and hold A clear of the count wheel until the locking notch of D is again reached, when a deep notch in the wheel will allow C to catch, as in Fig. 93, unless C is stopped by A falling on the rim of the wheel, as in Fig. 94.

One leaf, F, of the pinion of the locking arbor sticks out far enough to engage with the count wheel teeth and rotate

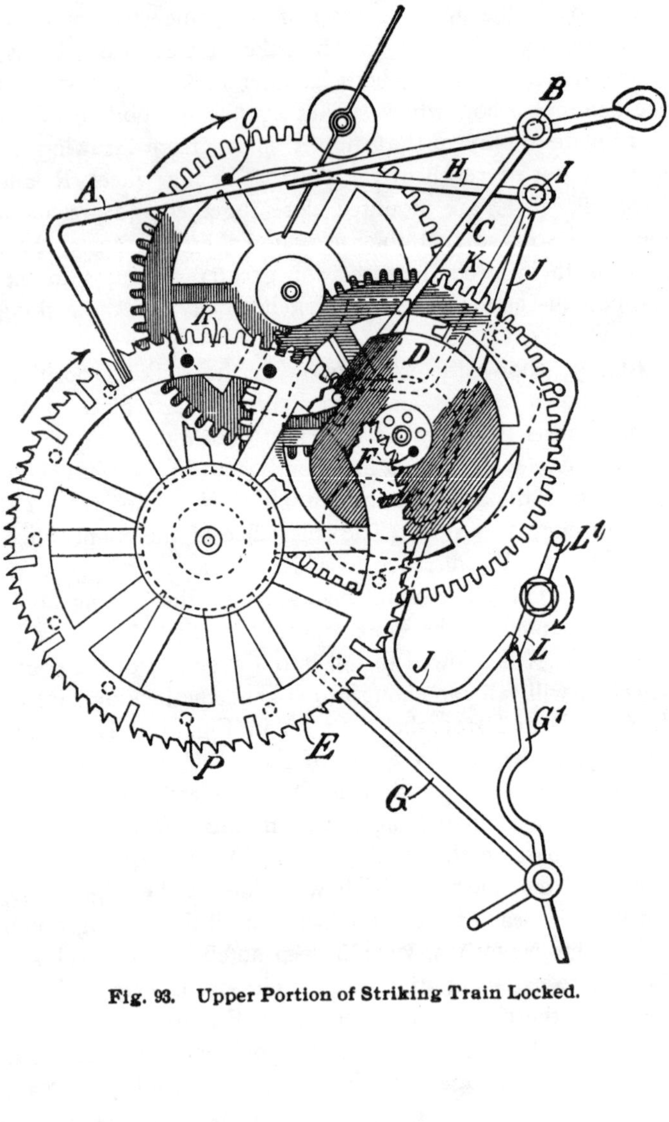

Fig. 93. Upper Portion of Striking Train Locked.

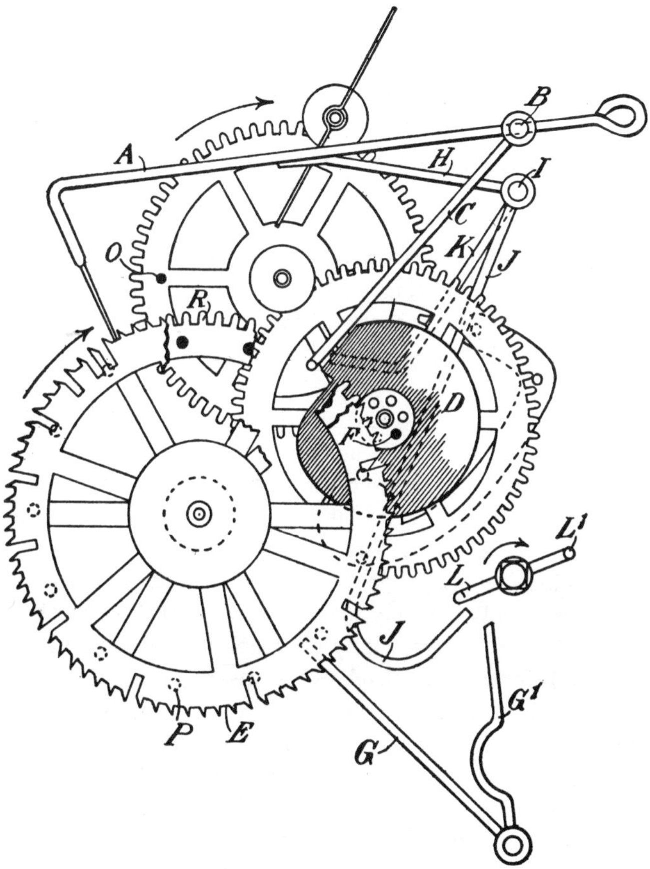

Fig. 94. Striking Train Unlocked and Running.

the wheel one tooth for each revolution of D, so that F forms a one-leaf pinion similar to that of a rack striking train. Here we have our counting mechanism; F and D go around together; F moves E one tooth every revolution. A holds C out of action (Fig. 94) until A reaches a deep slot, when C stops the train by engaging D (Fig. 93).

The count wheel, E, must have friction enough on its stud so that it will stay where the pin F leaves it, when F goes out of action and thus it will be in the right position to suitably engage F on the next revolution. Too much friction of the count wheel on its stud will use too much power for F to move it and thus slow the train; if there is too little friction here the count wheel may get in such a position that F will get stalled on the top of a tooth and stop the train.

The count hook, A, must strike exactly in the middle of the deep slots, without touching the sides of the slots in entering or leaving, as to do this would shift the position of the count wheel if the rubbing were sufficient, or it might prevent A from falling (as A and C are both very light) and the clock would go on striking. If the hook A does not strike the middle of the spaces between the teeth of the count wheel, it will gradually encroach on a tooth and push the wheel forward or back, thus disarranging the count. Many a clock has struck 13 for 12 in this way because the hook was a little out. This did not occur in the smaller numbers because the action was not continued long enough to allow the hook to reach a tooth. The pin, F, should also mesh fairly and freely in the teeth of the count wheel, or a similar defect is likely to occur.

When repairing or making new count hooks, A, Figs. 93 and 94, they must be of such a length that they will enter the slots on a line radial with the center of the wheel. The proper length and direction are shown at A, Fig. 95, while B and C are wrong. With hooks like either B or C you can set or bend the hook to strike right at one and as you turn the clock ahead the hook does not fall in far enough and at twelve it only strikes eleven. Then if you bend the same hook to strike right at twelve it will strike two at one and as you turn the clock ahead it will strike right at about five or seven. A, Fig. 95, being of the proper length and shape will give no trouble. Many of the count wheels of the older clocks were divided by hand and are not as accurate as they should be; when a wheel of this kind is found and a new wheel cannot be substituted (because the clock is an

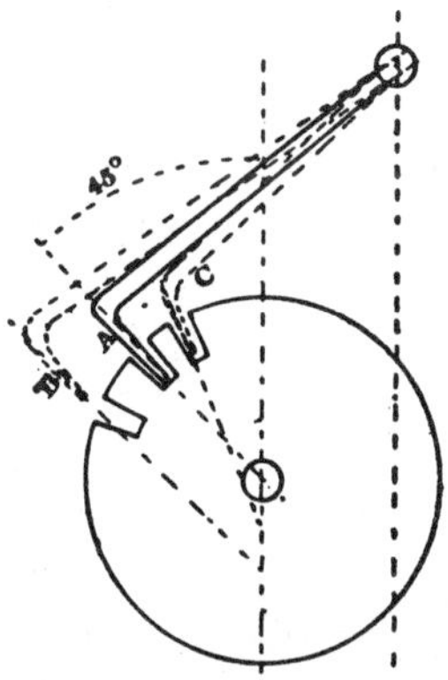

Fig. 95. The proper length of the count hook.

antique and must have the original parts preserved) it will sometimes require nice management of the hook A to obtain correct striking. A little manipulation of the pinion, F, Fig 93. is sometimes desirable also, if the count wheel is very bad.

The locking face of the cam, D, must also be on a line radial to its center, or it will either unlock too easily and go off on the slightest jar or movement of the clock, or the face will have too much draw and the hook C will not be unlocked when the clock is fully wound, and the spring pressure is greatest. In this case the clock will not strike when fully wound, but will do so when partly run down, and as the count wheel train strikes in rotation, without regard to the position of the hands, you will have irregular striking of a most puzzling sort. Repairs to this notch are sometimes required, when the corner has become rounded, and the best way to make them is to cut a new face on the cam with a sharp graver, being careful to keep the face radial with its center.

Because the count wheel strikes the hours in rotation, regardless of the position of the hands, if the hands are turned backwards past the figure 12 on the dial the striking will be thrown out of harmony with the hands. To remedy this the count hook, A, has an eye on its rear end and a wire, shown in Fig. 92, hangs down to where it can be reached with the hand when the dial is on. Pulling this wire will lift A and C and cause the clock to strike; by this means the clock may be struck around until the position of the striking train agrees with that of the hands. Where this wire is not present the striking is corrected by turning the hands back and forth between IX and XII until the proper hour is struck.

Now we come to the releasing mechanism, which causes the clock to strike at stated times. I, Figs. 93 and 94, is an arbor pivoted between the plates and carrying three levers, H, K and J, in different positions on the arbor. H is directly under the count hook, A, and lifts A and C whenever J is pushed far enough to one side by L on the center arbor, which revolves once an hour. Thus L, through J, H and A, C, unlocks the train once every hour. When C is thus lifted the train runs until the warning pin, O, Figs. 93 and 94, strikes against the lever K, which is on the same arbor with H and J. This preliminary run of the train makes a little noise and is called "warning," as the noise notifies us that the train is in position to commence striking. The lever K and the warning pin, O, then hold the train until L has been carried out of action with J and released it, when

O will push K out of its path at every revolution and the clock will strike.

The half hours are struck by L^1 pressing the short hammer tail, G^1, and thus raising and releasing the hammer once an hour.

In setting up the striking train after cleaning, place the pin wheel so that the hammer tail, G, may be about one-fourth of the distance from the next pin, as shown in Fig. 93; this allows the train to get well under way before meeting with any resistance and will insure its striking when nearly run down. If the hammer tail is too close to the pin, it might stop the train when there is but little power on.

Then place D in the locked position, with A in a deep slot of the count wheel and C in the notch of D. Next place the warning wheel with its pin, O, on the opposite side of its arbor from the lever K, see Fig. 93. This is done to make sure that when it is unlocked for "warning" the train will run far enough to get the corner of the lock, D, safely past C, so that it will not allow C to fall into the notch again and lock the train when J, K and H are released by L. This is the rule followed in assembling these clocks at the factories and is simple, correct and easily understood. A study of these points in Fig. 93 will enable any one to set up a train correctly before putting the front plate on.

If the workman gets a clock that has been butchered by some one who did not understand it (and there are many such), he may find that when correctly set up the clock does not strike on the 60th minute of the hour; in such a case a little bending of J, in or out as the case may be, will usually remedy the trouble. The same thing may have to be done to the hammer tails, G and G^1, or the stop on the hammer arbor. If both hammer tails are out of position, bend the stop; if one is right, let the stop alone and bend the other tail.

A rough, set or gummy spring will cause irregular striking. In such a case the clock will strike part of the blows and then stop and finally go on again and complete the number. Much time has been lost in examining the teeth of wheels and pinions in such cases when the trouble lay in the spring. Too strong a spring will make the movement strike too fast; too weak a spring will make it strike slow, especially in the latter part of the day or week, when it has nearly run down.

Too small a fan, or a fan that is loose on its arbor, will allow the clock to strike too fast. If this fan is badly out of balance it will prevent the train from starting when there is but little power on.

There is a class of clocks which have the count wheel tight on the arbor, outside the clock plate. Many of them are on much tighter than they should be. In such a case take an alcohol lamp and heat the wheel evenly, especially around the hub; the brass will expand twice as much as the steel and the wheel may then be driven off without injury.

Fig. 96 shows another typical American eight-day train, made by the Gilbert Clock Company, and striking the half hours from the train. Here we notice, on comparing with Fig. 92, that there are many points of difference. First the notches on the count wheel are twice as wide as they are in Fig. 92. This means that half hours are struck on the train; this will be explained later. Next there are two complete sets of notches on the wheel, which shows that the wheel turns only once in twenty-four hours, whereas the other makes two revolutions in that time. There are no teeth on the count wheel, so that it must be fast to its arbor, which is that of the great wheel and spring, while Fig. 92 has a separate stud and it is loose. The wheel being on the spring arbor and going once in 24 hours, there must be one turn of spring for each 24 hours which the train runs. There is no pin wheel in Fig. 96, but instead of this two pins are cut out of the locking cam to raise the hammer tail as they pass. There are also two locking notches in

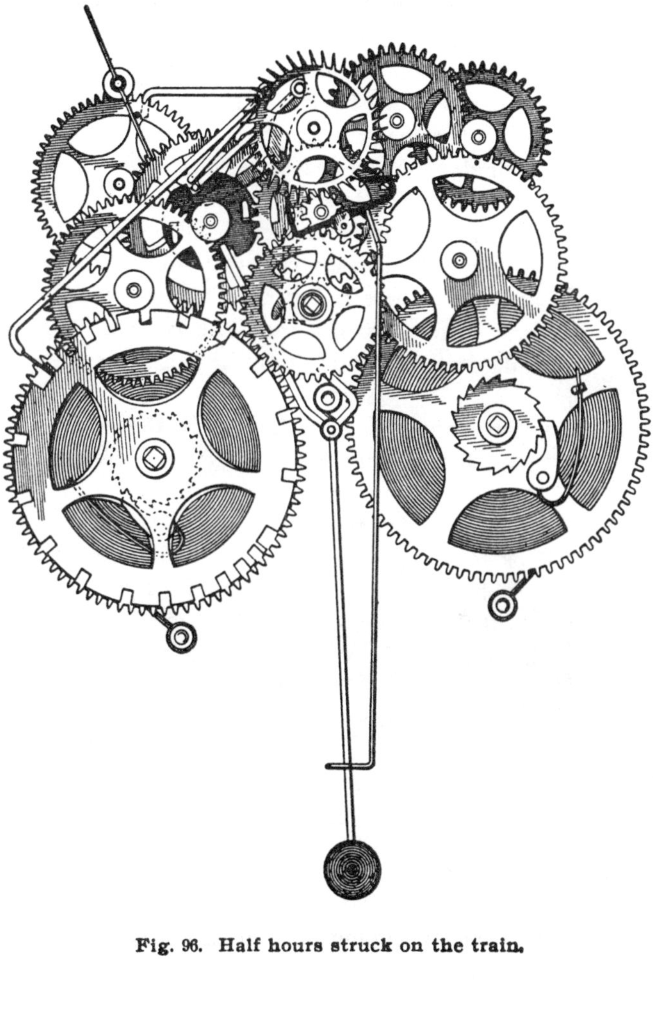

Fig. 96. Half hours struck on the train.

the locking cam. The cams on the center arbor are stamped out of brass sheet. while those of Fig. 92 were of wire.

Turning to the enlarged view in Fig. 97 and comparing it with Fig. 93, we find further differences. The levers K and J are here made of one piece of brass, while the others were separate and of wire. The lifting lever, H, is flattened at its outer end in Fig. 93, while in Fig. 97 it is bent at right angles and passed under the count hook, A. The hook, C, Fig. 97, is added to the arbor, B, as a safety device, in case the locking hook should fail to enter its slot in the cam, D. It is shown as having just stopped the warning pin in Fig. 96. There is but one hammer tail, G, and the hammer stop acts against the stud for the hammer spring, instead of against the bottom of the front plate, as in Fig. 92.

The first important difference here is in the position of the count hook, A. In Figs. 92 and 93 the hook must be exactly in the middle of the slot, or there will be trouble. In trains striking half hours from the train, we must never allow the hook to occupy the middle of the slot, or we will have more trouble than we ever dreamed of. In this instance the count hook must enter the slot close to (but not touching) the side of the slot when the clock stops striking; then when the half hour is struck the count wheel will move a little and the hook must drop back into the same slot without touching; this brings it close to the opposite side of the same slot and the next movement will land the hook safely on top of the wheel for the strokes of the hour. Fig. 96 shows its position after striking the half hour and ready to strike the hour of two. Fig. 97 shows it dropping back after striking two.

In setting up this train, see that the count hook, A, goes into the slot of the count wheel close to, but not touching, one side of the slot in the count wheel, and, after placing the intermediate, insert the locking cam, D, so that it engages the locking hook; then put in the warning wheel

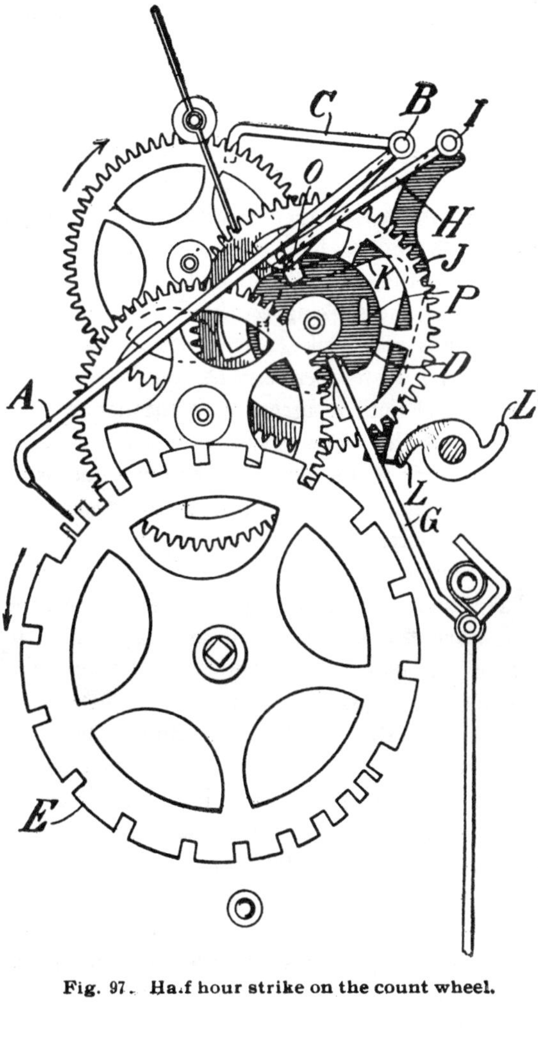

Fig. 97. Half hour strike on the count wheel.

with the warning pin, O, safely to the left of the hook C, Fig. 97, so that it cannot get past that hook after striking. Placing the wheel with its warning pin six or eight teeth to the left of the edge of the bottom plate is generally about right. The action of the levers, H, J, K, the hammer tail, G, and the cam, L, in striking the hours is the same as that already described in detail for Figs. 93 and 94, hence need not be repeated here. L^1 strikes the half hours by being enough shorter than L to raise the hooks for one revolution, but not quite so high as for the hours. The cams L, L^1 are friction tight on the center arbor and may be shifted on the arbor to register the striking on the 60th minute, if desired. When the hands and strike do not agree, turn the minute hand back and forward between IX and XII, thus striking the clock around until it agrees with the hands.

Sometimes, if the warning pin is not far enough away, an eight-day clock will strike all right for a number of days and then commence to gain or lose on the striking side. It either does not strike at some hours, or half hours, or it may strike sometimes both hour and half hour before stopping. Take the movement out of the case and put the hands on; then move the minute hand around slowly until the clock warns. Look carefully and be sure there is no danger of the clock striking when it warns. If this looks secure, then move the hand to the hour, making it strike; say it is going to strike 9 o'clock; when it has struck eight times, stop the train with your finger and let the wheels run very slow while striking the last one, and when the rod drops into the last notch stop the train again and hold it there.

For the striking part to be correct, the warning pin on the wheel wants to be about one-fourth of a revolution away from the rod when the clock has struck the last time, or as soon as this rod falls down far enough to catch the pin. The object of this is so there is no chance of the warning pin getting past the rod at the last stroke; this it is liable to do if the pin is too close to the rod when the rod drops. If you will examine the clock as above, not only when it strikes IX, but all the hours from I to XII, you will generally find the fault. Of course, if the pin is too close to the rod when the rod drops, you must lift the plates apart and change the wheel so that the warning pin and the rod will be as explained.

SHIP'S BELL STRIKING WORK.—Of all the count wheel striking work which comes to the watchmaker, the ship's bell is most apt to give him trouble. This generally arises from ignorance as to what the system of bells on shipboard consists of and how they should be struck. If he goes to some nautical friend, he hears of long and short "watches" or "full watches" and "dog watches." If he insists on details, he gets the information that a "watch" is not a horological mechanism, but a period of duty for a part of the crew. Then he is told of the "morning watch," "first dog watch," "afternoon watch," "second dog watch," "off watch," "on watch," etc. Now the ship's bell clock does not agree with these "watches" and was never intended to do so. As a matter of fact, it is simply a clock striking half hours from one to eight and then repeating through the twenty-four hours.

The striking is peculiarly timed and is an imitation of the method in which the hours are struck on the bell of the ship. As this bell is also used for other purposes, such as tolling in fogs, fire alarms, church services, etc., it will readily be seen that a different method of striking for each purpose is desirable to avoid misunderstanding of signals.

The method of striking for time is to give the blows in couples, with a short interval between the strokes of the couples and three times that interval between the couples. Odd strokes are treated as a portion of the next couple and separated accordingly, thus:

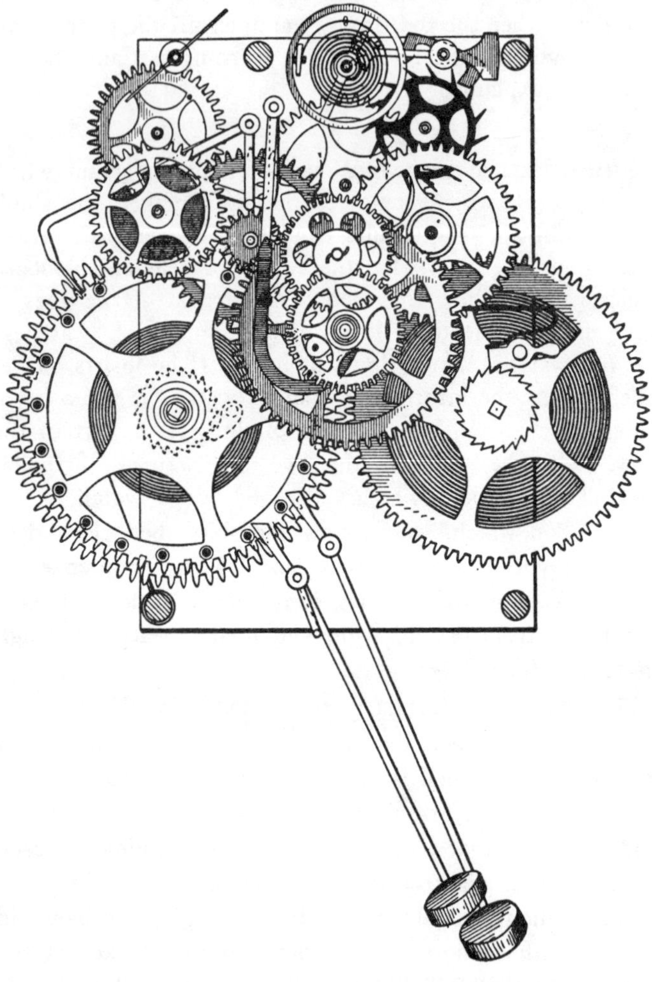

Fig. 98. Ships bell clock.

12:30 p. m.	One Bell,	O
1:00 p. m.	Two Bells,	O O
1:30 p. m.	Three Bells,	O O O
2:00 p. m.	Four Bells,	O O O O
2:30 p. m.	Five Bells,	O O O O O
3:00 p. m.	Six Bells,	O O O O O O
3:30 p. m.	Seven Bells,	O O O O O O O
4:00 p. m.	Eight Bells,	O O O O O O O O

After striking eight bells the clock repeats, although the ship's bell is generally struck in accordance with the two dog watches (which are of two hours' duration each) before commencing the evening watch (8 to 12 p. m.). It will thus be seen that the clock should strike eight at 12 m., 4 p. m., 8 p. m., 12 p. m., 4 a. m., and 8 a. m.

In order to strike the blows in pairs two hammers are necessary, see Fig. 98; these hammers are placed close together, but not in the same plane. The pin wheel has twenty

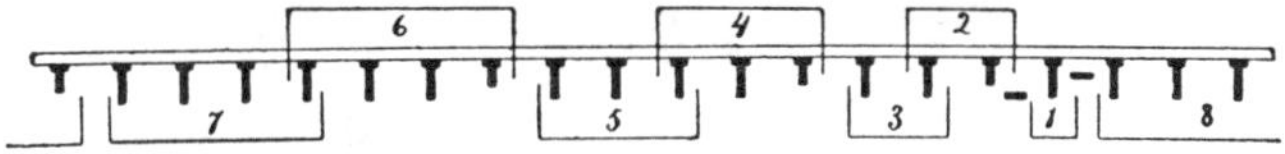

Fig. 100. The pins on the count wheel of the ships bell clock.

pins, see Figs. 98, 99, 100; some of these pins are shorter than the others, so that they do not operate one of the hammer tails. These are shown graphically in Fig. 100; where the two oblong marks at figure 1 represent the tops of the hammer tails shown in Fig. 99. It will be seen by studying Fig. 100 that with the wheel moving from left to right, the inside hammer tail will be operated for one blow, while the

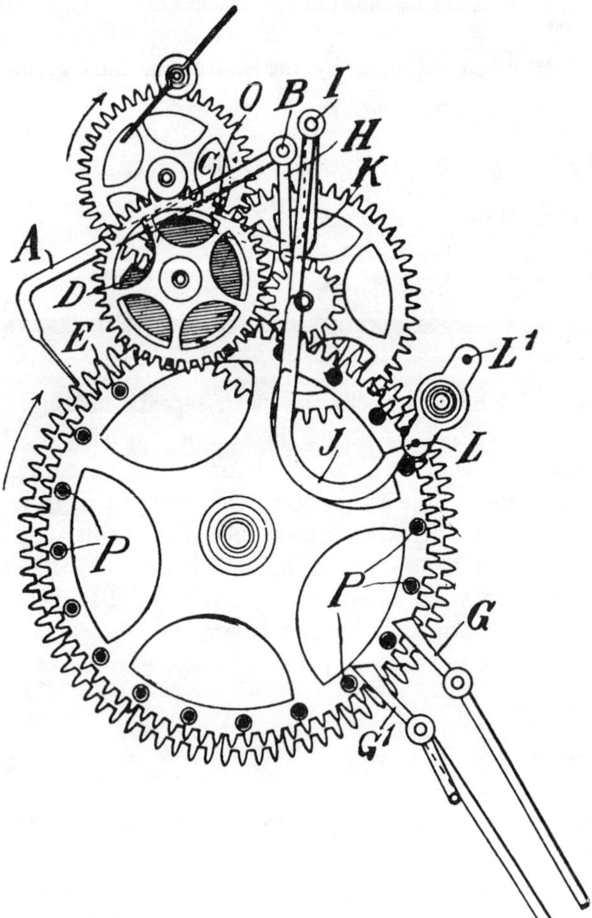

Fig. 99. Enlarged view of striking work, ships bell clock.

outer hammer tail will not be operated at all, thus giving but one blow, or "bell." At the next movement of the pin wheel, the outside hammer will be operated by the long pin and the inside hammer by the short pin, thus giving one blow of each hammer, or "two bells."

We now have these hammer tails advanced along the wheel so that the outside one is opposite the figure 3 in the drawing, while the other is opposite the figure 2, with one pin between them. The next movement of the pin wheel advances them so that the outside hammer will pass the next short pin and consequently that hammer will miss one blow and the pair will therefore strike three—one by the outside hammer and two by the inside. It thus goes on until the cycle is completed, eight blows being struck with the last four pins. The striking in pairs is effected by having the two hammer tails close together, so that the pins will operate both hammer tails quickly and there will then be an interval of time while the wheel brings forward the next pins. This is so spaced that the interval between pairs is three times that between the blows of a pair and the hammer tails should not be bent out of this position, or if found so they should immediately be restored to it. Tolling the bells, instead of striking them properly, is very bad form at sea and generally leads to punishment if persisted in, so that the jeweler will readily perceive that his marine customers are very particular on this point, and he should go any length to obtain the proper intervals in striking.

The pin wheel moves forward one pin for each couple of blows or parts of a couple, the odd blows being secured by the failure of the blow when the hammer tail passes the short pin. Thus it moves as far for one bell as for two bells; as far for three bells as for four, etc. The result is that the count wheel has no odd numbers on it, but instead two 2's, two 4's, two 6's and two 8's; the first two are counted on the count wheel, but only one is struck on the pin wheel, owing to the short pin; this is repeated at three, five and seven, when four, six and eight are counted on the wheel, but the last blow fails of delivery, owing to the short pin in the pin wheel at these positions.

The center arbor carries two pins, L and L^1, to unlock the train through the lever J, as it is really a half-hour-striking clock. The count hook, A; locking hook, C; count wheel, E; pins, P, and other parts have similar letters for similar parts as in the preceding figures and need not be further explained, as the mechanism is otherwise similar to the Seth Thomas movement shown in Fig. 92.

CHAPTER XVII.

CLEANING AND REPAIRING CUCKOO CLOCKS.

The cuckoos are in a class by themselves for several reasons, all of which have to do with their construction and should therefore be understood by the watchmaker. They are bought as timepieces by but two classes of people: those who were used to them in their former homes in Europe and buy them for sentimental reasons; and those who admire fine wood carvings as works of art and desire to possess a finely carved cuckoo clock for the reasons which govern in the purchase of paintings and statuary, bronzes, and other art objects. For this reason cuckoos have never been a success when attempts have been made to cheapen their production by the use of imitations of wood carving in composition or metal. The use of cuckoos in plain cases, with springs instead of weights, has also been attempted with the idea of thereby securing an inclosed movement, as in ordinary clocks; but while it offers advantages in cleanliness and protection of the movement, such clocks have never become popular, as they have lost their character as works of art by being enclosed in plain cases, or have become rather erratic in rate by the substitution of springs for weights.

The use of exposed weights and pendulum necessitates openings in the bottom of the case through which the dust enters freely and this makes necessary unusual side shake, end shake and freedom of depthing of the wheels and pinions and also the use of lantern pinions and an amount of driving weight in excess of that necessary for protected movements, as there must be enough weight to pull the cuckoo movement through obstructions which would stop the ordinary movement.

Repairers therefore should not attempt to close worn holes as snugly as in the ordinary movements, as when this is done the clock generally stops about three weeks after it has left the shop and a "comeback" is the result. Lightening the driving weights will have the same result, as the movement must have sufficient power to pull it through when dirty. As the plates and wheels are generally of cast metal, cutting of pivots from running dry is frequent in old clocks, and where it is necessary to close the holes care must be taken not to overdo it.

Another point where repairers fail is in not polishing the pivots. Many watchmakers seem to think that any kind of a pivot will do for a clock, although they take great care of them in their watchwork. Rough and dry pivots will cut the holes in a clock plate deep enough to wedge the pivots in the holes like a stuck reamer and stop a clock just after it has been repaired, when if they had been properly polished the job would not have come back.

The high prices of wood carving in America and the necessity for its genuineness, as explained above, has resulted in making it necessary to spend as little as possible for the movements; hence we ordinarily find a total lack of finish on the movements, and this, with the great freedom everywhere evident in its construction and the apparent excess of angular motion of the levers, combine to give it an appearance of roughness which surprises those who see them but rarely.

It has been frequently suggested by watchmakers that if the cases only were imported and the movements were made by the American factories better results should be obtained, in appearance at least. They forget that the bellows, pipes and birds, with their wires, are parts of the movements and the cost of having these portions made in this country is prohibitive, so that the whole movement is imported.

Arrangements are now being made by at least one firm to have the frames and wheels made of sheet metal by automatic machinery, instead of being cast and finished in the usual way, and when this is done the appearance of the movements will be greatly improved, so that American watchmakers will regard them with a more kindly eye. So far as is known to the writer all cuckoo movements are imported, although one firm is doing a large and constantly growing trade in such clocks with cases made in America.

There are a number of importing firms who sell to jobbers, large retailers and clock companies only, and as the large American clock manufacturers all list and carry cuckoos the clocks find their way to the consumer through many and devious channels. Probably more are sold in other ways than through the retailers for the reason that the average retailer does not understand the cuckoos and is reluctant to stock them, thereby deliberately avoiding a large amount of business from which he might make a handsome profit.

Under the general term Cuckoos are listed several kinds of movements, all having bellows, pipes and moving figures, such as the cuckoo, cuckoo and quail, trumpeter, etc., with or without the regular hammers and gongs of the ordinary movements.

Figs. 101 and 102 show front and back views of a time train in the center with quail strike train on the left and cuckoo strike train at the right. The positions of arbors, levers, depthings of trains, etc., are exact, but the movement plates have been left off for greater clearness, so that the arbors appear to be without support. The positions of the pillars are shown by the shaded circles above and below the trains in Fig. 101. The parts have the same letters in both Figs. 101 and 102, although as the movement is turned around to show the rear in 102, the quail train appears on the right side.

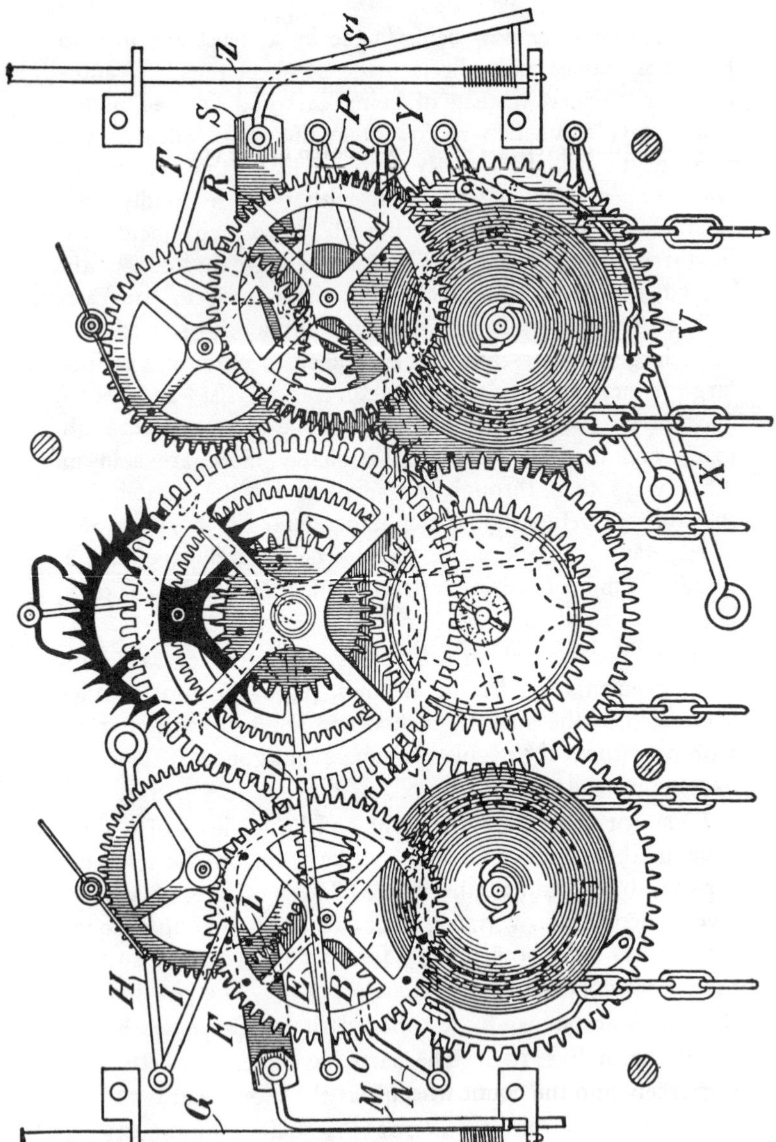

Fig. 101. Front View of Quail and Cuckoo Strike Movement.

NAMES OF PARTS.

A—Quail count wheel.
B—Quail striking cam.
C—Minute wheel.
D—Quail lifting lever.
E—Quail count hook.
F—Quail locking arm.
G—Quail bird stick; also called bird holder.
H—Quail bellows arm.
I—Quail bellows lifting lever.
J—Quail gong hammer.
K—Quail warning lever.
L—Quail lifting pin.
M—Quail bird stick lever.
N—Quail hammer lever.
O—Quail Lifting pin wheel.
P—Cuckoo lifting lever.
Q—Cuckoo warning lever.
R—Cuckoo lifting pin.
S—Cuckoo locking arm.
T—Cuckoo count hook.
U—Cuckoo striking cam.
V—Cuckoo lifting pin wheel.
W—Cuckoo count wheel.
X—Cuckoo bellows lifting lever.
Y—Cuckoo hammer.
Z—Cuckoo bird stick; also called bird holder.
S^1—Cuckoo bird stick lever.

In examining a movement the student discovers a peculiarity of cuckoo frames, which is that the pivot holes for several of the arbors of the striking levers have slots filed into them, reaching to the edges of the frames and narrower than the full diameter of the pivot holes. This is because such arbors have levers riveted into them which must function in front, between and at the rear of the plates and in setting up the movement the slots are necessary to allow the end levers to pass through the holes. Such arbors as have slots on the front plates are inserted and placed in their proper positions before setting the train wheels with which they function. The others are first inserted in the back plate and turned to position while putting on that plate.

Both quail and cuckoo trains are set up very simply and surely by observing the following points: In the quail train, when the quail bellows lever, H, is just released from a pin in the pin wheel, O, the locking lever, F, must just fall into the slot of the locking cam, B; the warning pin should then be near the fly pinion and the count hook, K, drop freely into the count wheel, A.

On the cuckoo side we find two levers, X; the upper one of these operates the low note of the cuckoo call and the lower one the high note. When this upper lever is released

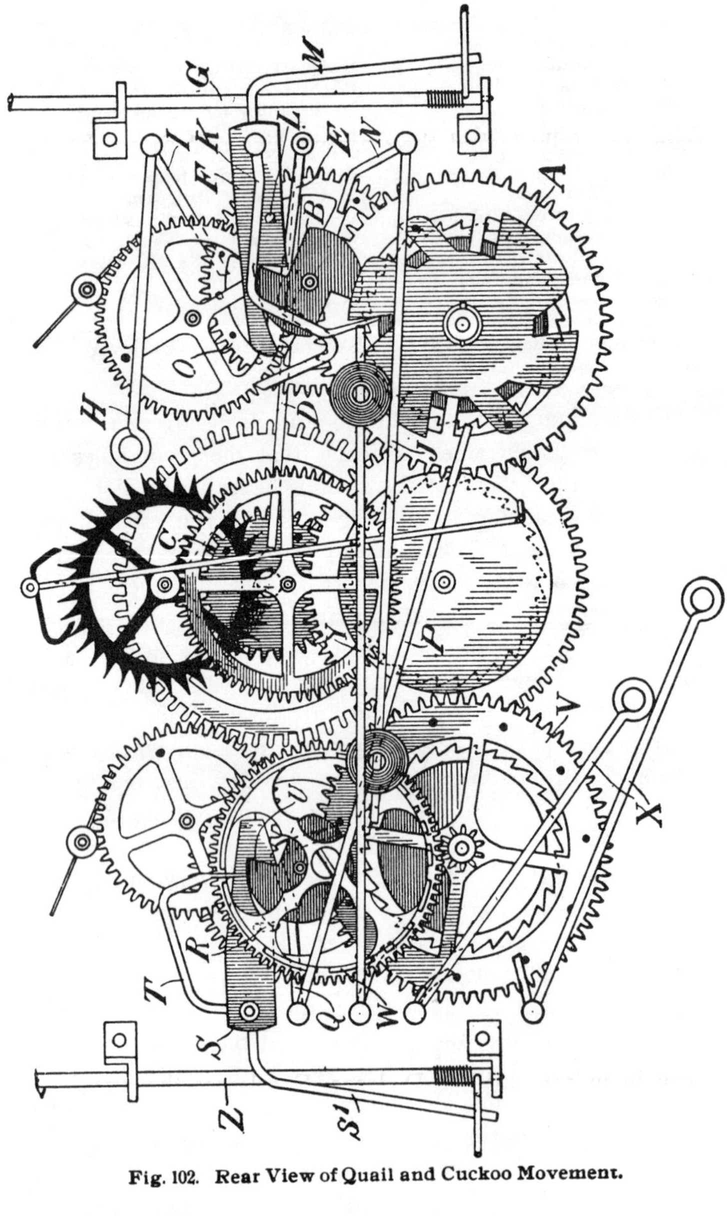

Fig. 102. Rear View of Quail and Cuckoo Movement.

from a pin in the pin wheel, the cuckoo locking lever, S. must drop into its locking cam, U, and the count hook, T, drop into its count wheel, while the warning pin must be near the fly pinion. After the run has stopped and the trains are fully locked the warning pins will be as shown in Fig. 102; but at the moment of locking they should be as described above.

The operation is as follows: Turning to Fig. 101, we find the minute wheel, C, has four pins projecting from its rear surface. This revolves once per hour and consequently the pins raise the lifting lever, D, every fifteen minutes. Here is a point that frequently is productive of trouble. The reader will readily see that if the hands of a cuckoo are turned backward the pins in the minute wheel will bend this wire, D, and derange the striking, as the warning lever is also attached to the same arbor. *Never push the hands backward* on a cuckoo clock; always push them forward. If the striking and hands do not register the same time, take off the weights of the striking trains; then push the hands forward until they register the hour which the trains struck last. As there is no power on the trains they will not be operated, the only action being the rising and falling of the lever, D, as the pins pass. When the hands point to the hour last struck by the trains, put on the striking weights again and push the hands *forward,* allowing time for each striking, until the clock has been set to the correct time.

Upon the lifting lever, D, being raised sufficiently the warning lever, E, on the same arbor is lifted into the path of the warning pin and at the same time unlocks the train by pressing against the lifting pin, L, in the locking lever, F. The locking lever, F, count hook, K, and the bird holder lever, M, are all on the same arbor and therefore work in unison. When D drops, E releases the warning pin and the train starts. The pin wheel has pins on both sides, the rear pins operate the gong hammer, N, J; the front pins operate the quail bellows, I, H. The rising and falling of the unlocking lever, F, operates the bird holder, G, through M and the wire in the bellows top tilts the tail of the bird and flutters the wings. When the fourth quarter has been struck, the pins shown in the quail count wheel, A, operate the hour lifting lever, P, and the action of that train becomes similar to that of the quarter train just described, with the difference that there are two bellows levers, X, for the high and low notes of the cuckoo, whereas there is but one for the quail.

There are several adjustments necessary to watch on these clocks. The wires to operate the bellows from the levers X and H may be so long that the bellows when stretched to its full capacity may not allow the tails of X and H to clear the pins of the pin wheels and thus stop the trains. The pins should clear safely with the bellows fully opened. The levers M and S′, which operate the bird holders, G and Z, may be turned in their arbors so as to be farther from or closer to the bird holder; this regulates the opening and closing of the doors and the appearance of the birds; if there is too much movement the birds may be sent so far out that they will not return, but will stay out and stop the trains. Moving S′ and M towards the bird holders, Z and G, will lessen the amount of this motion and the contrary movement will increase it.

Another important source of trouble—because generally unsuspected—is the fly. The fly on a cuckoo train must be tight; a loose fly will cause too rapid striking and allow the train to overrun, making wrong striking, or in a very bad case it will not stop until run down. When this happens turn your attention to the fly and make sure that it is tight before doing any bending of the levers, and also see to the position of the warning pin.

Sometimes the front of the case (which is also the dial) wili warp and cause pressure on the ends of the lever arbors and thus interfere with their proper working. Be sure that the arbors are free at both ends.

When replacing worn pins in the striking trains, care should be taken to get them the right length, as on account of the large amount of end shake in these movements they may slip past the levers without operation, if too short, or foul the other parts of the train if too long. For the same reasons bending the levers should only be done after exhausting the other sources of error and then be undertaken very slowly and cautiously.

The notes of a cuckoo are A and F, just below middle C; these should be sounded clearly and with considerable volume. If they are short and husky in tone it may be due to holes in the bellows, too short stroke of bellows, removal of the bellows weights, E, Fig. 103, dirt in the orifices of the pipes, or cracks in the pipes. Holes in the bellows, if small and not in the folds of the kid, may be mended by being glued up with paper or kid, or a piece of court plaster which is thin enough to not interfere with the operation of the bellows. If much worn a new bellows should be substituted. Cracks in the pipes may be mended with paper.

The orifice of the pipe, if dirty, may be cleaned with a piece of mainspring filed very thin and smooth and carefully inserted, as any widening or roughening of this slit will interfere with the tone. Sometimes a clock comes in which has been spoiled in this regard, then it becomes necessary to remove the outer portion or lip, A, Fig. 103, of the slot (which is glued in position) and make a new inner lip, B, or file the old one smooth again. The proper shape is shown in B, Fig. 103, while C and D show improper shapes which interfere with the tone.

Much time and money has been spent in trying to avoid the inherent defects of this portion of the clock; sometimes the lips will swell or warp and close the orifice; sometimes they will shrink and make it too wide; in either case a loss of purity of tone is the result. Brass tubes, if thin enough

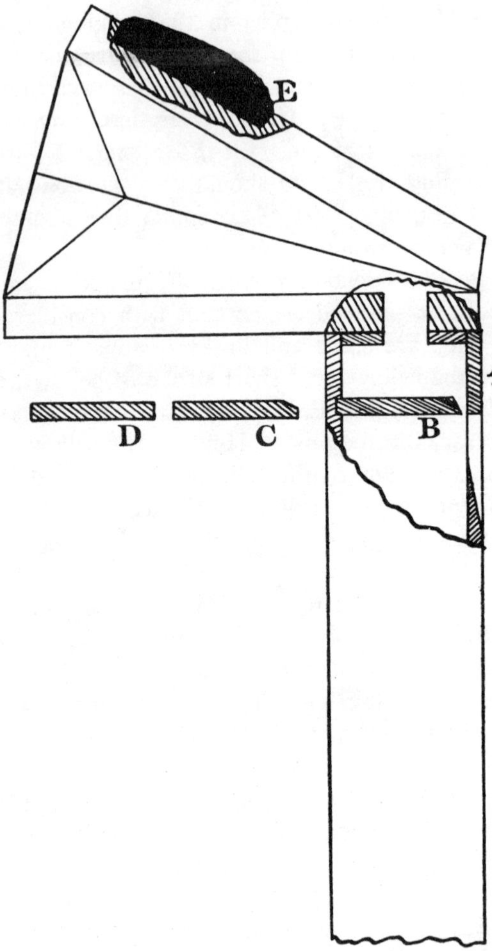

Fig. 103. Cuckoo bellows and pipe. A, outer lip; B, inner lip; C, D. incorrect forms of lip.

to be cheap, give a brassy tone to the notes; compositions of lead. tin and antimony (organ pipe metal) are readily cast, but give a softer, duller tone of less volume than the wood. Celluloid lips to a wooden tube were at first thought to be a great success, but were found to warp as they got older. Bone lips are costly; so there is nothing at present that seems likely to displace well seasoned wood, where discriminating lovers of music and art demand purity and correctness of tone, reasonably accurate time, artistic sculptural effects and durability, all in one article—a high class cuckoo clock.

When sending a clock home after repairing, each of the chains should be tied together with strings just outside the bottom of the case so that they will not slip off the sprockets and the customer should be instructed to hang the clock in its accustomed position before cutting the strings and attaching the weights.

CHAPTER XVIII.

SNAIL STRIKING WORK, ENGLISH, FRENCH AND AMERICAN.

While the majority of snail striking movements made in America are on the French system, because they are cheaper when made in that way, still this system is so condensed and so difficult to illustrate, with all its mechanism packed in a small space between the plates, that the student will gain a much better idea of the rack and snail and its principles by first making a study of an English snail striking clock, which has the whole of the counting and releasing levers placed outside the front plate, where they can occupy all the room that may be necessary. The calculation and planting of the striking train do not differ from those using the count wheel, up to and including the single toothed pinion or gathering pallet. The stopping of the train after striking is different and the counting is divided, being dependent upon four pieces acting in conjunction in an hour strike of the simplest order, which number may run to a dozen in a repeating clock.

As the count wheel system had the defect of getting out of harmony with the hands when the latter are turned backward, so the snail system has its defects, which are the displacement of the rack and failure to stop the striking in some clocks if the striking train runs down before the time side and is then rewound, and a most puzzling inaccuracy of counting, resulting from slight wear and inaccuracy of adjustment. We mention these things here because they have an influence on the construction of the clock and an advance knowledge of them will serve to make clearer some of the statements which follow.

HOUR AND HALF-HOUR SNAIL STRIKING WORK.—Fig. 104 is a view of the front plate of an English fusee striking clock, on the rack principle. The going train occupies the right and center and the striking train the left hand. The position of the trains is indicated in dotted lines, the trains having barrels and fusees as shown by the squared arbors, all the dotted work being between the clock plates, and that in full lines being placed on the outside of the front plate, under the dial. The connection between the going train and the striking work is by means of the motion wheel on the center arbor, and connection is made between the striking train and the counting work by the gathering pallet, F, which is fixed to the arbor of the last wheel but one of the striking train, and also by the warning piece, which is shown in black on the boss of the lifting piece, A. This warning piece goes through a slotted hole in the plate, and during the interval between warning and striking stands in the path of a warning pin in the last wheel of the striking train. The motion wheel on the center arbor, turning once in an hour, gears with the minute wheel, E, which has an equal number of teeth. There are two pins opposite each other and equidistant from the center of the minute wheel, which in passing raise the lifting piece, A, every half hour. Except for a few minutes before the clock strikes, the striking train is kept from running by the tail of the gathering pallet, F, resting on a pin in the rack, C. Just before the hour, as the boss of the lifting piece, A, lifts the rack hook B, the rack C, impelled by a spring in its tail, falls back until the pin in the lower arm of the rack is stopped by the snail, D. This occurs before the lifting piece, A, is released by the pin in the minute wheel, E, and in this position the warning piece stops the train. Exactly at the hour the pin in the minute wheel, E, gets past the lifting piece, A, which then falls, and the train is free. For every blow struck by the hammer the gathering pallet, F, which is really a one-toothed pinion, gathers up one tooth of the rack, C, which is then held, tooth by tooth, by the point of the hook, B. After the pinion, F, has gathered up the last tooth, its tail is caught by the pin in the rack, which stops and locks the train, and the striking ceases.

The snail, O, is mounted on a twelve-toothed star wheel, placed on a stud in the plate, so that a pin in the motion wheel on the center arbor moves it one tooth for each revolution of the motion wheel, and it is then held in position by the click and spring as shown. The pin, in moving the star wheel, presses back the click, which not only keeps the star wheel steady, but also completes its forward motion after the pin has pushed the tooth past the projecting center of the click. The steps of the snail are arranged so that at one o'clock it permits only sufficient fall of the rack for one tooth to be gathered up, and at every succeeding hour gives the rack an additional motion equal to one extra tooth. It will be seen that where a star wheel is used a cord or wire attached to A and run outside the case, so that A may be lifted, will cause the clock to repeat the hour whenever desired.

The lower arm of the rack, C, and the lower arm of the lifting piece, A, are made of brass, and thin, so as to yield when the hands of the clock are turned back; the lower extremity of the lifting piece, A, is a little wider, and bent to a slight angle with the plane of the arm, so as not to butt as it comes into contact with the pin when this is being done. If the clock is not required to repeat, the snail may be placed upon the center arbor, instead of on a stud with a star wheel as shown, and this is generally done with the cheaper class of hour striking clocks; but the position of the snail is not then so definite, owing to the backlash of the motion wheels, so that it will not repeat correctly, as the pin of the rack may fall on a slope of the snail and, besides, a smaller snail must be used, unless it is brought out to clear the nose of the minute wheel cock, or bridge if one be used.

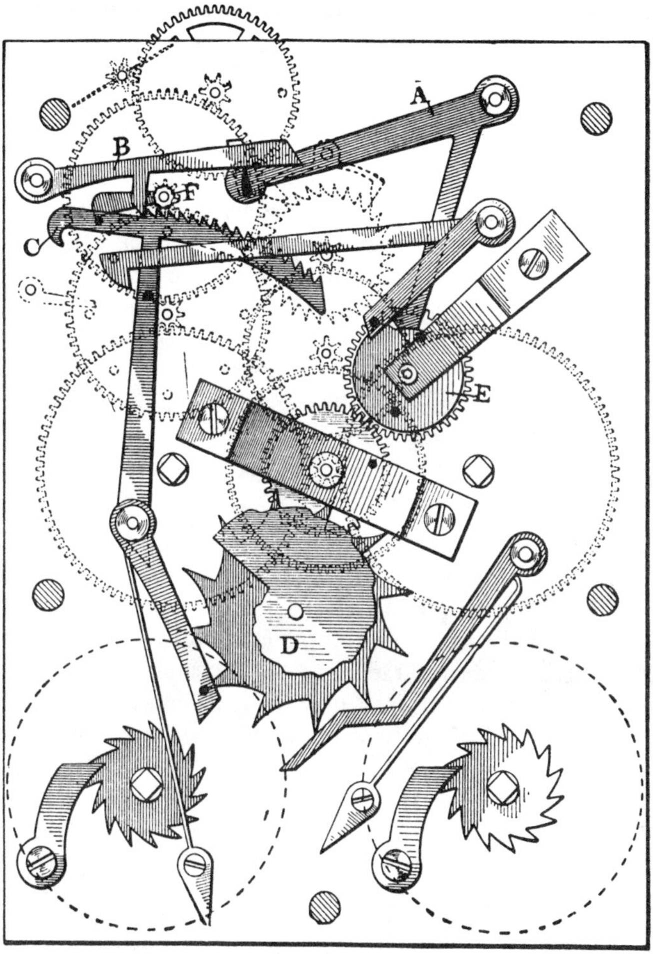

Fig. 104. Hour and half hour snail striking work with fusee train.

Half-Hour Striking.—The usual way of getting the clock to strike one at the half-hour, is by making the first tooth of the rack, C, lower than the rest, and placing the second pin in the minute wheel, E, a little nearer the center than the hour pin, so that the rack hook, B, is lifted free of the first tooth only at the half hour. But this adjustment is too delicate after some wear has occurred and the action is then liable to fail altogether or to strike the full hour, from the pin getting bent or from uneven wear of the parts. The arrangement shown in Fig. 104 is generally used in English work, as it is much safer. One arm of a bell crank lever rests on a cam fixed to the minute wheel, E. This arm is shaped so that just before the half-hour the other extremity of the bell crank lever catches a pin placed in the rack, C, and permits it to release the train and fall the distance of but one tooth. This is the position shown in Fig. 104. After the half-hour has struck, the cam carries the hook free from the pin in C.

Division of the Hour Snail.—The length of the rack tail, from the center of the stud hole in the rack to the center of the pin, should be equal to the distance between the center of the stud hole and the center of the snail. The difference between the radius of the top and the radius of the bottom step of the snail may be obtained by getting the angular distance of twelve teeth of the rack from center to pin. See A B, C D, E F, Fig. 105, which show the total distances for twelve steps of the snail for rack tails of different lengths. Divide the circumference of a piece of brass into twelve parts and draw radial lines as shown in Fig. 106. Each of these spaces is devoted to a step of the snail. Draw circles representing the top and bottom step. Divide the distance, A B or E F, Fig. 105, between these two circles, into eleven equal parts, and at each division draw a circle which will represent a step of the snail. The rise from one step to another should be sloped as shown, so as to raise the pin in the rack arm if the striking train has been allowed to run down, and it should be resting on the snail when it is desired to turn the hands back. The rise from the bottom to the top step is bevelled off, so as to push the pin in the rack arm on one side, by springing the thin brass of the arm and allow it to ride over the snail if it is in the way when the clock is going. It should also be curved to avoid interference with the pin. Clockmakers making new snails when repairing generally mark off the

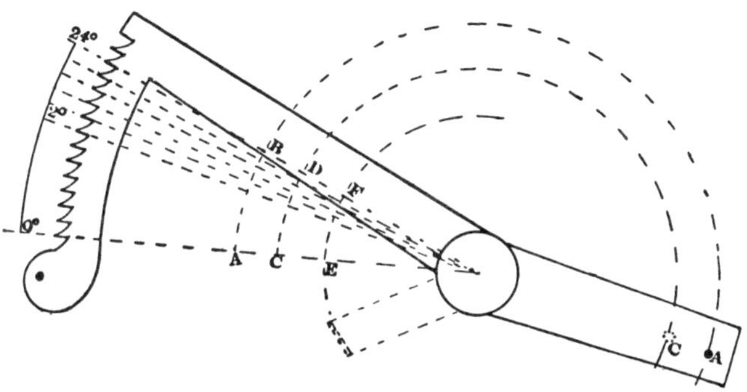

Fig. 105. Rack, showing method of getting sizes of snail steps according to distance from the rack center to the pin in the rack tail.

snail on the clock itself after the rest of the striking work is in position. A steel pointer is fixed in the hole of the lower rack arm, and the star wheel jumped forward twelve teeth (one at a time) by means of the pin in the motion wheel. After each jump a line is marked on the blank snail with the pointer in the rack arm by moving the rack arm. These twelve lines correspond to the twelve radial lines in Fig. 106. The motion wheel is then turned sufficiently to carry the pin in it free of the star wheel and leave the star wheel and blank snail quite free on their stud. The rack hook is placed in the first tooth of the rack, and while the pointer in the rack arm is pressed on the blank snail, the latter is rotated a little, so that a curve is traced on it. The rack hook is then placed in the second, and afterwards in the succeeding teeth consecutively, and the operation repeated till the twelve curves are marked. There is one advantage in marking off the snail in this way. Should there be any inaccuracy in the division of the teeth of the rack, the steps of the snail are thus varied to suit it. This frequently occurs in old clocks which have had new racks filed up by hand by some watchmaker.

Reference to the drawing, Fig. 105, will show that the rack is laid out as a segment of a wheel with teeth occupying two degrees each, with a few teeth added for safety. Fourteen to sixteen teeth are generally provided, for the following reasons: If the first tooth is used to strike the half hours, it may in time become worn so that it can no longer be stretched to its proper length. In such cases moving the pin two degrees nearer the rack teeth will allow us to use the teeth from the second to the thirteenth in striking twelve, which makes a cheap and easy repair, as compared to inserting a new tooth or making a new rack.

Weight driven snail clocks should have the weight cords of the striking side long enough so that the striking train will not run down before the time train, as in such a case the rack tail is pushed to one side by the progress of the snail (which is carried on the time train and is still running); then the rack will drop clear out of reach of the gathering pallet and when the striking train is wound that train will continue striking until it runs down, or the dial is removed and the rack replaced in mesh with the gathering pallet. This happens with short racks and with large, old-fashioned snails. By leaving a few more teeth in the rack the rack tail will strike the stud, or hour wheel sleeve, before the rack teeth get out of reach of the gathering pallet.

Many watchmakers put a stud or pin in the plate to stop the rack from falling beyond the twelfth step, to prevent troubles of this kind.

The rack tail is friction-tight on its arbor and should be adjusted so that the proper tooth shall come in mesh with the gathering pallet for each step of the snail, or irregular striking will result. Such a clock may strike one, two, three and four correctly and then strike six for five, or seven or nine for eight, or thirteen for twelve, or it may strike one or two hours wrong and the rest correctly. This is because the gathering pallet, F, Fig. 104, does not carry the

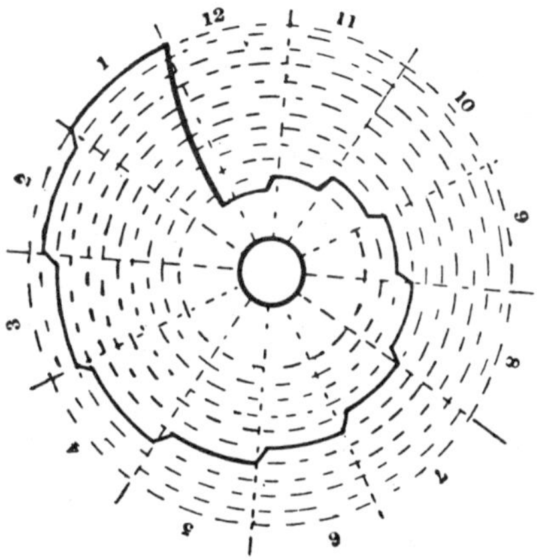

Fig. 106 Laying out steps of snail.

rack teeth safely past the edge of the rack hook, B, owing to the tail of the rack not being properly adjusted. The teeth should all be carried safely past the edge of the hook and then be dropped back a little as the hook engages; this is the more necessary to watch with hand-made racks and snails, or after putting in a new, and therefore larger, pin in the rack tail to replace one which is badly worn.

The snail should be put on so that the pin in the rack tail will strike the center of each step, or there is danger of irregular striking, or of failure to strike twelve, owing to the pin striking the surface of the cam midway between one and twelve and thus preventing the rack from falling the requisite number of teeth. When this occurs the clock will jam and stop.

The rack hook, B, Fig. 104, should be lifted far enough so that the rack will fall clear of the hook without the teeth catching and making a rattling noise as they pass the hook. In many old hour strikes the first tooth of the rack is left longer than the rest to ensure this freedom of passage when the rack is released.

The gathering pallet, F, is the weakest member of the system and will be very likely to be split or worn out in clocks brought in for repair. It should be squared on its arbor, or pinned, but many are not. If split, and the arbor is round, where the pallet is put on, it may cause irregular striking by opening on the arbor and permitting the train to run when the tail strikes the pin in the rack. A new one should be made so as to lift one tooth and a very little of the next one at each revolution. It is necessary to cause the gathering pallet to lift a little more than one tooth of the rack, and let it fall back again, to insure that one will always be lifted; because if such was not the case the clock would strike irregularly, and would also be liable sometimes to strike on continually till it ran down. If the striking part is locked by the tail of the gathering pallet catching on a pin in the rack, the tail should be of a shape that will best prevent the rack from falling back when the clock warns for striking the next hour; and of course the acting faces of the pallet must be perfectly smooth and polished.

The teeth of the rack may require dressing up in some cases and to allow this to be done the rack may be stretched a little at the stem, with a smooth-faced hammer, on a smooth anvil; or, if it wants much stretching, take the pene of the hammer and strike on the back, with the front lying on the smooth anvil. The point of the rack hook, B, will probably be much worn, and when dressing it up it will be safe to keep to the original shape or angle. The point of the rack hook is always broader than the rack, and the mark worn in it will be about the middle of the thickness; so enough will be left to show what the original shape or angle was.

After cleaning, particularly if it be French, look for dots on the rims of the wheels, and for pinions with one end of one leaf filed off slantingly. When putting it together, place the pin wheel (that is the one with the pins) and the pinion it engages with so that the leaf of the pinion (which you will find filed slanting at one extremity) enters between the two teeth of the wheel, opposite which you will find a countersunk mark, on the side of the wheel. See also that the gathering pallet, F, which lifts the rack, does so at the same time that the gong hammer falls. Then place the hour and minute wheels and cannon pinion so that the countersunk marks on each line with each other. Neglect of the marks on a marked train generally means that you will have to take the clock down again and set it up properly before it will run; therefore pay attention to these marks the first time.

Quarter Chiming Snail Strikes.—Fig. 107 shows the counting mechanism and trains of an English, fusee, quarter-strike work. The time train occupies the center, the hour striking train the left and the chiming train the right. All the train wheels are between the plates and are dotted in as in Fig. 104, while the counting mechanism is on the front plate, behind the dial and is drawn in full lines, to show that it is outside.

GOING TRAIN.

Fusee Wheel	96
Pinion	8
Center Wheel	84
Pinion	7
Third Wheel	78
Pinion	7

STRIKING TRAIN.

Fusee Wheel	84
Pinion	8
Pin Wheel, 8 pins in Pin Wheel	64
Pinion	8
Pallet Wheel	70
Pinion	7
Warning Wheel	60
Fly Pinion	7

CHIMING TRAIN.

Fusee Wheel	100
Pinion	8
Second Wheel	80
Pinion	8
Pallet Wheel	64
Pinion	8
Chiming Wheel	40
Warning Wheel	50
Fly Pinion	8

The reader will see a marked resemblance between the hour and time trains of Fig. 104 and the same trains of Fig. 107. The hour rack hook in 107, however, is hung from the center and the hour warning lever is raised by a spring instead of a lifting piece.

The minute wheel of Fig. 107 carries a snail of four steps, corresponding to the four teeth of the quarter rack, and the tail of the quarter rack is bent upwards towards the rack, to engage with the quarter snail. The quarter rack carries a pin which projects on both sides of the rack; one side of this pin stops the tail of the quarter gathering pallet and therefore locks the train as fully described in Fig. 104. The other side of the same pin acts on the tail of the hour warning lever, so that whenever the quarter rack falls the hour warning lever is released and its spring moves it into the path of the hour warning pin. This goes on whether the hour rack hook is released or not. Behind the quarter snail, there are four pins in the minute wheel; these pins

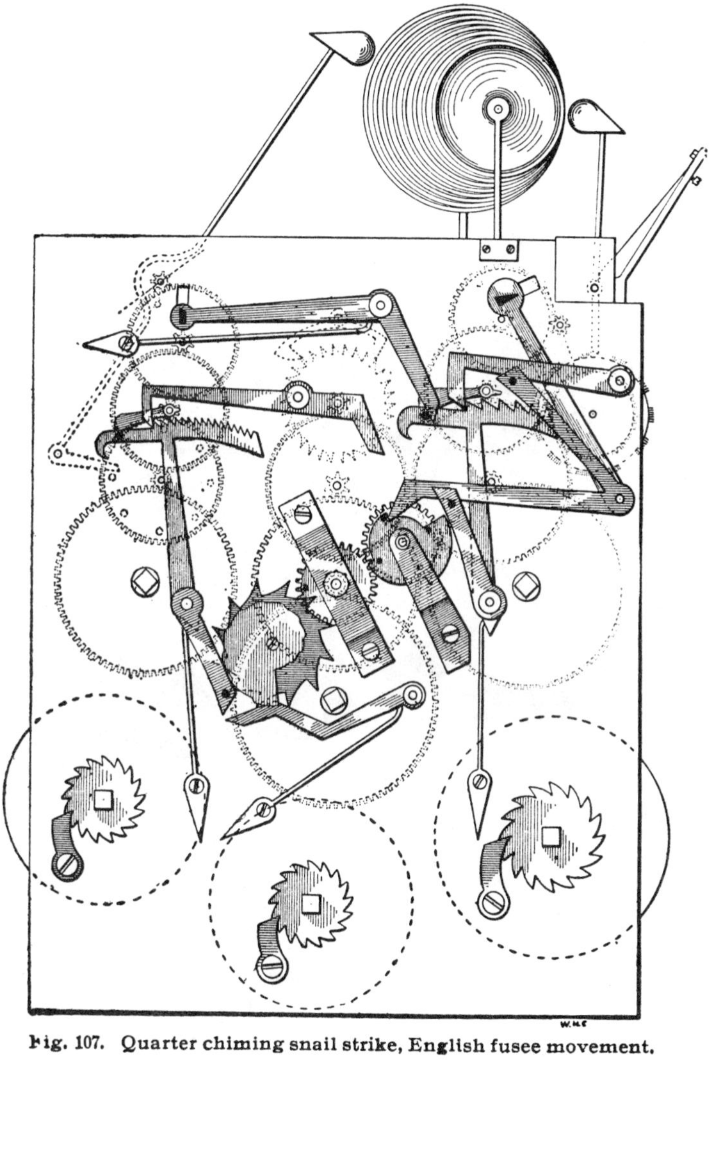

Fig. 107. Quarter chiming snail strike, English fusee movement.

raise the quarter lifting piece, which raises the quarter rack hook and the quarter warning lever at the same time, thus warning and dropping the quarter rack; as soon as the lifting piece drops, the warning lever and rack hook are released and the quarter train starts.

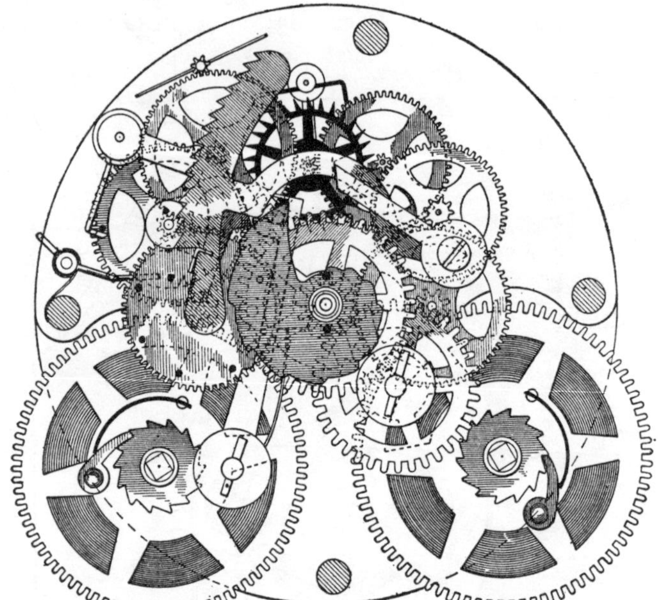

Fig. 108. Eight day snail half hour strike, French system, striking train locked.

One, two, three, or four quarters are chimed according to the position of the quarter snail, which turns with the minute wheel. At the time for striking the hour (when the quarter rack is allowed to fall its greatest distance), the pin in it falls against the bent arm of the hour rack hook, and releases the hour rack and hour warning lever. As the last tooth of the quarter rack is gathered up, the pin in the quarter rack pulls over the hour warning lever, and lets off the hour striking train. The position of the pieces in the drawing is as they would be directly after the hour was struck.

Figs. 108, 109 and 110 are three views of the New Haven eight-day snail strike, which is on the French system. As nearly all American strikes utilize this system and the work is between the plates, this may be considered a typical American snail strike.

As will be seen in Fig. 108, by the two pins at the center arbor, immediately behind the snail, this is a half-hour strike; and as the rack hook has for its lower step a little more than twice the depth of the other steps in the snail, it will readily be perceived that this rack hook may be pushed almost out and thus release the train without dropping the rack. This is the method pursued in striking half hours.

Figs. 109 and 110 show the parts more clearly than in 108. They are drawn a little larger than actual size and we will discover that the rack is the only portion of this system that works by gravity, all the others being spring operated. We see here the pins J K, which are used to push out the lever M sufficiently far so that the upper portion, which is bent at right angles to form a stop, will free the warning pin O and allow the train to run. The rack hook and the locking lever L are mounted on the same arbor and are kept in position by a coiled spring on the arbor until they are pushed out by the lower projection at the upper end of M for either the half-hour or hour strike.

As shown in Fig. 109, the lever M and the rack hook are pushed out by J far enough to pass the warning pin O and to unlock the train, which is normally locked by the pin N and the lever L. G is the gathering pallet, which is a long pin in a lantern pinion as in the ordinary count wheel strike. H is the hammer tail and P the pin wheel; R is the rack and T the rack tail. The rack arm is curved to pass the center arbor when dropping for twelve and the rack tail is bent toward the teeth in order that it may admit of a longer rack in a small movement, thus permitting of a large snail and consequently less liability of disarrangement. The same necessity of the proper adjustment of the rack tail T with the snail exists as has already been spoken of in regard to the English form of the snail strike.

In Fig. 110 will be seen the rack dropped clear with the tail resting clear of the snail at one stroke from the snail. In other words, the train is now in position to give eleven more strokes, having struck the first stroke of twelve. By comparison with Fig. 109, it will be seen that the spring actuated arm M has been thrown forward so that its dog is resting on the center arbor, after having been released from the hour pin K. This holds M out of the way of the warning pin O and the rack hook and allows the parts to operate as fully described with the English rack.

The gathering pallet G must have as many teeth as there are teeth between the pins in the pin wheel P. The train is locked by L coming in contact with N, the locking pin on the wheel on the same arbor as the gathering pallet. In setting this train up, it should stop so that the warning pin O should be near the fly.

As all the parts are operated by springs on the arbor, as shown by the hammer spring H, it will be seen that this strike mechanism will work in any position, while that which is operated by gravity must be kept upright. A loose fly will cause the clock to strike too fast and may cause it to strike wrong. Careless adjustment of the rack tail T with the snail will also induce wrong counting, although this is somewhat easier to adjust than the English form of strike. The hook should safely clear the rack teeth just as the gathering pallet G lets go of a tooth. If attention is paid to this point in adjusting the rack tail there will generally be little trouble.

The cam bearing the pins J K on the center arbor may be shifted with a pair of pliers to secure accurate register of hands and strike, as is the case with most American strikes. In putting in the pin wheel it should be set so that the pins may have a little run before striking the hammer tail, as

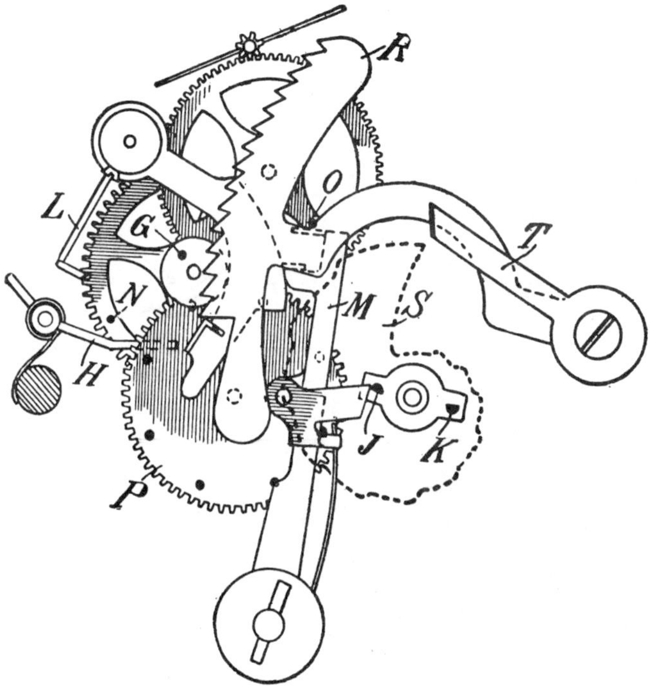

Fig. 109. Train about to strike the half hour; the hook L free of the train, which is held by the warning pin O; one stroke will be given when M drops.

this hammer tail is very short, and if the spring is strong the pins may not be able to lift the hammer tail without sufficient run to get the train thoroughly under motion. The half-hour strike should also be tested so that the pin J will release the warning pin O from the lever M without releasing the rack hook from the rack, as shown in Fig.

109. The parts of the train when at rest will be readily discerned in Fig. 108, where the hook L has locked the train by the pin N and the freedom between the pins and the hammer tail is about what it should be.

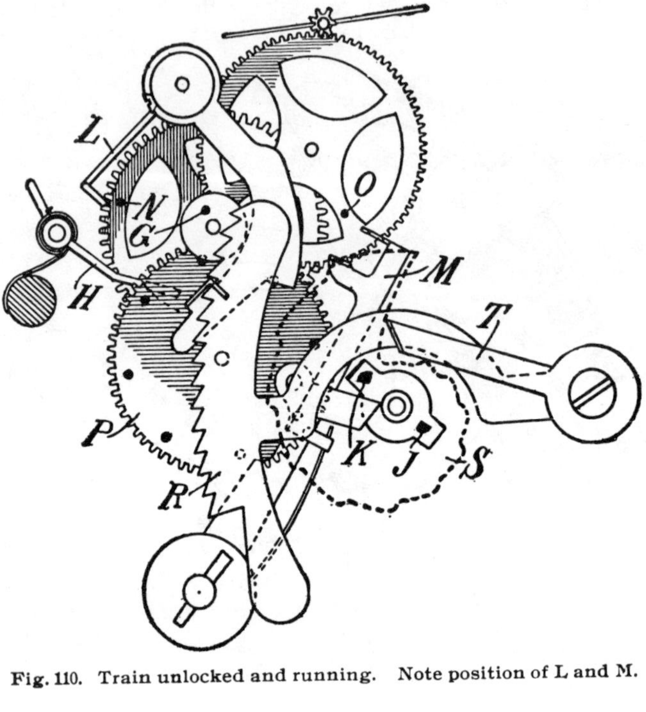

Fig. 110. Train unlocked and running. Note position of L and M.

The relative position of the locking lever L and the rack hook is also very clearly shown in Fig. 108; that is, when the rack hook is pressed clear home at the lower notch of the rack, the lever L should safely lock the train and the lever M be resting with its link against the center arbor.

CHAPTER XIX.

THE CONSTRUCTION OF SIMPLE AND PERPETUAL CALENDARS.

In taking up the study of calendar work the first thing that the student observes is the irregularity of motion of the various members. Every other portion of a clock has for its main object the attainment of the nicest regularity of motion, while the calendar must necessarily have irregular motion. The hand of the day of the month proceeds around its dial regularly from 1 to 28 and then jumps to 1 in February of some years, while it continues to 29 in others; sometimes it revolves regularly from 1 to 31 for several revolutions and then jumps from 30 to 1. What is the reason of this?

If the moon's phases are shown they do not agree with the changes of the month wheels, but keep gaining on them, while if an "equation of time" is shown, we have a hand that moves irregularly back and forth from the Figure XII at the center of its dial. What is the cause of this gaining and losing?

In order to understand this mechanism properly we shall have to first know what it is intended to show and this brings us to the study of the various kinds of calendar.

The earth revolves about its axis with a circular motion; it revolves about the sun with an elliptical motion. This means that the earth will move through a greater *angular* distance, measured from the sun's center, in a given time at some portions of its journey than it will do at others; at times the sun describes an arc of 57 minutes of the ecliptic; at other times an arc of 61 minutes in a day; hence the sun will be directly over a given meridian of the earth (noon), a little sooner at some periods than at others. Now the time at which the sun is directly over the given meridian is *apparent* noon, or solar noon. As before stated, this is irregular, while the motion of our clocks is regular, consequently the sun crosses the meridian a little before or a little after twelve by the clock each day, varying from 15 minutes before twelve to 15 minutes after twelve by the clock. The best we can do under these circumstances is to divide these differences of gaining or losing, take the average or *mean* of them and regulate the clock to keep *mean* time. Here then we have two times—the irregular *apparent* time and the regular *mean apparent* time. The amount to be added to or subtracted from the mean in order to get the solar or actual apparent time is called the *equation* of time and this is shown by the equation hand on an astronomical or perpetual calendar clock.

The moon revolves on its axis with a circular motion and it revolves about the earth with an elliptical motion, the earth being at one focus of the ellipse; as this course does not agree with that of the sun, but is shorter, it keeps gaining so that the lunar months do not agree with the solar.

Certain stars are so far away that they apparently have no motion of their own and are called *fixed;* hence in observing them the only motion we can discern is the circular motion of the earth. We can set our clocks by watching such stars and a complete revolution of the earth, measured by such a star, is called an *astronomical* or *siderial* day. This is the one used in computing all our time. It is shorter than the *mean solar* day by 3 minutes 56 seconds.

A *year* is defined as the period of one complete revolution of the earth about the sun, returning to the same starting point in the heavens. By taking different starting points we are led to different kinds of years. The point generally taken is the vernal equinoctial point, and when measured thus it is called the *tropical* year, which gives us the seasons. It is 20 minutes shorter than the siderial year.

A *siderial year* is the period of a complete revolution of the earth about the sun. This period is very approximately 365 days, 6 hours, 9 minutes, 9.5 seconds of mean time. Here we see an important difference between the siderial and the *civil* year of 365 days, and it is this difference, which must be accounted for somehow, that causes the irregularities in our calendar work.

For ordinary and business purposes the public demands that the year shall contain an exact number of days and that it should bear a simple relation to the recurrence of the seasons. For this reason the *civil* year has been introduced. The Roman emperor, Julius Caesar, ordered that three successive years should have 365 days each and the fourth year should have 366 days.

The fourth year, containing 366 days, is called a *leap* year, because it leaps over, or gains, the difference between the civil and siderial time of the preceding three years. For convenience the leap year was designated as any year whose number is exactly divisible by 4. This is called the Julian calendar.

But as a siderial year is 365 days, 6 hours, 9 minutes, 9.5 seconds of mean time, the addition of one day of twenty-four hours would not exactly balance the two calendars; therefore Pope Gregory XIII., in 1582, ordered that every year whose number is a multiple of 100 shall be a year of 365 days, unless the number of the year is divisible by 400, when it shall be a leap year of 366 days.

The calendar constructed in this way is called the Gregorian calendar, and is the one in common use. Its error is very small and will amount to only 1 day, 5 hours, 30 minutes in 4,000 years.

The revolution of the moon around the earth in relation to the *stars,* takes place in 27 days, 7 hours and 43 minutes, this is called a *siderial* month. But during this period the earth has advanced along the plane of its path about the sun and the moon must make up this distance in order to return to the same point in relation to the sun. This period is called a *synodic* month. Its average length is 29 days, 12 hours, 44 minutes, 2.9 seconds.

Having now understood these differences we shall be able to intelligently examine the various calendar mechanisms on the market and understand the reasons for their apparent departures from regular mechanical progression, as the equation of time gives us the difference between *real* and *mean* apparent, or solar time; we regulate our clocks by means of siderial time; the irregular procession of 30 and 31 days makes the civil calendar agree with the seasons, or the tropical year, and the remainder of the discrepancy between civil and siderial time is made up in February at the period when it is of the least consequence.

SIMPLE CALENDAR WORK.—Fig. 111 shows the American method of making a simple calendar, the example shown being drawn from a movement of the Waterbury Clock Company as a typical example. No attempt is made here to show the day of the week or the month. The days of the month are shown by a series of numbers from 1 to 31, arranged concentrically with the time dial and the current day is indicated by a hand of different color, carried on a pipe outside the pipe of the hour hand on the center arbor.

In order to accomplish this the motion work for the hands is mounted inside the frames, the hour pipe and center arbor being suitably lengthened. In the Figure A is the cannon pinion; B, the minute wheel; C, the minute pinion; D, the hour wheel at the rear end of the hour pipe; this pipe projects through the frame and forms a bearing in the frame for the center arbor. Friction-tight on the hour pipe, in front of the front plate, is the pinion E, which drives a wheel F of twice as many teeth. This wheel F is mounted loosely on a stud and has a pin which meshes with the teeth of a ratchet wheel G. G is carried at the bottom end of a pipe which fits loosely on

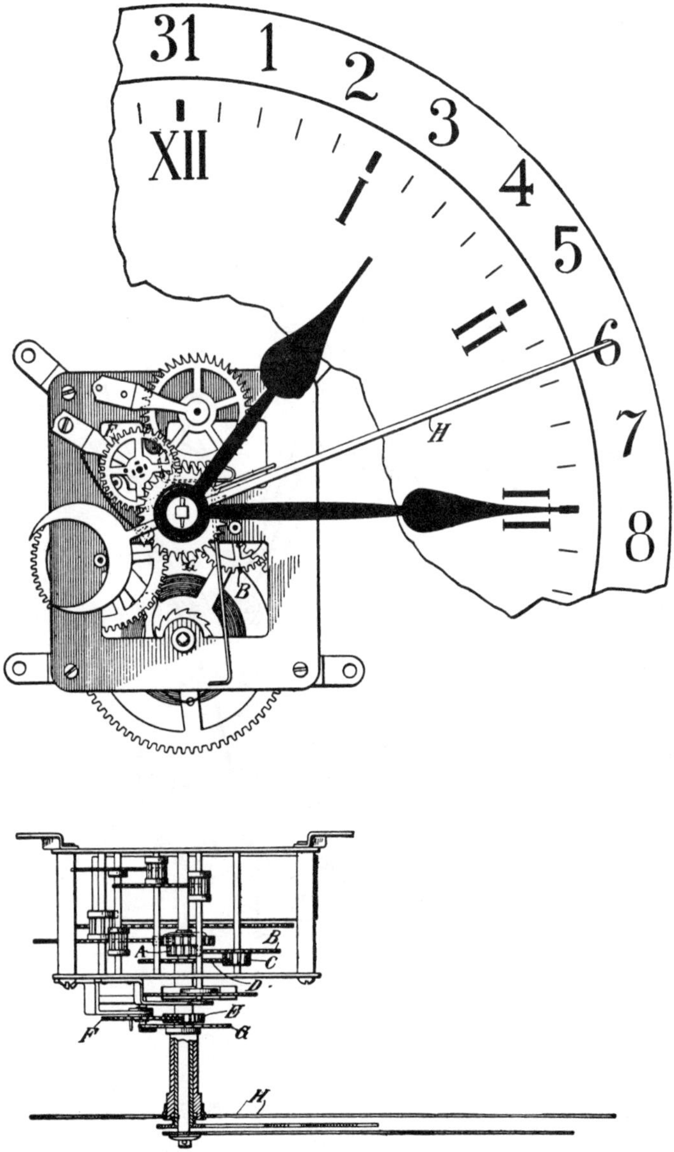

Fig. 111. Simple calendar on time train.

the hour pipe and carries the calendar hand H under the hour hand and close to the dial. The pinion on the hour pipe revolves once in twelve hours. The wheel E has twice

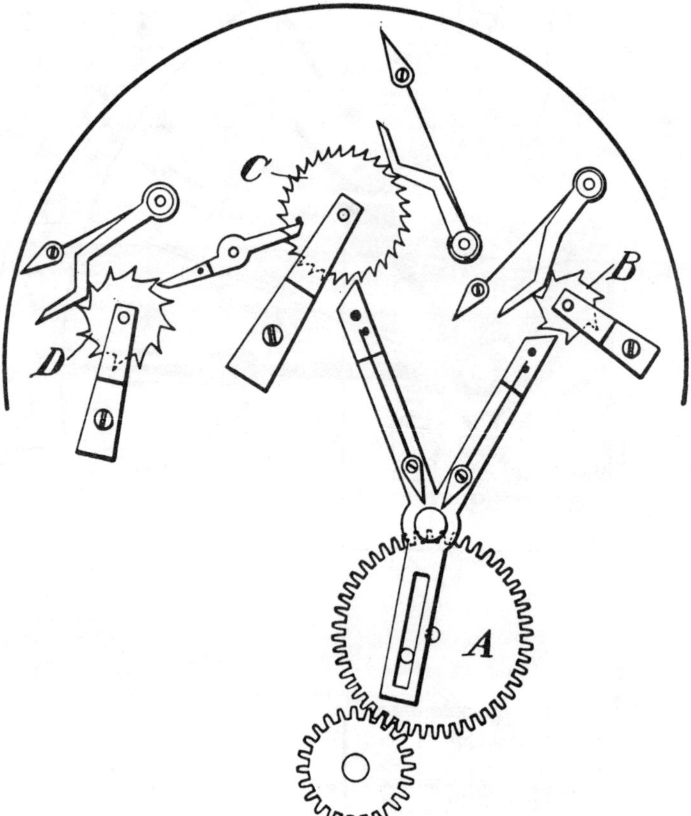

Fig. 112. Calendar work for grandfather clocks.

as many teeth and will therefore revolve once in twenty-four hours. It moves the ratchet G one tooth at each revolution; therefore the hand H moves one space every twenty-four hours. There are 31 teeth, so that the hand must be set forward every time it reaches the 28th and 29th of February and the 30th of April, June, September and November. This is the simplest and cheapest of all the calendars, occupies the least space and is frequently attached to nickel alarm clocks for that reason.

A simple calendar work often met with in old clocks of European origin is shown in Fig. 112. Gearing with the hour wheel is a wheel, A, having twice its number of teeth, and turning therefore once in twenty-four hours. A three-armed lever is planted just above this wheel; the lower arm is slotted and the wheel carries a pin which works in this slot, so that the lever vibrates to and fro once every twenty-four hours. The three upper wheels, B, C and D in the drawing, represent three star wheels. B has seven teeth, corresponding to the days of the week; C has 31 teeth, for the days of the month; and D has 12 teeth, for the months of the year. Each carries a hand in the center of a dial on the other side of the plate. Every time the upper arms of the lever vibrate they move forward the day of the week, B, and the day of the month, C, wheels each one tooth. The extremities of the two upper levers are jointed so as to yield on the return vibration, and are brought into position again by a weak spring. There is a pin in the wheel, C., which, by pressing on a lever once every revolution, actuates the month of the year wheel, D. This last lever is also jointed, and is pressed on by a spring so as to return to its original position. Each of the star wheels has a click kept in contact by means of a spring. For months with less than 31 days, the day of the month hand has to be shifted forward.

Perpetual Calendar Work.—Figs. 113, 114, 115, show a perpetual calendar which gives the day of the week, day of the month and the month, making all changes automatically at midnight, and showing the 31 days on a dial beneath the time dial, by means of a hand, and the days of the week and the month by means of cylinders operating

Fig. 113. Perpetual Calendar Movement.

behind slots in the dial on each side of the center. This is also a Waterbury movement.

A pinion on the hour pipe engages a wheel, A, having twice the number of teeth and mounted on an arbor which projects through both plates. The rear end of this arbor carries a cam, B, on which rides the end of a lever, C, which is pivoted to the rear frame. The lever is attached to a wire, D, which operates a sliding piece, E, which is weighted at its lower end. The cam, which, of course, revolves once in twenty-four hours, drops its lever at midnight and the weight on E pulls it down. E bears a spring pawl, F, which on its way down, raises the spring actuated retaining click, H, and then moves the 31-toothed wheel G one notch. This wheel is mounted on the arbor which carries the hand and, of course, advances the hand.

Lying on top of the wheel, G, is a cam, I, pivoted to G near its circumference and having an arm reaching toward the months cylinder and another reaching towards the right leg of the pawl, H, while it is cut away in the center, so as to clear the center arbor carrying the hand. Trace this cam, I, carefully in Figs. 113 and 114, as its action is vital. The lower arm of this cam is shown more clearly in Fig. 114. It projects above the wheel and engages the long teeth, J, and the cam, K, mounted on the year cylinder arbor; where the lower arm of I strikes one of these teeth it shoves the upper arm outward, so that it strikes the retaining end of the pawl, H, and holds it up, and the descending pawl, F, may then push the wheel, G, forward for *more* than one tooth. The upper end of I is broad enough to cover three teeth of the wheel, G, when pushed outward, and the slot in E is long enough so that F may descend far enough to push G forward three teeth at once, unless it is stopped by H falling into a tooth, so that the position of I, when it is holding up H and the extra drop thus given to E serve to operate the jumps of 30 to 1, 28 to 1 and 29 to 1 of the hand on the dial. The teeth, J, Fig. 114, operate for two notches,

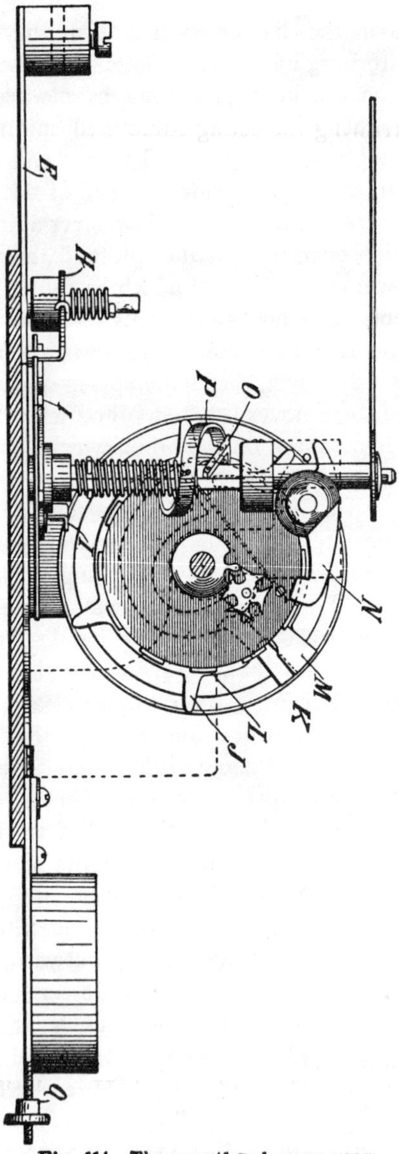

Fig. 114. The months change gear.

thus making the changes from 30 to 1. The wide tooth, M, and cam, K, acting together, make the change for February from 28 to 31. The 29th day is added by the movement of the cam, K, narrowing the acting surface once in four years, as follows:

Looking at Fig. 114 we see an ordinary stop works finger, mounted on the months arbor and engaging a four-armed maltese cross on the wheel. Behind the wheel is a circular cam (shown dotted in) with one-fourth of its circumference cut away; the pivot holds the cam and cross rigidly together while permitting them to revolve loosely in the wheel. The cam, K, lies close to the wheel and is pressed against the cam on the cross by a spring, so that ordinarily the full width of M and K act as one piece on the end of the cam, I, which thus is pressed against the retaining pawl, H, during the passage of three teeth, making the jump from 28 to 1 each of these three years.

The fourth revolution of the maltese cross brings the cut portion of its cam to operate on K and allows K to move behind M, thus narrowing the acting surface so that I only covers two teeth (30 and 31) for every fourth revolution of the month's cylinder, thus making the leap year every fourth year.

The months cylinder is kept in position by the two-armed pawl, N, engaging the teeth, L, which stand at 90 degrees from the wheel, as shown in Fig. 113. Attached to the bearing for the week cylinder (not shown) is one revolution of a screw track, or worm, surrounding the arbor for the hand. Attached to the arbor is a finger, O, held taut by a spring and engaging the track, P. The revolution of the arbor raises O on P until it slips off, when O, drawn downward by its spring, raises the pawl, N, drops on one of the teeth, L, and revolves the cylinder one notch.

Q is a shifter for raising the pawl, H, and allowing the hand to be set.

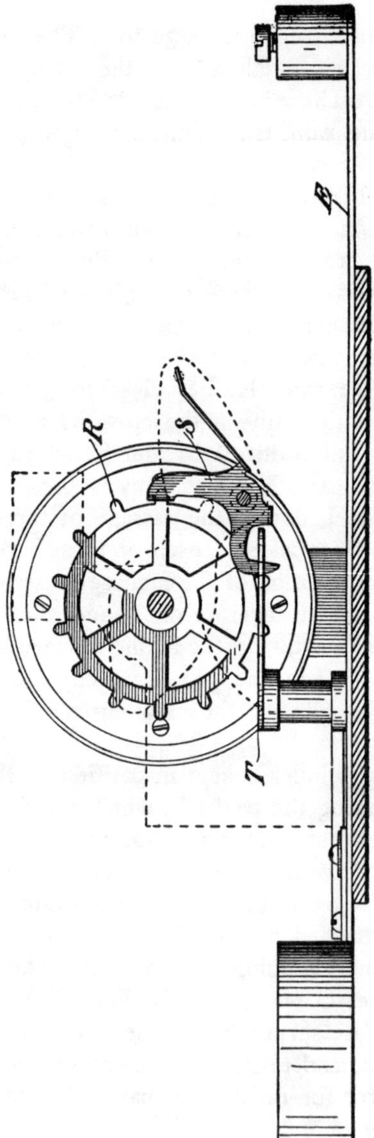

Fig. 115. The weeks change gear.

Fig. 115 shows the inner end of the cylinder for the days of the week. There are two sets of these and fourteen teeth on the sprocket, R, so as to get the two cylinders approximately the same size (there being 14 days and 12 months on the respective cylinders). S is a pawl whose upper end is forked so as to embrace a tooth and hold the cylinder in position. T is a hook, carried on the sliding piece, E, which swings outward in its upward passage as E is raised and on its downward course raises the pawl, S, and revolves the sprocket, R, one tooth, thus changing the day of the week at the same time the hand is advanced.

To set the calendar, raise the pawl, N, and revolve the year cylinder until M and K are at their narrowest width; that is, a leap year. Then give the year cylinder as many additional turns as there are years since the last leap year, stopping on the current month of the current year. For instance, if it is two years and four months since the 29th of February last occurred, give the cylinder 2 and 4/12 turns which should bring you to the current month, raise the shifter, Q, and set the hand to the current day. Then raise the pawl, S, and set the week cylinder to the current day. Place the hour hand on the movement so that the cam will drop E at midnight.

Fig. 116 shows the dial of Brocot's calendar work, which, with or without the equation of time and the lunations, is to be met with in many grandfather, hall and astronomical clocks. We will assume that all of these features are present, in order to completely cover the subject. It consists of two circular plates of which the front plate is the dial and the rear plate carries the movement, arranged on both sides of it. All centers are therefore concentric and we have marked them all with the same letters for better identification in the various views as the inner plate is turned about to show the reverse side, thus reversing the position of right to left in one view of the inner plate.

Fig. 117 shows the wheel for the phases of the moon, which is mounted on the outside of the inner plate immediately behind the opening in the dial. The dark circles have the same color as the sky of the dial and the rest is gilt, white or cream color to show the moon as in Fig. 116.

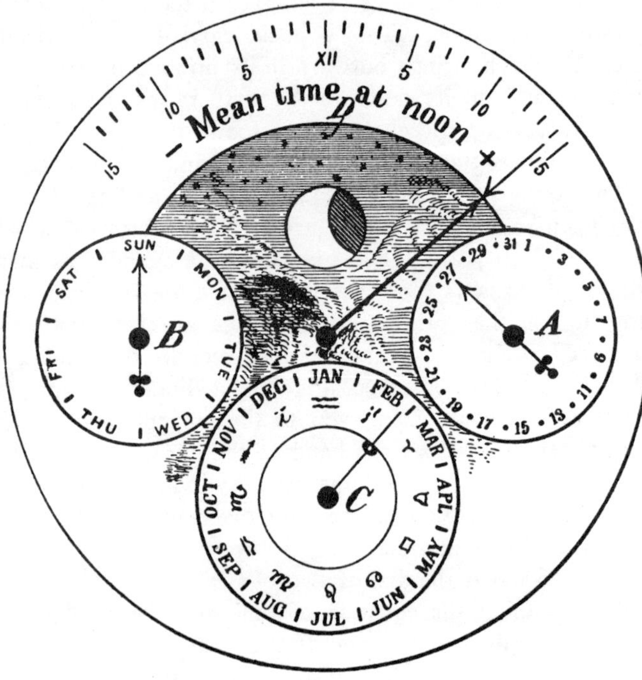

Fig. 116. Dial of Brocot's Calendar.

The position of this plate is also shown in Fig. 120. By the dotted circles, about the center D.

The inner side containing the mechanism for indicating the days of the week and the days of the month is shown in Fig. 118. The calendar is actuated by means of a pin, C, fixed to a wheel of the movement which turns once in twenty-four hours in the manner previously described with

Fig. 113. Two clicks, G and H, are pivoted to the lever, M. G, by means of its weighted end, see Fig. 119, is kept in contact with a ratchet wheel of 31 teeth, and H with a ratchet wheel of 7 teeth. As a part of these clicks and wheels is concealed in Fig. 118, they are shown separately in Fig. 119.

When the lever, M, is moved to the left as far as it will go by the pin, e, the clicks, G and H, slip under the teeth; their beaks pass on to the following tooth; when e has moved out of contact the lever, M, falls quickly by its own weight, and makes each click leap a tooth of the respective wheels, B of 7 and A of 31 teeth. The arbors of these wheels pass through the dial (Fig. 116), and have each an index which, at every leap of its own wheel, indicates on its special dial the day of the week and the day of the month. A roll, or click, kept in position by a sufficient spring, keeps each wheel in its place during the interval of time which separates two consecutive leaps.

This motion clearly provides for the indication of the day of the week, and would be also sufficient for the days of the month if the index were shifted by hand at the end of the short months.

To secure the proper registration of the months of 30 days, for February of 28 during three years, and of 29 in leap year, we have the following provision: The arbor, A, of the day of the month wheel goes through the circular plate, and on the other side is fixed (see Fig. 120) a pinion of 10 leaves. This pinion, by means of an intermediate wheel, I, works another wheel (centered at C) of 120 teeth, and consequently turning once in a year. The arbor of this last wheel bears an index indicating the name of the month, C, Fig. 116. The arbor, C, goes through the plate, and at the other end, C, Fig. 118, is fixed a little wheel gearing with a wheel having four times as many teeth, and which is centered on a stud in the plate at F. This wheel is partly concealed in Fig. 118 by a disc V, which is fixed to it, and with the wheel makes one turn in four years. On this disc, V, are made 20 notches, of which the 16 shallowest correspond to the months of 30 days; a deeper notch corresponds to the month of February of leap year, and the last three deepest to the month of February common years in each quarternary period. The uncut portions of the disc correspond to the months of 31 days in the same period. The wheel, A, of 31 teeth, has a pin (i) placed before the tooth which corresponds to the 28th of the month. On the lever, M, is pivoted freely a bell-crank lever (N), having at

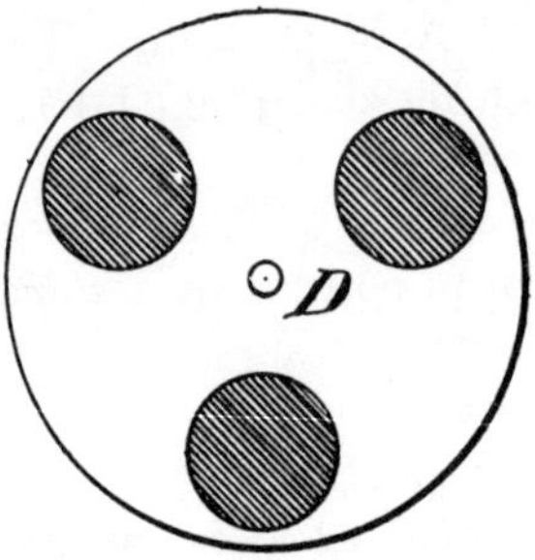

Fig. 117. Dial of Moon's Phases.

the extremity of the lower arm a pin (o) which leans its own weight upon the edge of the disc, V, or upon the bottom of one of the notches, according to the position of the month, and the upper arm of N is therefore higher or lower according to the position of the pin, o, upon the disc.

It will be easy to see that when the pin, o, rests on the contour of the disc the upper arm, N, of the bell-crank lever is as high as possible, and out of contact with the pin as it is dotted in the figure, and then the 31 teeth of the month wheel will each leap successively one division by the action of the click, G, as the lever, M, falls backward till the 31st day. But when the pin, o, is in one of the shallow notches of the plate, V, corresponding to the months of 30 days, the upper arm, N, of the bell-crank lever will take

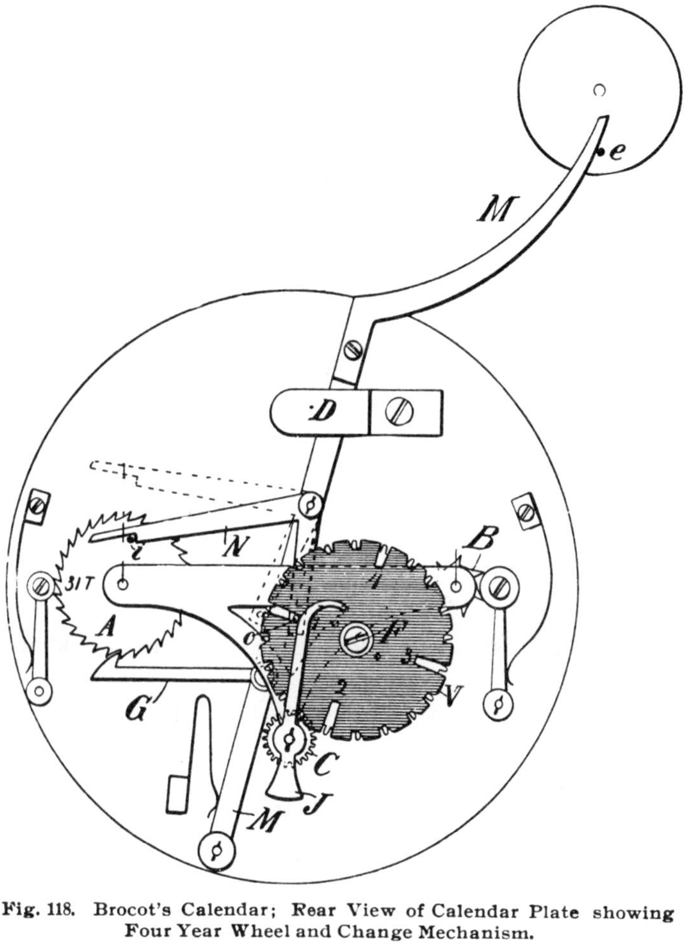

Fig. 118. Brocot's Calendar; Rear View of Calendar Plate showing Four Year Wheel and Change Mechanism.

a lower position, and the inclination that it will have by the forward movement of the lever, M, will on the 30th bring the pin, i, in contact with the bottom of the notch, just as the lever, M, has accomplished two-thirds of its forward movement, so the last third will be employed to make the wheel 31 advance one tooth, and the hand of the dial by consequence marks the 31st, the quick return of the lever, M, as it falls putting this hand to the 1st by the action of the click, G. If we suppose the pin, o, is placed in the shal-

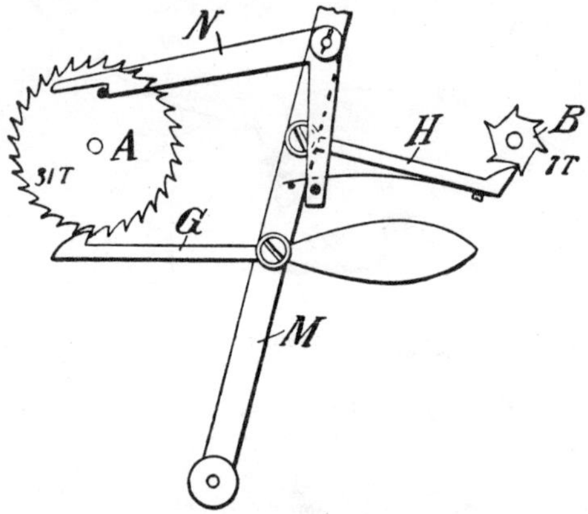

Fig. 119. Change Mechanism behind the Four Year Wheel in Fig. 118

lowest of the four deep notches, that one for February of leap year, the upper end of the arm, N, will take a position lower still, and on the 29th the pin, i, will be met by the bottom of the notch, just as the lever has made one-third of its forward course, so the other two-thirds of the forward movement will serve to make two teeth of the wheel of 31 jump. Then the hand of the dial, A, Figs. 116 and 118, will indicate 31, and the ordinary quick return of the lever, M, with its detent, G, will put it to the 1st. Lastly, if, as it is represented in the figure, the pin, o, is in one of the three deepest notches, corresponding to the months of February in ordinary years, the pin will be in the bottom of the notch on the 28th just at the moment the lever begins its movement, and three teeth will pass before the return of the lever makes the hand leap from the 31st to the 1st.

The pin, o, easily gets out of the shallow notches, which, as will be seen, are sloped away to facilitate its doing so. To help it out of the deeper notches there is a weighted finger (j) on the arbor of the annual wheel. This finger, having an angular movement much larger than the one of the disc, V, puts the pin, o, out of the notch before the notch has sensibly changed its position.

Phases of the Moon.—The phases of the moon are obtained by a pinion of 10, Fig. 120, on the arbor, B, which gears with the wheel of 84 teeth, fixed on another of 75, which last gears with a wheel of 113, making one revolution in three lunations. By this means there is an error only of .00008 day per lunation. On the wheel of 113 is fixed a plate on which are three discs colored blue, having between them a distance equal to their diameter, as shown in Fig. 117, these discs slipping under a circular aperture made in the dial, produce the successive appearance of the phases of the moon.

Equation of Time.—On the arbor of the annual wheel, C, Figs. 116, 118, 120, is fixed a brass cam, Y, on the edge of which leans the pin, s, fixed to a circular rack, R. This rack gears with the central wheel, K, which carries the hand for the equation. That hand faces XII the 15th of April, 14th of June, 1st of September and the 25th of December. At those dates the pin, s, is in the position of the four dots marked on the cam, Y. The shape of the cam, Y, must be such as will lead the hand to indicate the difference between solar and mean time, as given in the table of the Nautical almanac.

To set the calendar first see that the return of the lever, M, be made at the moment of midnight. To adjust the hand of the days of the week, B, look at an almanac and see what day before the actual date there was a full or new moon. If it was new moon on Thursday, it would be necessary, by means of a small button fixed at the back, on the arbor of the hand of the wheel, B, of the week, to make as many returns as requisite to obtain a new moon, this hand

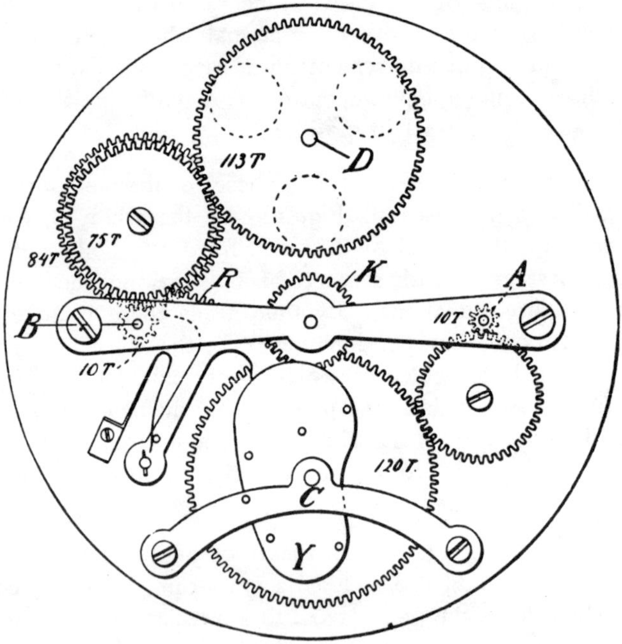

Fig. 120. Brocot's Calender: Wheels and Pinions under the Dial with their Number of Teeth.

pointing to a Thursday; afterward bring back the hand to the actual date, passing the number of divisions corresponding to the days elapsed since the new moon. To adjust the hand of the day of the month, A, see if the pin, o, is in the proper notch. If for the leap year, it is in the month of February in the shallowest of the four deep notches (o); if for the same month of the first year after leap year, then the pin should be, of course, in the notch, 1, and so on.

CHAPTER XX.

HAMMERS, GONGS AND BELLS.

Just as the tone of a piano depends very largely upon the condition of the felts on the hammers which strike the wires, so does the tone of a clock gong or bell depend on its hammer action. The deep, soft, resonant tone in either instance depends on the vibration being produced by something softer than metal. Ordinarily this condition is reached by facing the hammer with leather. The second essential is that the hammer shall immediately rebound, clear of the bell, so as not to interfere with the vibrations it has set up in the bell, wire or tube. As the leather gets harder the tone becomes harsher and "tinny," sometimes changing to another much higher tone and entirely destroying the harmony. The remedy is either to oil the leather on the hammers, or if they are much worn to substitute new and thicker leathers until the tone is sufficiently mellowed, so that a vigorous blow will still produce a mellow tone of sufficient carrying power. A piece of round leather belting will be found very convenient for this purpose.

The superiority of a chiming clock lies in its hammer action. If this mechanism is not perfect, only inferior results can be obtained. The perfect hammer is the one that acts with the smallest strain and is operated with the least power. Heavy weights create a tremendous strain on the mechanism and bring disastrous results when one of the suspending cords break. The method of lifting the hammer is one of importance, and the action of the hammer spring is but seldom right on old clocks brought in for repairs, especially if it be a spring bent over to a right angle at its point. If there are two springs, one to force the hammer down after the clock has raised it up, and another shorter one, fastened on to the pillar, to act as a counter-spring and prevent the hammer from jarring on the bell, there will seldom be any difficulty in repairing it; and the only operation necessary to be done is to file worn parts, polish the acting parts, set the springs a little stronger, and the thing is done. But if there is only one spring some further attention will be necessary, because the action of the one spring answers the purpose of the two previously mentioned, and to arrange it so that the hammer will be lifted with the greatest ease and then strike on the bell with the greatest force, and without jarring, requires some experience. That part of the hammer-stem which the spring acts on should never be filed or bent beyond the center of the arbor, as is sometimes done, because in such a case the hammer-spring has a sliding motion when it is in action, and some of the force of the spring is thereby lost. The point of the spring should also be made to work as near to the center of the arbor as it is possible to get it, and the flat end of the spring should be at a right angle with the edge of the frame, and that part of the hammer-stem that strikes against the flat end of the spring should be formed with a curve that will stop the hammer in a particular position and prevent it jarring on the bell. This curve can only be determined by experience; but a curve equal to a circle six inches in diameter will be nearly right.

The action of the pin wheel on the hammer-tail is also of importance. The acting face of the hammer-tail should be in a line with the center of the pin-wheel, or a very little above it, but never below it, for then it becomes more difficult for the clock to lift the hammer, and the hammer-tail should be of such a length as to drop from the pins of the pin-wheel, and when it stops be about the distance of two teeth of the wheel from the next pin. This allows the wheel-work to gain a little force before lifting the hammer, which is sometimes desirable when the clock is a little dirty or nearly run down. We might also mention that in setting the hammer-spring to work with greater force it is always well to try and stop the fly with your finger when the clock is striking, and if this can be done it indicates that the hammer spring is stronger than the striking power of the clock can bear, and it ought to be weakened, because the striking part will be sure to stop whenever the clock gets the least dirty.

Gong wires are also the cause of faulty tones. In the factories these are made by coiling wires of suitable lengths and sections on arbors in a lathe. They are then heated to a dull red and hardened by dipping in water or oil. After cooling they are trued in the round and the flat like a watch hairspring and then drawn to a blue temper. The tone comes with the tempering, and if they are afterwards bent beyond the point where they will spring back to shape the tone is interfered with. Many repairers, not being aware of this fact, have ruined the tone of a gong wire while trying to true it up by bending with pliers. When the owner is particular about the tone of the clock, a new gong should always be put in if the old one is badly bent.

The wires are soldered to their centers and if they are at all loose they should be refastened in the same manner if it can be done without drawing the temper of the wire. When this cannot be done a plug of solder may be driven in between the wire and the side of the hole so as to stop all vibration or the solder already in place may be driven down so as to make all tight, as any vibration at this point will interfere with the tone.

TUNING THE BELLS.—Bells only very slightly out of tone offend the musical ear, and they may easily be corrected to the extent of half a tone. To sharpen the tone make the bell shorter by turning away the edge of it if it be a shell, or by cutting off if it be a rod or tube; to flatten the

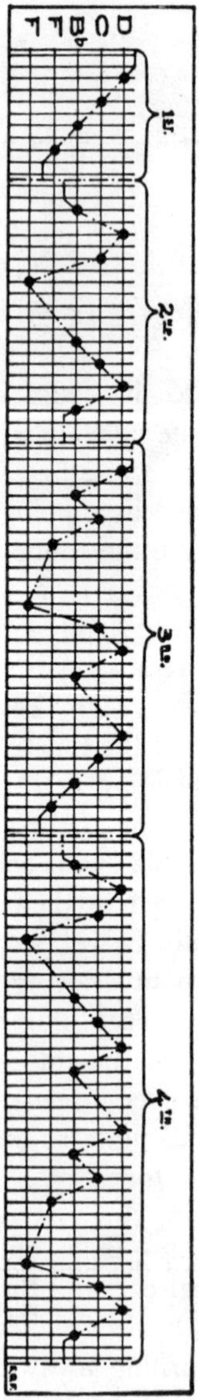

Fig. 121. The pins in the chiming barrels.

tone, thin the back basin-shaped part of the bell by turning some off the outside. Bells which are cracked give a poor sound because the edges of the crack interfere with each other when vibrating. They may be repaired by sawing through the crack to the end of it, so that the edges will not touch each other when vibrating. If there is danger of the crack extending further into the bell, first drill a round hole in the solid metal just beyond the end of the crack, and then saw through into the hole; this will generally prevent any further trouble.

Marking the Chime Barrel.—The chime barrel in small clocks is of brass and should be as large in diameter as can be conveniently got in. To mark off the positions of the pins for the Cambridge chimes, first put the barrel in the lathe and trace circles round the barrel at distances apart corresponding to the positions of the hammer tails. There are five chimes of four bells each for every rotation of the barrel, and a rest equal to two or three notes between each chime. Assuming the rest to be equal to three notes, divide the circumference of the barrel into thirty-five equal parts by means of an index plate, and draw lines at these points across the barrel with the point of the tool by moving it with the slide rest screw. Call the hammer for the highest note D, and that for the lowest note F. Then the first pin is to be inserted where one of the lines across the barrel crosses the first circle; the second pin where the next line crosses the second circle; the third pin where the third line crosses the third circle and the fourth pin where the fourth line crosses the four circle, because the notes of the first chime are in the order, D, C, Bb, F. Then miss three lines for the rest. The first note of the second chime is Bb and the pins for it will consequently be inserted where the first line after the rest crosses the third circle, and so on. Where two or more notes on the same bell come so close as to make it difficult to strike them properly, it is usual to put in another hammer, as it shown in Fig. 121, where there are two Fs. In fine clocks the pins are of varying lengths so as to strike the hammers on the bells with varying force and thus give more expression to the music.

The following gives the Cambridge Chimes, which are used in the Westminster Great Clock. They are founded on a phrase in the opening symphony of Handel's air, "I

Fig. 122. Westminster chimes.

know that my Redeemer liveth," and were arranged by Dr. Crotch for the clock of Great St. Mary's, Cambridge, in 1793.

In Europe these chiming clocks are sometimes very elaborate, as the following description of a set of bells in Belgium will show:

"So far as the experience of the writer goes the Belgian carillons are invariably constructed on one prevailing plan, with the exception that the metal used for the cylinder is generally brass; here, however, it is of steel, and consists of a large barrel measuring 4 feet 2 inches in width and 3 feet 6 inches in diameter, its surface being pierced with horizontal lines of small square holes about ⅜ inch square. There are lines of 60 of these in the width of the barrel, while there are 120 lines of them round the circumference, making a total of 7,200 holes. The drilling of these, of course, takes place when the cylinder is made, and, so far as this part is concerned, the barrel is complete before it is brought to the tower.

"Into these square holes are fixed the 'pins,' adjusted on the inside of the cylinder by nuts.

"The pins are of steel of finely graduated sizes, corresponding with the value of the notes of music. Some idea of the precision obtainable may be gathered by the fact, as the carillonneur told the writer, that there were no less than 24 grades of pins, so as to insure the greatest accuracy of striking the bells.

"Over the cylinder are 60 steel levers with steel nibs; these are lifted by the 'pins' and, connected by wires with the hammers, strike the bells.

"The 35 bells are furnished with 72 hammers, which are fixed as ordinary clock-hammers outside of the bells; three of the bells (in the ring of eight) have a single hammer only, the limited space in the 'cage' making it impossible to put more, while others are supplied with two or three apiece for use in rapidly repeating notes of the music. On a visit some years ago to the carillon at Malines, the writer noticed that some of the bells there had no less than five hammers apiece.

"Obviously, though there are 72 hammers in connection with the carillon, only 60, corresponding with the number of levers, can be used at one time; these are selected according to the requirement of the tune; in case of new tunes, the wires can easily be adjusted so as to bring other hammers and bells into use.

"The feature of the Belgian carillons is that instead of the single notes of the air being struck as with the old familiar 'chimes,' harmonized tunes of great intricacy are rendered with chords of three, four or even five bells striking at one time.

"The cylinder here is capable of 120 'measures' of music, but as a matter of fact it is subdivided so that half a revolution plays every hour.

"A march is, as a rule, played at the odd hours, and the national air at the even, but the bells are silent after 9 p. m. and start again at 8 a. m.

"The motive power is supplied by a weight of 8 cwt., and is controlled by a powerful fly of four fans artistically formed to represent swans. It may be mentioned that the keyboard for hand-playing consists of thirty-five keys of wood and eleven pedals; these, as indeed the whole apparartus of this part, are entirely separate from the automatic carillon; in this instance the keys connect with the clappers of the bells and have no association with the hammers. The pedals are connected with the eleven largest bells and are supplementary to the hour key."

TUBULAR CHIMES are tubes of bell metal, cut to the proper lengths to secure the desired tones and generally, but not always, nickel plated. As they take up much room in the clock, they are generally suspended from hooks at the top of the back board of the case, being attached to the hooks by loops of silk or gut cords, passed through holes drilled in the wall of the tubes near the top ends. The hour tube, being long and large, generally extends nearly to the bottom of a six-foot case, while the others range upwards, shortening according to the increase of pitch of the notes which they represent.

This makes it necessary to place the movement on a seat board and hang the pendulum from the front plate of the movement, so that such clocks have, as a rule, comparatively light pendulums. On account of the position and the great spread of the tubes, the chiming cylinder and hammers are placed on top of the movement, parallel with the plates, and operated from the striking train by means of bevel gears or a contrate wheel. The hammers are placed vertically on spring hammer stalks and connected with the chiming cylinder levers by silken cords. This gives great freedom of hammer action and results in very perfect tones. The hammers must of course be each opposite its own tube and thus they are rather far apart, which necessitates a long cylinder. This gives room for several sets of chimes on the same cylinder if desired, as a very slight horizontal movement of the cylinder would move the pins out of action with the levers and bring another set into action or cause the chimes to remain silent.

Practically all of the manufacturers of "hall" or chiming clocks import the movements and supply American cases, hammers and bells. The reason is that there is so little sale for them (from a factory standpoint) that one factory could supply the world with movements for this class of clocks without working overtime, and therefore it would be useless to make up the tools for them when they can be bought without incurring that expense.

CHAPTER XXI.

ELECTRIC CLOCKS AND BATTERIES.

Electric clocks may be divided into three kinds, or principal divisions. Of the first class are those in which the pendulum is driven directly from the armature by electric impulse, or by means of a weight dropping on an arm projecting from the pendulum. In this case the entire train of the clock consists of a ratchet wheel and the dial work.

The second class comprises the regular train from the center to the arbor. This class has a spring on the center arbor, wound more or less frequently by electricity. In this case the aim is to keep the spring constantly wound, so that the tension is almost as evenly divided as with the ordinary weight clock, such as is used in jewelers' regulators.

The third system uses a weight on the end of a lever connected with a ratchet wheel on the center arbor and does away with springs. One type of each of these clocks will be described so that jewelers may comprehend the principles on which the three types are built.

In the Gillette Electro-Automatic, which belongs to the class first mentioned, the ordinary clock principle is reversed. Instead of the works driving the pendulum, the pendulum drives the train, through the medium of a pawl and ratchet mechanism on the center arbor. The pendulum is kept swinging by means of an impulse given every tenth beat by an electro-magnet. This impulse is caused by the weight of the armature as it falls away from the magnet ends, the current being used solely to pull back and re-set the armature for the next impulse. Any variation in the current, therefore, does not affect the regulation of the

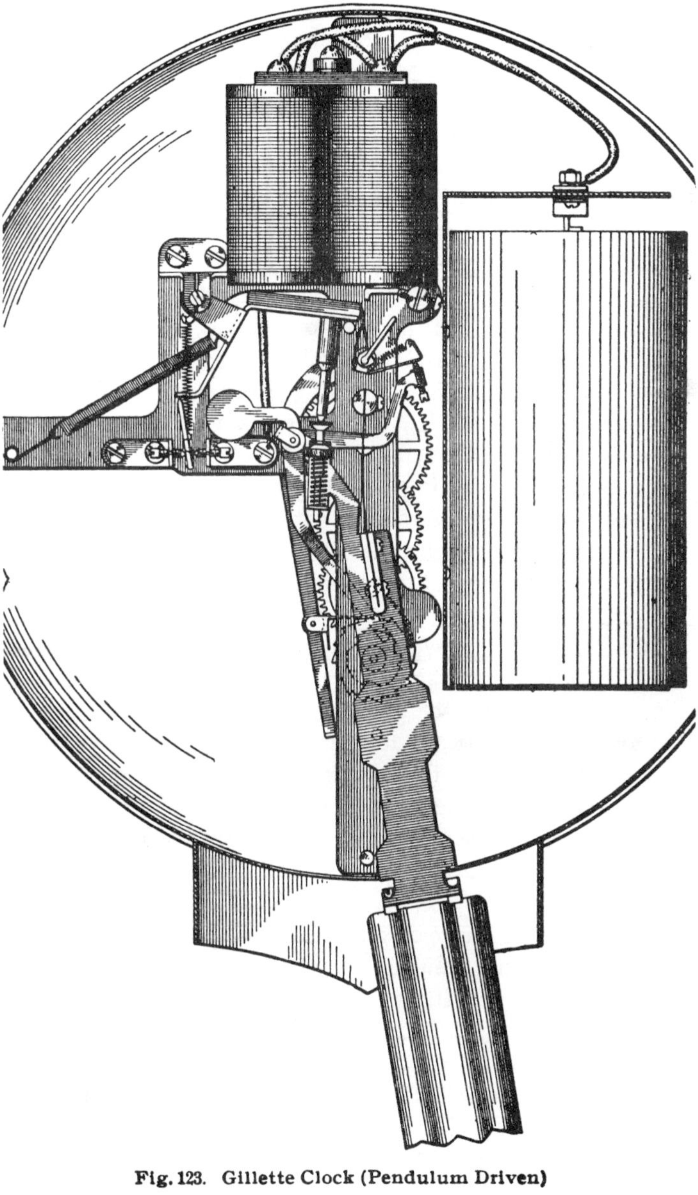

Fig. 123. Gillette Clock (Pendulum Driven)

clock, as the power is obtained from gravity only, by means of the falling weight. Referring to the drawings, Figs. 123 and 124, it is seen that each time the pendulum swings the train is pushed one tooth forward. A cam is carried by the ratchet (center) arbor in which a slot is provided at a position equivalent to every fifth tooth of the ratchet. Into this slot drops the end of a lever, releasing at its other end the armature prop. Thus at the next beat of the pendulum the armature is released and in its downward swing impulses the pendulum, giving it sufficient momentum to carry it over the succeeding five swings.

The action of the life-giving armature is entirely disconnected and independent of the clock mechanism. It acts on its own accord when released every tenth beat and automatically gives its impulse and re-sets itself. It is provided with a double-acting contact spring (see Fig. 125) which "flips" a contact leaf from one adjustable contact screw to the other as the action of the armature causes the spring to pass over its dead center. Thus, when the armature reaches the lowest point in its drop (Figs. 126 and 127) the leaf snaps against the right contact screw, the circuit is completed, the magnet energized and the armature drawn up. As the armature rises above a certain point, the dead center of the flipper spring is again crossed and the leaf snaps back against the post at the left. In the meantime, however, the armature prop has slipped under the end of the armature and retains it until the time comes for the next impulse.

In adjusting the mechanism of this type of clock the increasing pendulum swing should catch and push the ratchet before the buffer strikes and lifts the armature from the prop. The adjustment of the "flipper" contact screws (with 1-32 inch play) should be such that as the armature falls the contact leaf will be thrown and the armature drawn up at a point just beyond the half-way position in the swing of the pendulum. The power of the impulse can

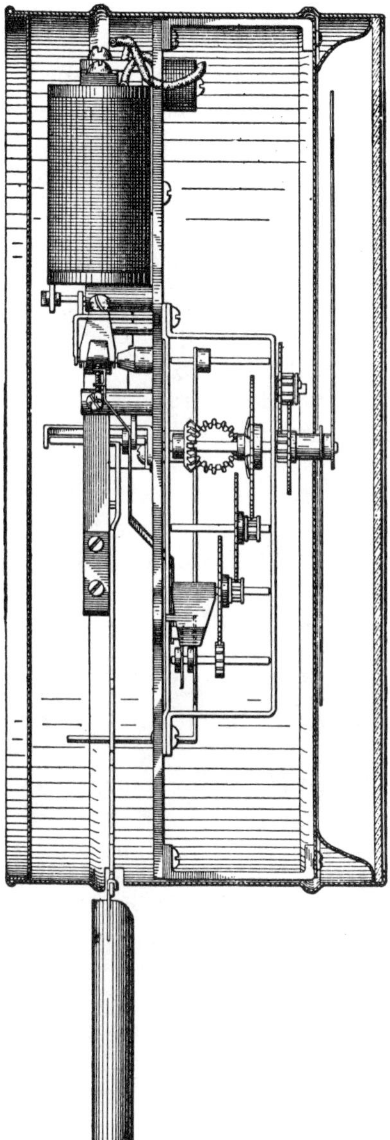

Fig. 124. Side View.

be regulated by turning the adjusting post with pliers, thus varying the tension of the armature spring, the pull of which reinforces the weight of the armature. Care should be taken, however, that the tension is not beyond the "quick action" power of the electro-magnet. It is much better to ease up the movement in other ways before putting too great a load on the life of the battery.

The electrical contacts on the leaf and screw are platinum tipped to prevent burning by the electric sparking at the "make" and "break." This sparking is also much reduced by means of a resistance coil placed in series connection with the magnet coil, Fig. 127, to reduce the amount of current used. If this coil is removed or disconnected the constant sparking and heat would soon burn out the contact tips.

Care should be taken to see that the batteries are dated and the battery connections are clean at the time of sliding in a new battery. The brush which makes connection with the center or carbon post of the battery is insulated with mica from the framework of the case. The other connection is made from the contact of the uncovered zinc case of the battery with the metal clock case surrounding it. The contact points should be bright and smooth to insure good contacts.

These clocks need but little cleaning of the works as *no oil whatever is used,* except at one place, viz., the armature pivot. Oil should *never* be used on the train bearings, or other parts. This clock ran successfully on the elevated railway platforms of the loop in Chicago where no other pendulum clock could be operated on account of the constant shaking.

In considering the electrical systems of these clocks, let us commence with the batteries. While undoubtedly great improvements have been made in the present form of dry battery they are still very far from giving entire satisfaction. Practically all of them are of one kind, which is that which produces electricity at 1½ volts from zinc, carbon and sal-ammoniac, with a depolarizer added to the elements to absorb the hydrogen. The chemical action of such a battery is as follows:

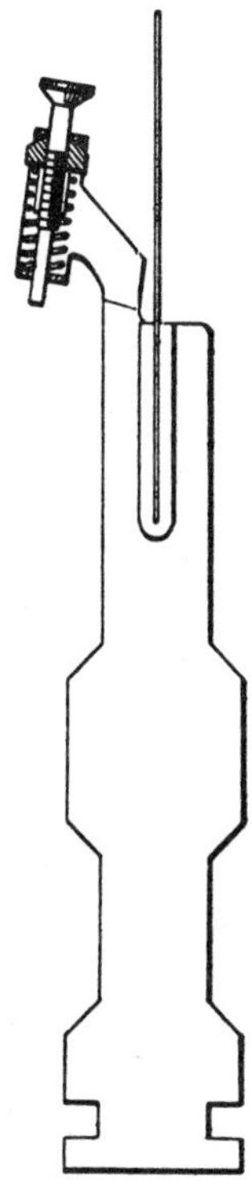

Fig. 125.

The water in the electrolite comes in contact with the zinc and is decomposed thereby, the oxygen being taken from the water by the zinc, forming oxide of zinc and leaving the hydrogen in the form of minute bubbles attached to the zinc. As this, if allowed to stand, would shut off the water from reaching the zinc, chemical action would therefore soon cease and when this happens the battery is said to be polarized and no current can be had from it.

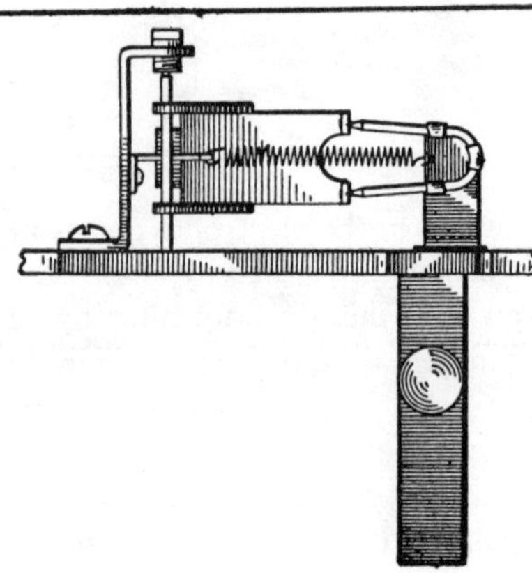

Fig. 126. The Electric Contact.

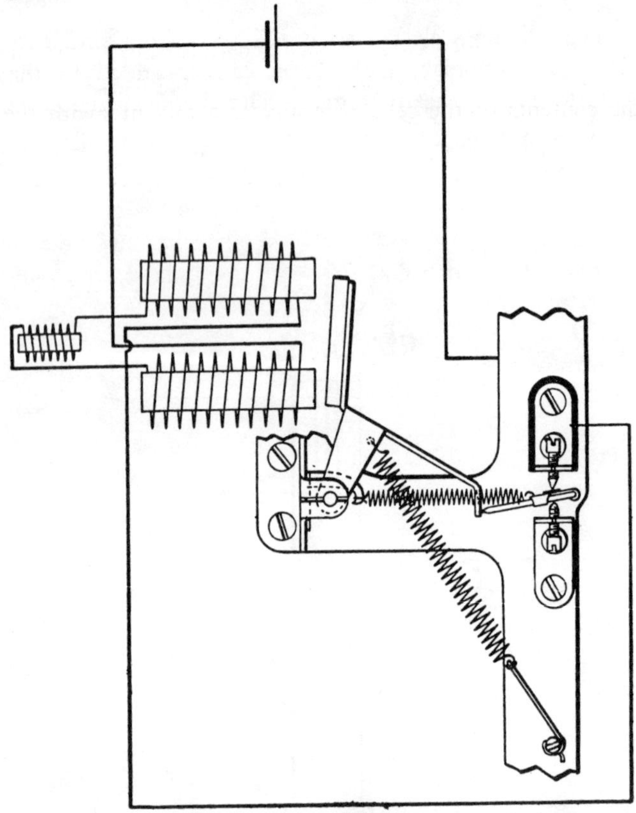

Fig. 127. The Wiring System.

In order to take care of the hydrogen and thus insure the constant action of the battery, oxide of manganese is added to the contents of the cell, generally as a mixture with the carbon element. Manganese has the property of absorbing oxygen very rapidly and of giving it off quite easily. Therefore while the hydrogen is being formed on the zinc, it becomes an easy matter for it to leave the zinc and take its proper quantity of oxygen from the manganese and again form water, which is again decomposed by the zinc. As long as this cycle of chemical action takes place the battery will continue to give good satisfaction, and usually when a battery gives out it is because the depolarizer is exhausted, for the reason that the carbon is not affected at all and the zinc element forming the container is present in sufficient quantity to outlast the chemical action of the total mass.

There are great differences in the various makes of batteries; also in the methods of their construction. It would seem to be an easy matter for a chemist to figure out exactly how much depolarizer would serve the purpose for a given quantity of zinc and carbon and therefore to make a battery which should give an exact performance that could be anticipated. In reality, however, this is not the case, owing to the various conditions. There are three qualities of manganese in the market; the Japanese, which is the best and most costly; the German, which comes second, and the American, which is the cheapest and varies in quality so much as to be more or less a matter of guesswork. We must remember that in making batteries for the price at which they are now sold on the market we are obliged to take materials in commercial quantities and commercial qualities and cannot depend upon the chemically pure materials with which the chemists' theories are always formulated. This therefore introduces several elements of uncertainty.

In practice the Japanese manganese will stand up for a far longer time than any other that is known and it is used in all special batteries where quality and length of life are considered of more importance than the price. The German manganese comes next. Then comes a mixture of American and German manganese, and finally the American manganese, which is used in making the cheaper batteries which are sorted afterwards, as we shall explain farther on. These batteries are sealed after having been made in large quantities, say five thousand or ten thousand in the lot, and kept for thirty days, after which they are tested. The batteries which are likely to give short-life will show a local action and consequent reduction of output in thirty days. They are, therefore, sorted out, much as eggs are candled on being received in a storage warehouse, for the reason that after a cell has been made and put together it would cost more to find out what was the matter with it and remedy that than it would to make a new cell. Many of the battery manufacturers, therefore, make up their batteries with an attempt to reach the highest standard. They are sorted for grade in thirty days and those which have attained the point desired are labeled as the factories' best battery and are sold at the highest prices. The others have been graded down exclusively and labeled differently until those which are positively known to be short-lived are run out and disposed of as the factories' cheapest product under still another label.

When buying batteries always look to see that the tops are not cracked, as if the seal on the cell is broken, chemical action induced from contact with the air as the battery dries out, will rapidly deteriorate the depolarizer and sulphate the zinc, both of which are of course a constant draft on the life of the battery, which contains only a stated quantity of energy in the beginning. Always examine the terminal connections to see that they are tight and solid.

Batteries when made up are always dated by the factory, but this does the purchaser little good, as the dates are in codes of letters, figures, or letters and figures, and are constantly changed so that even the dealers who are handling thousands of them are unable to read the code. This is done because many people are prone to blame the battery for other defects in the electrical system and many who are using great quantities would find an incentive to switch the covers on which the dates appear if they knew what it meant. This is perhaps rather harsh language, but a good many men would be tempted to send back a barrel of old batteries every now and then with the covers showing that they had not lasted three months, if they could read these signatures.

Practically the only means the jeweler has of obtaining a good cell, with long life, is to buy them of a large electrical supply house, paying a good price for them and making sure that that house has trade enough in that battery to insure their being continuously supplied with fresh stock.

The position of the battery also has to do with the length of life or amount of its output. Thus a battery lying on its side will not give more than seventy-five per cent of the output of a battery which is standing with the zinc and carbon elements perpendicular. Square batteries will not give the satisfaction that the round cell does. It has been found in practice by trials of numerous shapes and proportions that the ordinary size of 2½ x 6 inches will give better satisfaction than one of a different shape—wider or shorter, or longer and thinner; that is for the amount of material which it contains. The battery which has proved most successful in gas engine ignition work is 3¾ x 8 inches. That maintains the same proportions as above, or very nearly so, but owing to local action it will give on clock work only about fifty per cent longer life than the smaller size.

It has been a more or less common experience with purchasers of electric clocks to find that the batteries which came with the clock from the factory ran for two or three years (three years not being at all uncommon) and that they were then unable to obtain batteries which would stand up to the work for more than three weeks, up to six months. The difference is in the quality and freshness of the battery bought, as outlined above.

In considering the rest of the electrical circuit, we find three methods of wiring commonly used and also a fourth which is just now coming into use. The majority of electric clocks are wound by a magnet which varies in size from three to six ohms; bridged around the contact points,

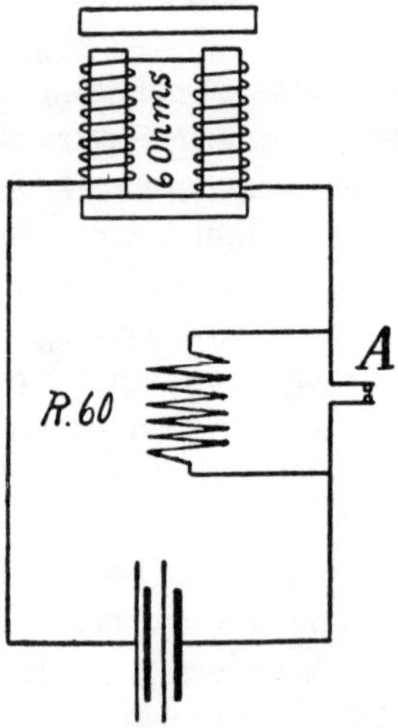

Fig. 128.

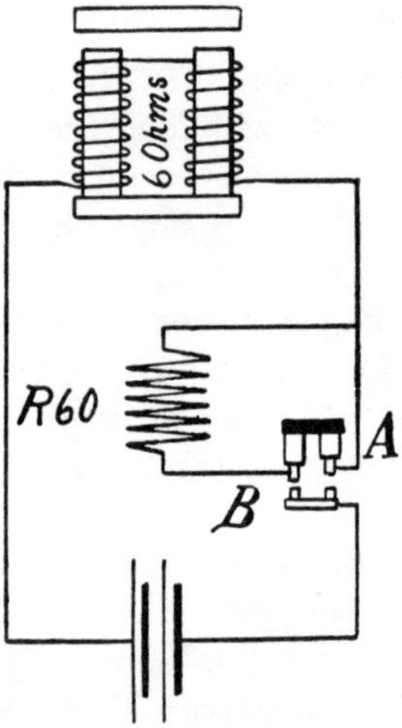

Fig. 129.

there has generally been placed a resistance spool which varies in size from ten to twenty-five times the number of ohms in the armature magnets. See Fig. 128. This practically makes a closed circuit on which we are using a battery designed for open circuit work.

If we use an electro-magnet with a very soft iron core, we will need a small amount of current, but every time we break the contact, we will have a very high counter electromotive force, leaping the air gap made while breaking the contact and therefore burning the contact points. If our magnet is constructed so as to use the least current, by very careful winding and very soft iron cores, this counter electro-motive force will be at its greatest while the draft on the battery is at its smallest. If the magnet cores are made of harder iron, the counter electro-motive force will be much less; but on the other hand much more current will be needed to do a given quantity of work with a magnet of the second description; and the consequence is that while we save our contact points to some extent, we deplete the battery more rapidly.

If we put in the highest possible resistance—that of air—in making and breaking our contacts, we use current from the battery only to do useful work; but we also have the spark from the counter electro-motive force in a form which will destroy our contact points more quickly. If we reduce the resistance by inserting a German silver wire coil of say sixty ohms on a six-ohm magnet circuit, we have then with two dry batteries (the usual number) three volts of current in a six-ohm magnet during work and three volts of current in a sixty-six ohm circuit while the contacts are broken, Fig. 128. Dividing the volts by the ohms, we find that one twenty-second of an ampere is constantly flowing through such a circuit. We are therefore using a dry battery (an open circuit battery) on closed circuit work and we are drawing from the life of our battery constantly in order to save our contact points.

It then becomes a question which we are going to sacrifice, or what sort of a compromise may be made to obtain the necessary work from the magnet and at the same time get the longest life of the contact points and the batteries. Most of the earlier electric clocks manufactured have finally arranged such a circuit as has been described above.

The Germans put in a second contact between the battery and the resistance with a little larger angular motion than the first or principal contact, so that the contact is then first made between the battery and resistance spool, B, Fig. 129, then between the two contact points of the shunt, A, Fig. 129, to the electro-magnet, and after the work is done they are broken in the reverse order, so that the resistance is made first and broken after the principal contact. This involves just twice as many contact points and it also involves more or less burning of the second contact.

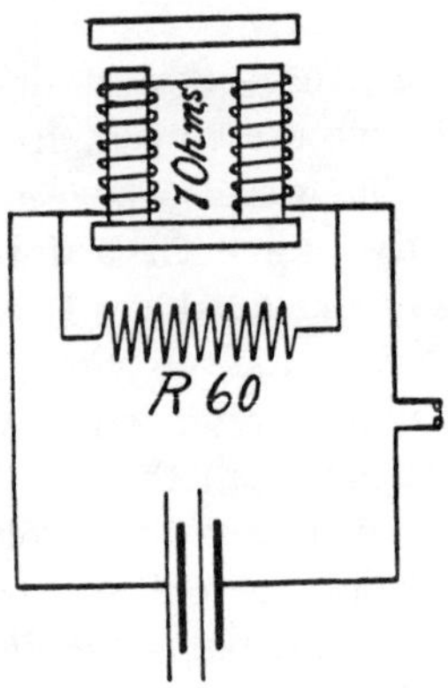

Fig. 130.

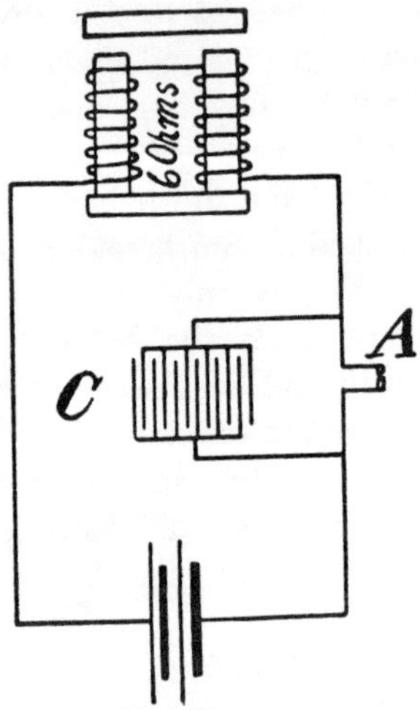

Fig. 131.

The American manufacturers seem to prefer to waste more or less current rather than to introduce additional contact points, as they find that these become corroded in time with even the best arrangements and they desire as few of them as possible in their movements, preferring rather to stand the draft on the battery.

One American manufacturer inserts a resistance spool of 60 ohms in parallel with a magnet of seven ohms (3½ ohms for each magnet spool) as in Fig. 130. He states that the counter electro-motive force is thus dissipated in the resistance when the contact is broken, as the resistance thus becomes a sort of condenser, and almost entirely does away with heating and burning of the contacts, while keeping the circuit open when the battery is doing no work.

It has been suggested to the writer by several engineers of high attainments and large experience that what should be used in the above combination is a condenser in place of a resistance spool, as there would then be no expenditure of current except for work. One of the clocks changed to this system just before the failure of its manufacturers, but as less than four hundred clocks were made with the condensers (Fig. 131), the point was not conclusively demonstrated.

It should also be borne in mind that the condenser has been vastly improved within the last twelve months. With the condenser it will be observed that there is an absolutely open circuit while the armature is doing no work and that therefore the battery should last that much longer, Figs. 130 and 131. As to the cost of the condensers as compared with resistance spools, we are not informed, but imagine that with the batteries lasting so much longer and the clock consequently giving so much better satisfaction, a slight additional cost in manufacture by changing from resistance to condensers would be welcomed, if it added to the length of life and the surety of operation.

Electric clocks cost more to make than spring or weight clocks and sell for a higher price and a few cents additional per movement would be a very small premium to pay for an increase in efficiency.

The repairer who takes down and reassembles one of these clocks very often ignorantly makes a lot of trouble for himself. Many of the older clocks were built in such a way that the magnets could be shifted for adjustment, instead of being put in with steady pins to hold them accurately in place. The retail jeweler who repairs one of these clocks is apt to get them out of position in assembling. The armature should come down squarely to the magnets, but should not be allowed to touch, as if the iron of the armature touches the poles of the magnet it will freeze and retain its magnetism after the current is broken. Some manufacturers avoid this by plating their armatures with copper or brass and this has puzzled many retailers who found an electro-magnet apparently attracting a piece of metal which is generally understood to be non-magnetic.

The method offers a good and permanent means of insulating the iron of the armature from the magnet poles while allowing their close contact and as the strength of a magnet increases in proportion to the square of the distance between the poles and the armature, it will be seen that allowing the armature to thus approach as closely as possible to the poles greatly increases the pull of the magnet at its final point. If when setting them up the magnet and armature do not approach each other squarely, the armature will touch the poles on one side or another and soon wear through the copper or brass plating designed to maintain their separation and then we will have freezing with its accompanying troubles.

A very good test to determine this is to place a piece of watch paper, cigarette paper or other thin tissue on the poles of the magnet before the naked iron armature is drawn down. Then make the connection, hold the armature and see if the paper can be withdrawn. If it cannot the armature and poles are touching and means should be taken to separate them. This is sometimes done by driving a piece of brass into a hole drilled in the center of the pole of the magnet; or by soldering a thin foil of brass on the armature. As long as the separation is steadily maintained the object sought is accomplished, no matter what means is used to attain it.

Another point with clocks which have their armatures moved in a circular direction is to see that the magnet is so placed as to give the least possible freedom between the armatures and the circular poles of the magnet, but that there must be an air-gap between the armature and magnet poles.

In those clocks which wind a spring by means of a lever and ratchet working into a fine-toothed ratchet wheel, or are driven by a weighted lever, there is an additional point to guard against. If the weight lever is thrown too far up, either one of two things will happen. The weight lever may be thrown up to ninety degrees and become balanced if the butting post is left off or wrongly replaced; the power will then be taken off the clock, if it is driven directly by weight, so that a butting post should meet the lever at the highest point and insure that it will not go beyond this and thus lose the efficiency of the weight.

In the cases where a spring on the center arbor is interposed between the arbor and the ratchet wheel, it should be determined just how many teeth are necessary to be operated when winding, as if a clock is wound once an hour and the aim is to wind a complete turn (which is the amount the arbor has run down) if the lever is allowed to vibrate one or two teeth beyond a complete turn, it will readily be seen that in the course of time the spring will wind itself so tightly as to break or become set. This was a frequent fault with the Dulaney clock and has not been guarded against sufficiently in some others which use the fine ratchet tooth for winding.

When such a clock is found the proper number of teeth should be ascertained and the rest of the mechanism adjusted to see that just that number of teeth will be wound. If less is wound there will come a time when the spring will run down and the clock will stop. If too much is wound the spring will eventually become set and the clock will stop. Therefore such movements should be examined to see that the proper amount of winding occurs at each operation. Of course where a spring is wound and there are but four notches in the ratchet wheel and the screw stop is accurately placed to stop the action of the armature, over action will not harm the spring, provided it will not go to another quarter, as if the armature carries the ratchet wheel further than it should, the smooth circumference between the notches will let it drop back to its proper notch.

There are a large number of clocks on the market which wind once per hour. These differ from the others in that they do not depend upon a single movement of the armature for an instantaneous winding. Thus if the batteries are weak it may take twenty seconds to wind. If the batteries are strong and new it may wind in six seconds. In this respect the clock differs radically from the others, and while we have not personally had them under test, we are informed that on account of winding once per hour the batteries will last very much longer than would be expected proportionately from those which wind at periods of greater frequency. The reason assigned is that the longer period allows the battery to dispose of its hydrogen on the zinc and thus to regain its energy much more completely between the successive discharges and hence can give a more effective quantity of current for hourly discharge than those which are discharged several times a minute, or even several times an hour. It is only proper to add that the manufacturers of clocks winding every six or seven minutes dispute this assertion.

Another point is undoubtedly in the increased length of life of the contacts; but speaking generally the electric clock may be said now to be waiting for further improvements in the batteries. Those who have had the greatest experience with batteries, as the telephone companies, telegraph companies and other public service corporations, have generally discarded their use in favor of storage batteries and dynamos wherever possible and where this is not possible they have inspected them continuously and regularly.

In this respect one point will be found of great service. When putting in a new set of batteries in any electrical piece of machinery, write the date in pencil on the battery cover, so that you, or those who come after you, some time later, will know the exact length of time the battery has been in service. This is frequently of importance, as it will determine very largely whether the battery is playing out too soon, or whether faults are being charged to the battery which are really due to other portions of the apparatus.

Never put together any piece of electrical apparatus without seeing that all parts are solidly in position and are clean; always look carefully to connections and see that the insulation is perfect so that short circuits will be impossible.

All contacts must be kept smooth and bright and contact must be made and broken without any wavering or uncertainty.

Fig. 132 shows the completely wired movement of the American Clock Company's weight-driven movement, which may be accepted as a type of this class of movements—weight-driven, winding every seven minutes.

The train is a straight-line time train, from the center arbor to the dead beat escapement, with the webs of the wheels not crossed out. It is wired with the wire from the battery zinc screwed to the front plate H and that from the battery carbon to an insulated block G.

Fig. 133 shows an enlarged view of the center arbor. Upon this arbor are secured (friction tight) two seven-notched steel ratchets, E, and carried loosely between them are two weighted levers pivoted loosely on the center arbor. Each lever is provided with a pawl engaging in the notches of the nearest ratchet, as shown. The weighted lever has a circular slot cut in it, concentric with the center hole and also has a portion of its circumference at the arbor cut away, thus forming a cam. Between these two levers is a connecting link D with a pin in its upper end, which pin projects into the circular slots of the weight levers.

The lever F is pivoted to the front plate of the clock and carries at right angles a beveled arm which projects over the ratchets E, but is ordinarily prevented from dropping into the notches by riding on the circumferences of the

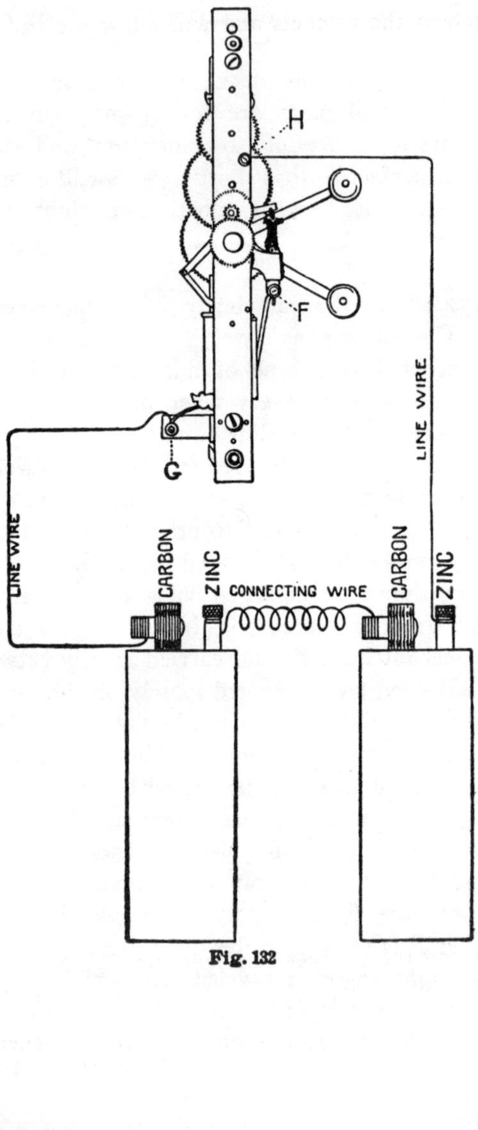
H
F
G
LINE WIRE
LINE WIRE
CARBON
ZINC
CONNECTING WIRE
CARBON
ZINC

Fig. 132

weighted levers. When one lever has dropped down and the other has reached a horizontal position the cut portions of the circumferences of these levers will be opposite the upper notch of the ratchets and will allow the bar project-

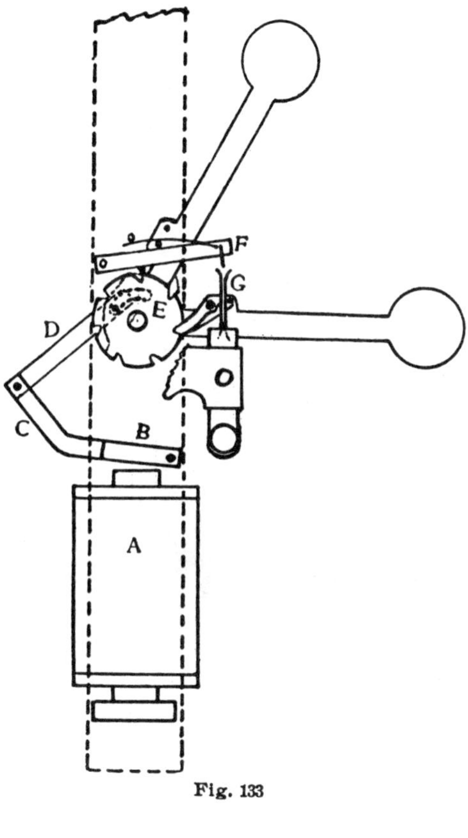

Fig. 133

ing from F to drop into the notches. This allows F and G to connect and the magnet A is energized, pulls the armature B, the arms C D, and thus lifts the lever through the pin in D pulling at the end of the circular slot. As the lever flies upward, the cam-shaped portion of its circumference raises the arm out of the notches, thus separating F and G and breaking the circuit. A spring placed above F keeps its arms pressed constantly upon E in position to drop. The wiring of the magnets is shown in Fig. 130.

The upper contact (carried in F) is a piece of platinum with its lower edge cut at an angle of fifteen degrees and beveled to a knife-edge. The lower point of this bevel comes into contact first and is the last to separate when breaking connection, so that any sparking which may take place will be confined to one edge of the contacts while the rest of the surface remains clean. (See Fig. 134.) Ordi-

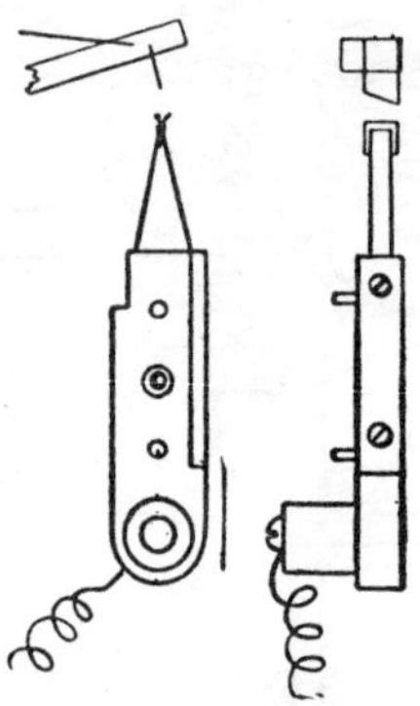

Fig. 134

narily there is very little corrosion from burning and this is constantly rubbed off by the sliding of the surfaces upon each other. The lower contact, G, consists of a brass block mounted upon an insulating plate of hard rubber. The block is in two pieces, screwed together, and each piece carries a platinum tipped steel spring. These springs are so set as to press their platinum tips against each other directly beneath the upper contact. The upper and lower platinum tips engage each other about one-sixteenth inch at the time of making contact. The lower block being in two pieces, the springs may be taken apart for cleaning, or to adjust their tension. The latter should be slight and should in no case exceed that which is exerted by the spring in F, or the upper knife-edge will not be forced between the two lower springs. The pin on which F is pivoted and that bearing on the spring above it must be clean and bright and *never be oiled,* as it is through these that the current passes to the upper contact in the end of F. The contacts are, of course, never oiled.

The two weighted levers should be perfectly free on the center arbor and their supporting pawls should be perfectly free on the shoulder screws in the levers. Their springs should be strong enough to secure quick action of the pawls. This freedom and speed of action are important, as the levers are thrown upward very quickly and may rebound from the butting post without engaging the ratchets if the pawls do not work quickly.

The projecting arm, C, of the armature, B, has pivoted to it, a link, D, which projects upward and supports at its upper end a cross pin. The link should not be tight in the slot of C, but should fit closely on the sides, in order to keep the cross pin at the top of D parallel with the center staff of the clock. This cross pin projects through D an equal distance on either side, each end respectively passing through the slot of the corresponding lever, the total length of this pin being nearly equal to the distance between the ratchets. When the electric circuit is closed, and the magnets energized, B, C and D are drawn downward; the weighted end of one of the levers which runs the clock, being at this time at the limit of its downward movement, see Fig. 135, the opposite or slotted end of said lever, is then at its highest point, and the downward pull in the slot by one end of the above described crosspin which enters it will throw the weighted end of the said lever upward. The direct action of the magnets raises the lever nearly to the horizontal position, and the momentum acquired carries it the remainder of the distance. By this arrangement of stopping the downward pull of the pin when the ascending lever reaches the horizontal, all danger of disturbing the other lever A is avoided. The position is such that the top of the ascending lever weight is about even with the center of the other weight when the direct pull ceases.

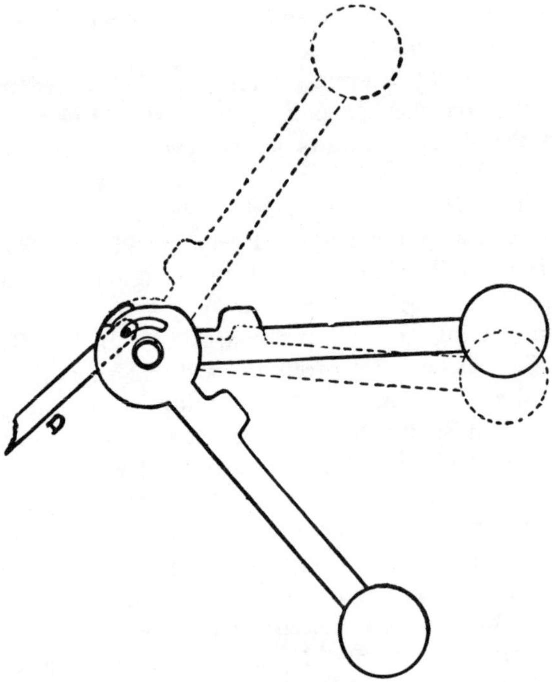

Fig. 135

Before starting the clock raise the lever weights so that one lever is acting upon a higher notch of the ratchet than the other. They are designed to remain about forty-five degrees apart, so as to raise only one lever at each action of the magnet. This maintains an equal weight on the train, which would not be the case if they were allowed to rise and fall together; keeping the levers separated also reduces the amount of lift or pull on the battery and uses less current, which is an item when the battery is nearly run down. If these levers are found together it indicates that the battery is weak, the contacts dirty, making irregular winding, or the pawls are working improperly. See that the levers rise promptly and with sufficient force. After one of them has risen stop the pendulum and see that the butting post is correctly placed, so that there is no danger of the lever wedging under the post and sticking there, or causing the lever to rebound too much. The butting post is set right when the clock leaves the factory and seldom needs adjustment unless some one has tinkered with it.

The time train should be oiled as with the ordinary movements, also the pawls on the levers. The lever bushings should be cleaned before oiling and then well oiled in order to avoid friction on the center arbor from the downward pull of the magnets when raising the levers. In order to clean the levers drive out the taper pin in the center arbor and remove the front ratchet, when the levers will slip off. In putting them back care should be used to see that the notches of the ratchets are opposite each other. Oil the edges of the ratchets and the armature pins. Do not under any circumstances oil the contact points, the pins or springs of the bar F, as this will destroy the path of the current and thus stop the clock. These pins must be kept clean and bright.

Hourly Winding Clocks.—There are probably more of these in America than of all other electric kinds put together (we believe the present figures are something like 135,000), so that it will not be unreasonable to give considerable space to this variety of clocks. Practically all of them are made by the Self Winding Clock Company and are connected with the Western Union wires, being wound by independent batteries in or near the clock cases.

Three patterns of these clocks have been made and we will describe all three. As they are all practically in the same system, it will probably be better to first make a simple statement of the wiring, which is rigidly adhered to by the clock company in putting out these goods. All wires running from the battery to the winding magnets of the movement are brown. All wires running from the synchronizing magnet to the synchronizing line are blue. Master clocks and sub-master clocks have white wires for receiving the Washington signal and the relay for closing the synchronizing line will have wires of blue and white plaid.

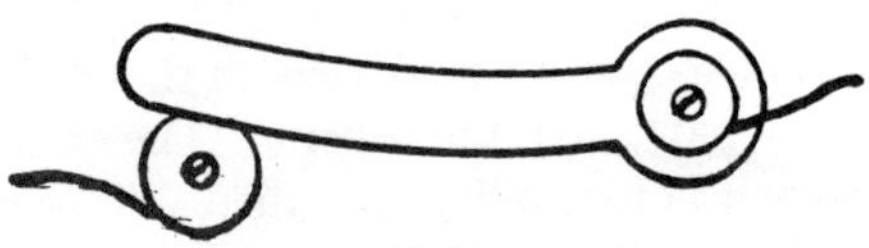

Fig. 136

By remembering this system it is comparatively easy for a man to know what he is doing with the wires, either inside or outside of the case. For calendar clocks there are, in addition, two white wires running from the calendar to the extra cell of battery. There is also one other peculiarity, in that these clocks are arranged to be wound by hand whenever run down (or when starting up) by closing a switch key, shown in Fig. 136, screwed to the inside of the case. This is practically an open switch, held open by the spring in the brass plate, except when it is pressed down to the lower button.

The earliest movement of which any considerable number were sent out was that of the rotary winding from a three-pole motor, as shown in Fig. 137. Each of these magnet spools is of two ohms, with twelve ohms resistance, placed in parallel with the winding of each set of magnet spools, thus making a total of nine spools for the three-pole motor.

On the front end of the armature drum arbor is a commutator having six points, corresponding to the six arma-

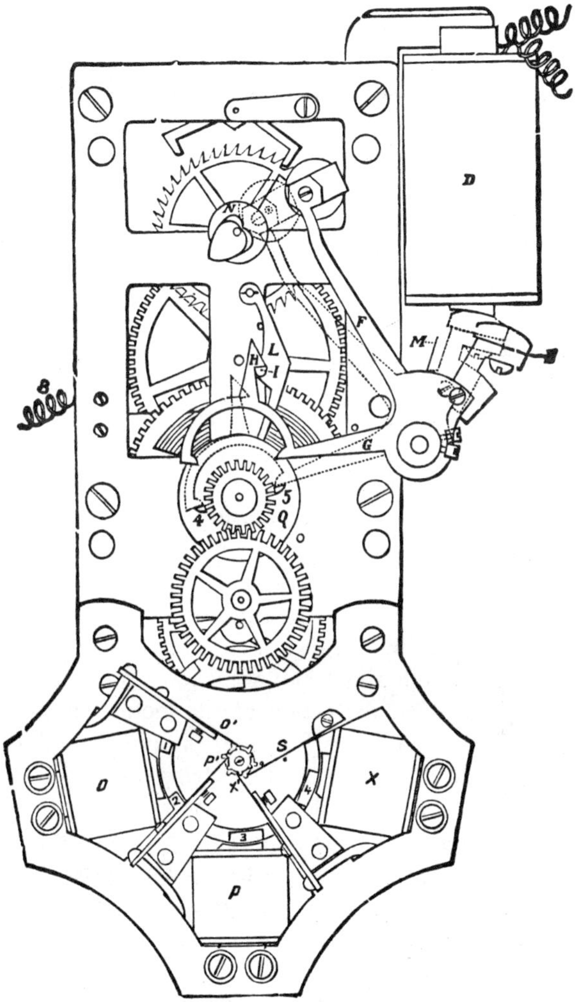

Fig. 137

tures in the drum. There are three magnets marked O, P and X; each magnet has its own brush marked O′, P′ and X′. When an armature approaches a magnet (see Fig. 137) the brush makes contact with a point of the commutator, and remains in contact until the magnet has done its work and the next magnet has come into action. When properly adjusted the brush O′ will make contact when armatures 1 and 2 are in the position shown, with No. 2 a little nearer the core of the magnet than No. 1; and it will break contact when the armature has advanced into the position shown by armature No. 3, the front edge of the armature being about one-sixteenth of an inch from the corner of the core, armature No. 4 being entirely out of circuit, as brush X′ is not touching the commutator.

The back stop spring, S, Fig. 137, must be adjusted so that the brush O′ is in full contact with a point of the commutator when the motor is at rest, with a tooth of the ratch touching the end of the spring, S.

Sometimes the back stop spring, S, becomes broken or bent. When this occurs it is usually from overwinding. It must be repaired by a new spring, or by straightening the old one by burnishing with a screwdriver. Set the spring so that it will catch about half way down the last tooth.

Having explained the action of the motor we come now to the means of temporarily closing the circuit and keeping it closed until such time as the spring is wound a sufficient amount to run the clock for one hour; as the spring is on the center arbor this requires one complete turn.

This is the distinguishing feature of this system of clocks and is not possessed by any of the others. It varies in construction in the various movements, but in all its forms it maintains the essential properties of holding the current on to the circuit until such time as the spring has been wound a sufficient quantity, when it is again forcibly broken by the action of the clock. This is termed the "knock away," and exists in all of these movements.

To start the motor the circuit is closed by a platinum tipped arm, A, Fig. 138, loosely mounted on the center arbor, and carried around by a pin projecting from the center wheel until the arm is upright, when it makes contact with the insulated platinum tipped brush, B. A carries in its front an ivory piece which projects a trifle above the platinum top, so that when B drops off the ivory it will make contact with the platinum on A firmly and suddenly. This contact then remains closed until the spring barrel is turned a full revolution, when a pin in the barrel cover brings up the "knock away," C, which moves the arm, A, forward from under the brush, B, and breaks the circuit. The brush, B, should lie firmly on its banking piece, and should be so adjusted that when it leaves the arm, A, it will drop about one-thirty-second of an inch. Adjusted in this way it insures a good, firm contact.

The angle at the top of the brush, B, must not be too abrupt, so as to retard the action of the clock while the contact is being made. Wire No. 8 connects the spring contact, B, to one of the binding plates at the left-hand side of the case; and wire No. 6 connects the motor, M, to the other. To these binding plates are attached brown wires that lead one to each end of the battery.

When the clock is quite run down, it is wound by pressing the switch key, Fig. 136, from which a wire runs to the plate. The switch key should *not* be permanently connected to its contact screw, J. See that all wires are in good condition and all connections tight and bright. The main spring is wound by a pinion on the armature drum arbor, through an intermediate wheel and pinion to the wheel on the spring barrel.

At stated times—say once in eighteen months or two years—all clocks should be thoroughly cleaned and oiled, and at the same time inspected to be sure they are in good order.

Never let the self-winding clocks run down backward, as the arm, A, Fig. 138, will be carried back against the brush, B, and bend it out of adjustment.

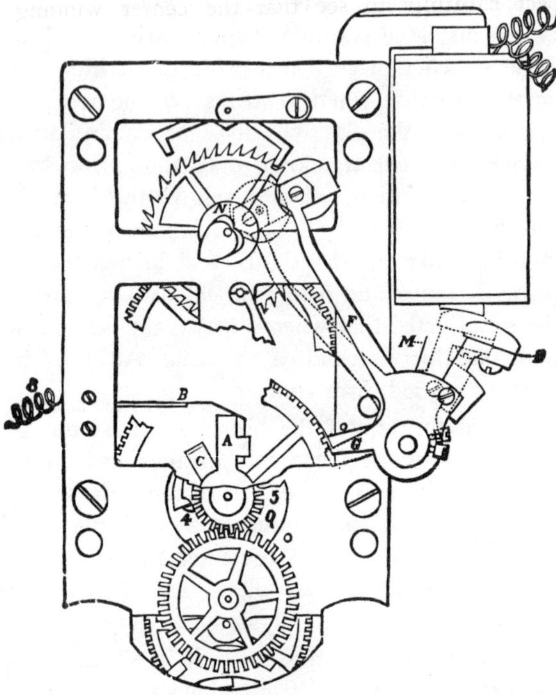

Fig. 138

To clean the movement, take it from the case, take out the anchor and allow it to run down gently, so as not to break the pins, then remove the motor. Take off the *front* plate and separate all the parts. Never take off the back plate in these clocks. Wash the plates and all parts in a good quality of benzine, pegging out the holes and letting them dry thoroughly before reassembling. The motor must not be taken apart, but may be washed in benzine, by using a small brush freely about the bearings, commutator and brushes. Put oil in all the pivot holes, but not so much that it will run. The motor bearings and the pallets of the anchor should also be oiled.

Inspect carefully to see that the center winding contact is right and that the motor is without any dead points. Dust out the case and put the movement in place. Before putting on the dial try the winding by means of the switch, Fig. 136, to be sure that it is right; also see that the disc on the cannon socket is in the right position to open the latch at the hour, and after the dial and hands are on move the minute hand forward past the hour and then backward gently until it is stopped by the latch. This will prove that the hand is on the square correctly.

On account of the liability of the motor to get out of adjustment and fail to wind, from the shifting of the springs and brushes, under careless adjustment, various attempts have been made to improve this feature of these clocks and the company is now putting out nearly altogether one of the two vibrating motors, shown in Figs. 139 and 140.

In Style C, Fig. 139, the hourly contact for winding is the same as in the clock with the three-magnet motor, as shown in Fig. 138. The magnet spools are twelve ohms and the resistance coil is eighty ohms, placed in parallel, as described in Fig. 130.

The vibrating motor, Fig. 139, is made with a pair of magnets and a vibrating armature. The main spring is wound by the forward and backward motion of the armature, one end of the connecting rod, 8, being attached to a lug of the armature, 2, and the other to the winding lever, 10. This lever has spring ends, to avoid shock and noise. As the winding lever is moved up and down, the pawl, 9, turns the ratch wheel, 11, and a pinion on the ratch wheel arbor turns the spring barrel until the winding is completed.

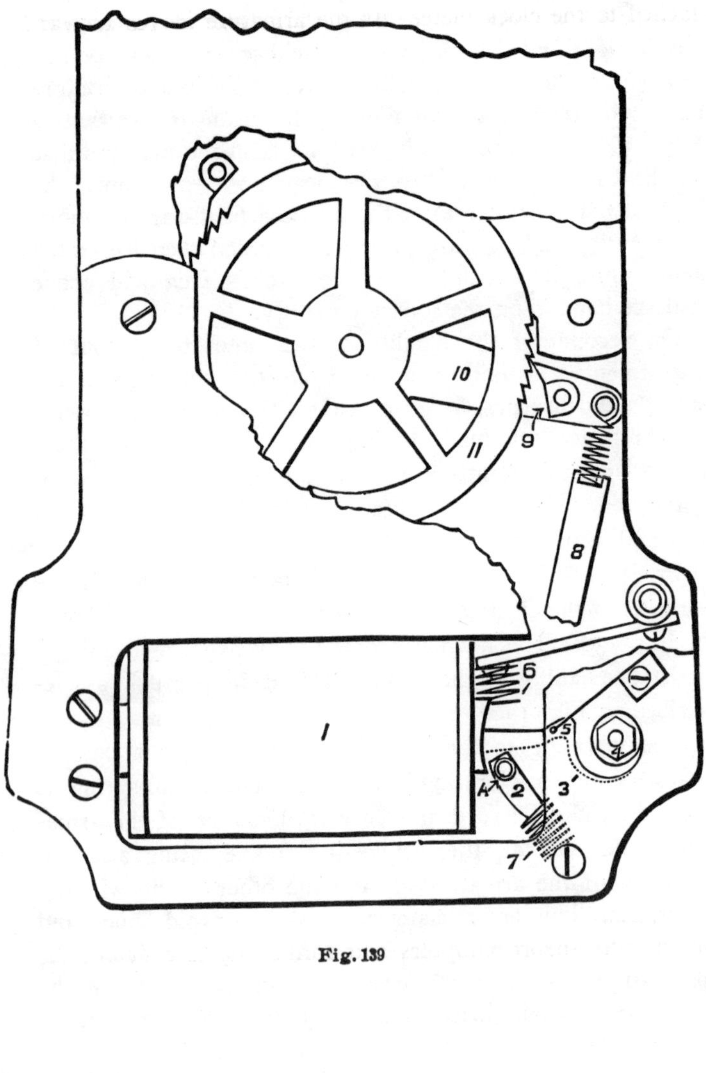

Fig. 139

The contact for operating the motor is made by the brass spiral spring, 3, which is attached to the insulated stud, 4, and the platinum pin, 5, which is carried on a spring attached to the clock plate. As the armature moves forward the break pin, A, in the end of the armature lifts the contact spring, 3, thus breaking the circuit. The acquired momentum carries the armature forward until it strikes the upper banking spring, 6, when it returns rapidly to its original position, banking on spring 7, by which time contact is again made between springs 3 and 5 and the vibration is repeated until the clock is wound one turn of the barrel and the circuit is broken at the center winding contact.

Fig. 140, Style F, is a similar motor so far as the vibrating armature and the winding is concerned, but the winding lever is pivoted directly on the arbor of the winding wheel and operates vertically from an arm and stud on the armature shaft, working in a fork of the winding lever, 8, Fig. 140. It will be seen that the train and the motor winding mechanism are combined in one set of plates. The motor is of the oscillating type and its construction is such that all its parts may be removed without dissembling the clock train.

CONSTRUCTION OF THE MOTOR.—The construction of the motor is very simple, having only one pair of magnets, but *two sets* of make and break contacts, one set of which is placed on the front and the other on the back plate of the movement, thus ensuring a more reliable operation of the motor, and reducing by fifty per cent the possibility of its failing to wind.

The center winding contact also differs from those used in the three-magnet motors and former styles of vibrating motor movements. The center winding contact piece, 13, has no ivory and no platinum. The hourly circuit is not closed by the current passing through this piece, but it acts by bringing the plate contact spring, 16, in metallic connection with the insulated center-winding contact spring, 17, both of which are platinum tipped. It will thus be seen that no accumulation of dirt, oil or gum around the center arbor or the train pivots will have any effect in preventing the current from passing from the motor to the hourly circuit closer.

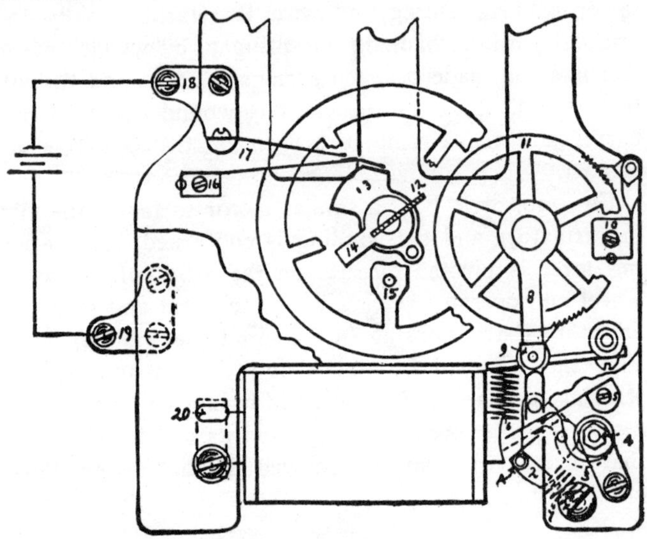

Fig. 140

The operation is as follows: As the train revolves, the pin, 12, securely fastened to the center arbor, in its hourly revolution engages a pin on the center winding contact piece, 13. This piece as it revolves pushes the plate contact spring, 16, upward, bringing it in metallic connection with the center winding contact spring, 17, which is fastened to a stud on an insulated binding post, 18, thereby closing the hourly circuit. The current passes from the binding post, 18, through the battery (or any other source of current supply) to binding post 19, to which is connected one end of the motor magnet wire. The current passes through these magnets to the insulated stud, 4. To this stud the spiral contact spring, 3, is fastened and the current passes from this spring to the plate contact spring, 5, thence through the movement plate to plate contact spring, 16, and from there through spring, 17, back to the battery.

The main spring is wound by the forward and backward motion of the armature, 2. To this armature is connected the winding lever, 8. As the winding lever is oscillated, the pawl, 9, turns the ratchet wheel, 11, and a pinion on the ratchet wheel arbor turns the winding wheel until the pin, 15, connected to it engages the knock-away piece, 14, revolving it until it strikes the pin on the center winding contact piece, 13, and pushes it from under the plate contact spring, thereby breaking the electric circuit and completing the hourly winding.

The proper position of the contact springs is clearly indicated in Fig. 140. The spring, 16, should always assume the position shown thereon. When the center winding contact piece, 13, comes in metallic connection with the plate contact spring, 16, the end of this spring should stand about one-thirty-second of an inch from the edge of the incline. The center winding contact spring, 17, should always clear the plate contact spring one-thirty-second of an inch. When the two springs touch they should be perfectly parallel to each other.

ADJUSTMENTS OF THE ARMATURE.—In styles C and F, when the armature, 2, rests on the banking spring, 7, its front edge should be in line with the edge of the magnet core. The upper banking spring, 6, must be adjusted so that the front edge of the armature will be one-sixteenth of an inch from the corner of the magnet core when it touches the spring.

When the contact spring, 3, rests on the platinum pin, 5, it should point to about the center of the magnet core, with the platinum pin at the middle of the platinum piece on the spring.

To adjust the tension of the spiral contact spring, 3, take hold of the point with a light pair of tweezers and pull it gently forward, letting it drop under the pin. It should take the position shown by the dotted line, the top of the spring being about one-thirty-second of an inch below the platinum pin. If from any cause it has been put out of adjustment it can be corrected by carefully bending under the tweezers, or the nut, 4, may be loosened and the spring removed. It may then be bent in its proper shape and replaced.

The hole in the brass hub to which the spring is fastened has a flat side to it, fitting a flat on the insulated contact stud. If the contact spring is bent to the right position it may be taken off and put back at any time without changing the adjustment, or a defective spring may readily be replaced with a new one. When the armature touches the upper banking spring the spiral contact spring, 3, should clear the platinum pin, 5, about one-sixteenth of an inch. Both contacts on front and back plates in style F are adjusted alike. The circuit break pins "A" on the armature should raise both spiral contact springs at the same instant.

If for any reason the motor magnets have become displaced they may readily be readjusted by loosening the four yoke screws holding them to the movement plates. Hold the armature against the upper banking spring, move the magnets forward in the elongated slot, 20, until the ends of the magnet cores clear the armature by one-sixty-fourth of an inch, then tighten down the four yoke screws. Connect the motor to the battery and see that the armature has a steady vibration and does not touch the magnet core. The adjustment should be such that the armature can swing past the magnet core one-eighth to three-sixteenths of an inch.

Description of Synchronizer.—At predetermined times a current is sent through the synchronizer magnet, D′, Fig. 141, which actuates the armature, E, to which are attached the levers, F and G, moving them down until the points on the lever, G, engage with two projections, 4 and 5, on the minute disc; and lever F engages with the heart-shaped cam or roll on the seconds arbor sleeve, causing both the minute and second hands to point to XII. These magnet spools are wound to twelve ohms, with an eighty-ohm resistance in parallel.

On the latch, L, is a pin, I, arranged to drop under the hook, H, and prevent any action of the synchronizing levers, except at the hour. A pin in the disc on the cannon socket unlocks the latch about two minutes before the hour and closes it again about two minutes after the signal. This is to prevent any accidental "cross" on the synchronizing line from disturbing the hands during the hour.

M is a light spring attached to the synchronizing frame to help start the armature back after the hands are set. The wires from the synchronizing magnet are connected to binding plates at the right-hand side of the clock and from these binding plates the blue wires, Nos. 9 and 10, pass out at the top of the case to the synchronizing line.

If the clock gets out of the synchronizing range it generally indicates very careless regulation. The clock is regulated by the pendulum, as in all others, but there is one peculiarity in that the pendulum regulating nut has a check nut.

If the clock gains time turn the large regulating nut under the pendulum bob slightly to the left.

If the clock loses time turn the nut slightly to the right.

Loosen the small check nut under the regulating nut before turning the regulating nut, and be *sure to tighten* the check nut after moving the regulating nut.

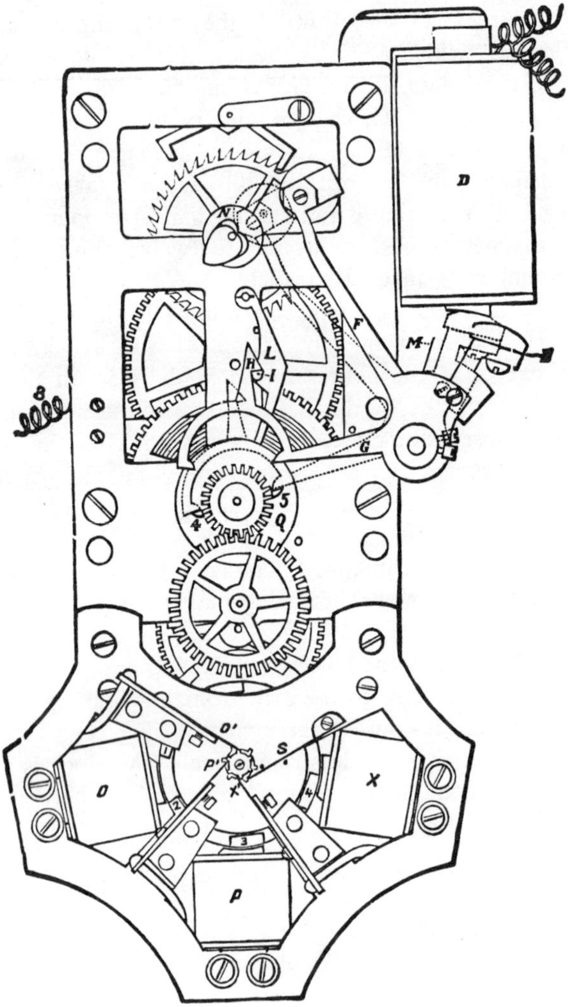

Fig. 141

The friction of the seconds hand is very carefully adjusted at the factory, being weighed by hanging a small standard weight on the point of the hand. If it becomes too light and the hand drives or slips backward, losing time, it can be made stronger by laying it on a piece of wood and rubbing the inner sides of the points with a smooth screw driver, and if too heavy and the clock will not set when the synchronizing magnets are actuated, the points of the spring in the friction may be straightened a little.

If the seconds hand sleeve does not hold on the seconds socket, pinch it a little with pliers. If the seconds hand is loose on the sleeve put on a new one or solder it on the under side.

In style F the synchronizing lever, heart-shaped seconds socket and cams on the cannon sockets are the same as in the old style movements, shown in Fig. 141. The difference is in the synchronizing magnets and the way they operate the synchronizing lever. The magnet has a flat ended core instead of being eccentric like the former ones. The armature is also made of flat iron and is pivoted to a stud fastened to the synchronizing frame. The armature is connected to the synchronizing lever by a connecting rod and pitman screws. A sector has an oblong slot, allowing the armature to be lowered or raised one-sixteenth of an inch. The synchronizing lever is placed on a steel stud fastened to the front plate and held in position by a brass nut. The synchronizing magnets are 12 ohms with 80 ohms resistance and are fastened to a yoke which is screwed to the synchronizing frame by four iron screws. The holes in the synchronizing frame are made oblong, allowing the yoke and magnets to be raised or lowered one-sixteenth of an inch. The spring on top of the armature is used to throw it back quickly and also acts as a diamagnetic, preventing the armature from freezing to the magnets. A screw in the stud is used to screw up against the magnet head, preventing any spring that might take place on the armature stud. Binding posts are screwed to the synchronizing frame and the ends of the magnet coils are fastened thereto with metal clips.

The blue wires in the clock case are coiled and have a metal clip soldered to them. They connect direct by these clips to the binding posts, thus making a firm connection, and are not liable to oxidize. With the various points of adjustment a pair of magnets burned out or otherwise defective may readily be replaced in from five to ten minutes.

When replacing a pair of synchronizing magnets proceed as follows: Remove the old pair and then loosen all four screws in the yoke, pushing it up against the tops of the oblong holes, then tighten down lightly. Fasten the new pair of magnets to the yoke with the inner ends of the coils showing at the outside of the movement. Press the armature upward until the synchronizing lever locks tightly on the cannon socket and the heart-shaped cams, then loosen the magnet yoke screws and press the magnets down on the spring on top of the armature. Then tighten the yoke screws on the front plate and see that the back of the magnets clears the armature by one-hundredth of an inch (the thickness of a watch paper), when the screws in the back of the yoke can be set down firmly. The adjustment screw may then be turned up until it presses lightly against the magnet head. When current is passed through the magnets and held there the armature must clear the magnets without touching. The magnet coils must then be connected to their respective binding posts by slipping the metal clips soldered to them under the rubber bushing, making a metallic connection with the binding plates. Fasten these screws down tight to insure good connections.

THE MASTER CLOCK.—Is a finely finished movement with mercurial pendulum that beats seconds and a Gerry gravity escapement. At the left and near the center of the movement is a device for closing the synchronizing circuit

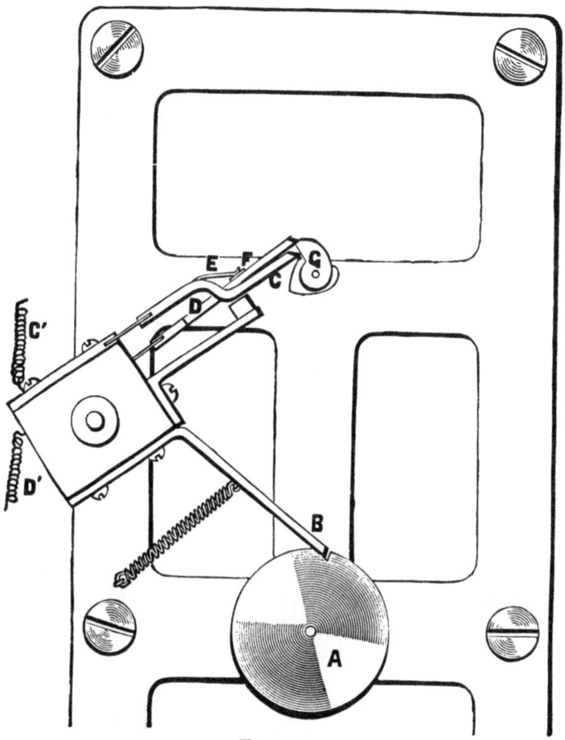

Fig. 142

once each hour. The device consists of a stud on which is an insulator having two insulated spring fingers, C and D, one above the other, as shown in Fig. 142, except at the points where they are cut away to lie side by side on an insulated support. On these fingers, and near the insulator, are two platinum pieces, E and F, so adjusted as to be held apart, except at the time of synchronizing.

A projection, B, from the insulator rests on the edge of a disc on the center arbor. At ten seconds before the hour, a notch in this disc allows the spring to draw the support downward, leaving the points of the fingers, C and D, resting on the raised part of the rubber cam on the escape arbor. The end of the finger, C, is made shorter than that of D, and at the fifty-ninth second, C drops and closes the circuit by E striking F. At the next beat of the pendulum the long finger D drops and opens the circuit again.

The winding is the same as in the regular self-winding clocks, the motor wire and seconds contact being connected to the binding plates at the left, from which brown wires lead up to the battery. Two wires from the synchronizing device are connected to the binding plates at the left, from which blue wires run out to the line.

Before connecting the clock to the line it must be run until it is well regulated, and also to learn if the contacts are working correctly. Regulate at first by the nut at the bottom of the rod until it runs about one second slow in 24 hours (a full turn of the nut will change the rate about one-half minute per day). The manufacturers send with each clock a set of auxiliary pendulum weights, the largest weighing one gram, the next in size five decigrams and the smallest two decigrams; these weights are to make the fine regulations by placing one or more of them on the little table that is fastened about the middle of the pendulum rod. The five decigram weight will make the clock gain about one second per day, and the other weights in proportion. Care must be taken not to disturb the swing of the pendulum, as a change of the arc changes the rate.

To start the clock after it is regulated, stop it, with the second hand on the fiftieth second; move the hands forward to the hour at which the signal comes from the observatory; then press the minute hand back gently until it is stopped by the extension on the hour contact, Fig. 142, and beat the clock up to the hour. This ensures the hour contact being in position to send the synchronizing signal.

A good way to start it with observatory time is with all the hands pointing to the "signal" hour; hold the pendulum to one side and when the signal comes let it go. With a little practice it can be started very nearly correct.

Clocks not lettered in the bottom of the case must be wound before starting the pendulum. To do this press the switch shown in Fig. 136, which is on the left side of the case and under the dial.

Continue the pressure until the winding ceases. Then set the hands and start the pendulum in the usual way. If the bell is not wanted to ring, bend back the hammer.

SECONDARY DIALS.—One of the most deceptive branches of clock work is the secondary dial, or "minute jumper." Ten years ago it was the rule for all manufacturers of electric clocks to put out one or more patterns of secondary dials. Theoretically it was a perfect scheme, as the secondary dial needed no train, could be cheaply installed and could be operated without trouble from a master clock, so that all dials would show exactly the same time. Practically, however, it proved a very deceptive arrangement. The clocks were subject to two classes of error. One was that it was extremely difficult to make any mechanical arrangement in which the hands would not drive too far or slip backward when the mechanism was released to advance the minute hand. The second class of errors arose from faulty contacts at the master clock and variation in either quantity or strength of current. Another and probably the worst feature was that all such classes of apparatus record their own errors and thereby themselves provide the strongest evidence for condemnation of the system. Clocks could be wound once an hour with one-sixtieth of the chance of error of those wound once per minute, and they could be wound hourly and synchronized daily with 1-1440th of the line troubles of a minute system.

The minute jumpers could not be synchronized without costing as much to build and install as an ordinary self-winding clock, with pendulum and time train, and after trying them for about ten years nearly all the companies have substituted self-winding time train clocks with a synchronizing system. They have apparently concluded that, since it seems too much to expect of time apparatus that it will work perfectly under all conditions, the next thing to do is to make the individual units run as close to time as is commercially practicable and then correct the errors of those units cheaply and quickly from a central point.

It is for these reasons that the secondary dial has practically disappeared from service, although it was at one time in extensive use by such companies as the Western Union Telegraph Company, the Postal Telegraph and the large buildings in which extensive clock systems have been installed.

Fig. 143 shows one form of secondary dial which involves a screw and a worm gear on the center arbor, which, it will be seen, is adapted to be turned through one minute intervals without the center arbor ever being released from its mechanism. This worm gear was described in the AMERICAN JEWELER about fifteen years ago, when patented by the Standard Electric Time Company in connection with their motor-driven tower clocks, and modifications of it have been used at various times by other companies.

The worm gear and screw system shown in Fig. 143 has the further advantage that it is suitable for large dials, as the screw may be run in a box of oil for dials above four feet and for tower clocks and outside work. This will readily be seen to be an important advantage in the case of large hands when they are loaded with snow and ice, requiring more power to operate them.

All secondaries operate by means of an electromagnet raising a weight, the weight generally forming the armature; the fall of the weight then operates the hands by gravity.

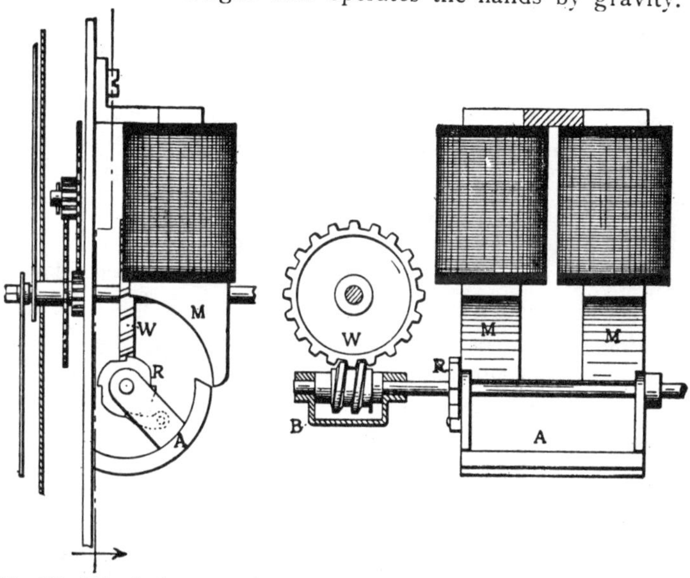

Fig. 143. Minute jumper. A, armature; M, magnets; W, worm gear on center arbor; B, oil box for worm; R, four toothed ratchet.

Direct action of the current in such cases is impracticable, as the speed of starting with an electric current would cause the machine to tear itself to pieces.

This screw gear is the only combination known to us that will prevent the hands from slipping or driving by and reduces the errors of the secondary system to those of one class, namely, imperfections in the contact of the master clock, insufficient quantity or strength of current, or accidental "crosses" and burnings.

The series arrangement of wiring secondaries was formerly greatly favored by all of the manufacturers, but it was found that if anything happened to one clock it stopped the lot of them; and where more than fifty were in series, the necessary voltage became so high that it was impracticable to run the clocks with minute contacts. The modern system, therefore, is to arrange them in multiples, very much after the fashion of incandescent lamps, then if one clock goes wrong the others are not affected. Or if the current is insufficient to operate all, only those which are farthest away would go out of time.

Very much smaller electromagnets will do the work than are generally used for it, and the economy of current in such cases is worth looking after, as with sixty contacts per hour batteries rapidly play out if the current used is at all excessive. Where dry batteries are used on secondaries care should be taken to get those which are designed for gas engine ignition or other heavy work. Wet batteries, with the zincs well amalgamated, will give much better satisfaction as a rule and if the plant is at all large it should be operated from storage cells with an engineer to look after the battery and keep it charged, unless current can be taken from a continuously charged lighting main. This can be readily done in such instances as the specifications call for in the new custom house in New York, namely, one master clock and 160 secondary dials.

ELECTRIC CHIMES.—There have lately come into the market several devices for obtaining chimes which allow the separation of the chimes and the timekeeping apparatus, connection being made by means of electricity. In many respects this is a popular device. It allows, for instance, a full set of powerful tubular chimes, six feet or more in length, to be placed in front of a jewelry store, where they offer a constant advertisement, not only of the store itself, but of the fact that chiming clocks may be obtained there. It also allows of the completion by striking of a street clock which is furnished with a time train and serves at once as timepiece and sign. Many of these have tubular chimes in which the hour bell is six feet in length and the others correspondingly smaller. They have also been made with bells of the usual shape, which are grouped on posts, or hung in

Fig. 144. Chimes of bells in rack.

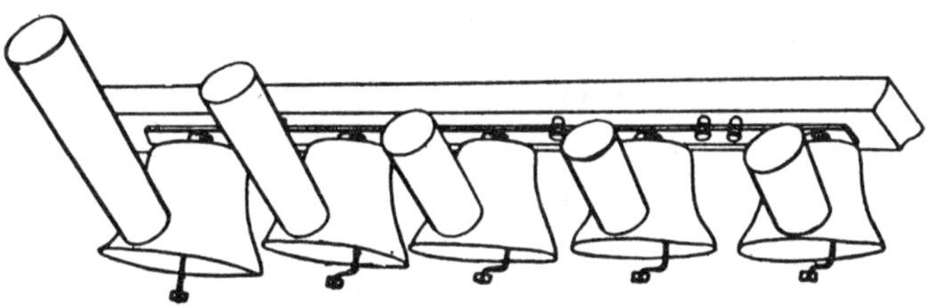

Fig. 145. Chimes of bells with resonators.

racks and operated electrically. It may also be used as a ship's bell outfit by making a few minor changes in the controller.

Fig. 144 shows a peal of bells in which the rack is thirty-six inches long and the height of the largest bell is eight inches, and the total weight thirty pounds. This, as will readily be seen, can be placed above a doorway or any other convenient position for operation; or it may be enclosed in a lattice on the roof, if the building is not over two stories in height. The lattice work will protect the bells from the weather and at the same time let out the sound.

Fig. 145 shows the same apparatus with resonators attached. These are hollow tubes which serve as sounding boards, largely increasing the sound and giving the effect of much larger bells. Fig. 146 shows a tubular chime and the electrical connections from the clock to the controller and to the hammers, which are operated by electro-magnets, so that a heavy leaden hammer strikes a solid blow at the tops of the tubes.

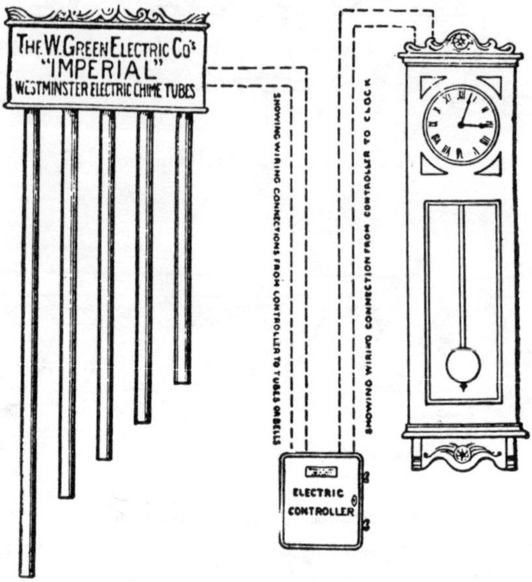

Fig. 146, Tubular electric chimes.

The dials of such clocks contain electrical connections and the minute hand carries a brush at its outer end. The contact is shown in enlarged view in Fig. 147, by which it will be seen that the metal is insulated from the dial by means of hard rubber or other insulating material, so that the brush on the minute hand will drop suddenly and firmly from the insulator to the metallic contact when the minute hand reaches fifteen, thirty, forty-five or sixty minutes. There is a common return wire, either screwed to the frame of the clock, or attached to the dial, which serves to close the various circuits and to give four strokes of the chimes at the quarter, eight at the half, twelve at the three-quarter, and sixteen at the hour, followed by the hour strike. The friction on the center arbor is of course adjusted so as to carry the minute hand without slipping at the contacts.

By this means a full chime clock may be had at much less cost than if the whole apparatus had to be self-contained and the facilities of separation between the chimes and the time-keeping apparatus, as hinted above, gives many advantages.

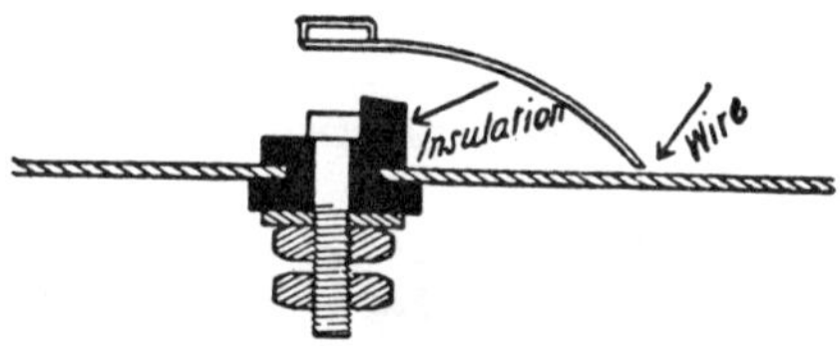

Fig. 147. Enlarged view of connections on dial.

For instance, the same clock and controller may operate tubes inside the room and bells outside, or vice versa. These are operated by wet or dry batteries purchased at local electrical supply houses, and the wiring is done with plain covered bell wire, or they may be operated by current from a lighting circuit, suitably reduced, if the current is constantly on the mains. As a full chime with sixteen notes at the hour strikes more than a thousand times a day, considerable care should be taken to obtain only the best batteries where these are used, as after the public gets used to the chimes the dealer will be greatly annoyed by the number of people asking for them if they are stopped temporarily.

There has lately developed a tendency to avoid the set tunes, such as the Westminster and the Whittington chimes, and to sound the notes as complete full notes, such as the first, third and fifth of the octave for the first, second and third quarters, followed by the hour strike. This allows them to be struck in any order and for a smaller chime reduces the cost considerably. The tubes used are rolled of bell metal and vary in pitch with the manufacture, so that the only way to obtain satisfactory tones is to cut your tubes a little long and then tune them by cutting off afterwards,

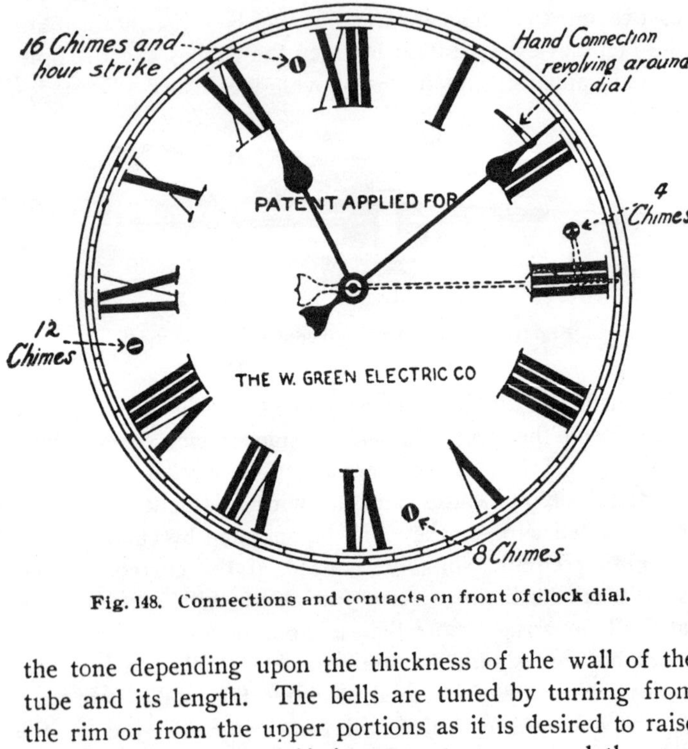

Fig. 148. Connections and contacts on front of clock dial.

the tone depending upon the thickness of the wall of the tube and its length. The bells are tuned by turning from the rim or from the upper portions as it is desired to raise or lower the tone, and if the resonators are used they are tuned in unison with the bells.

Of the ordinary bells, Fig. 144, the dimensions run: First, height four inches, diameter 5½; second, height four inches, diameter 5¼ inches; third, height 4½ inches, diameter 5⅝ inches; fourth, height 4½ inches, diameter 5⅝ inches; fifth, height 4⅝ inches, diameter 6½ inches. For the tubes the approximate length is six feet for the longest tube and the total weight of the chimes is 43 pounds. For the controller the size is nine by eleven by six inches,

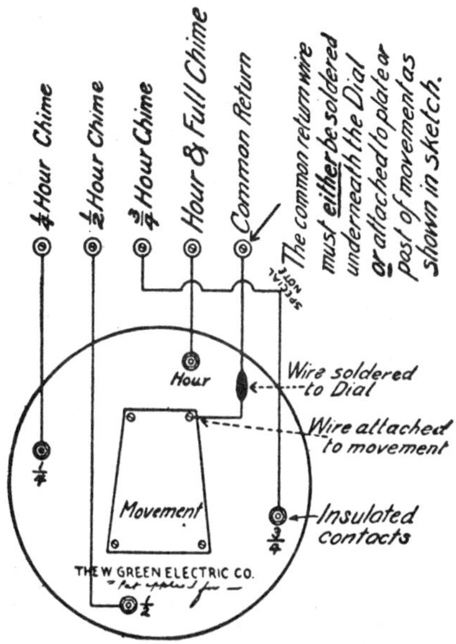

Fig. 149. Connections and wiring on back of clock dial.

with a weight of ten pounds. The hour strike may be had separately from the chimes if desired.

This makes an easily divisible system and one that is becoming very popular with retail jewelers and to some extent with their customers.

CHAPTER XXII.

THE CONSTRUCTION AND REPAIR OF DIALS.

Probably no portion of the clock is more important than the dial and it is apparently for this reason that we find so little variation in the marking. The public refuses to accept anything in the way of ornamentation which interferes with legibility and about all that may be attempted is a little flat ornament in light colors which will not obscure the sight of the hands, as it is in reality the angle made by the two hands which is read instead of the figures. In proof of this may be cited the many advertising dials in which one letter takes the place of each character upon the dial and of the tower clocks in which the hours are indicated merely by blackened characters, being nothing less than an oblong blotch on the dial. Thousands of people will pass such a dial without ever noticing that the regular characters do not appear. Various attempts have been made to change the colors and the sizes and shapes of the characters but comparatively few are successful. A black dial with gold characters and hands is generally accepted, or a cream dial with black hands, but any further experiments are dangerous except in the cases of tower clocks, which may have gold hands on any light colored dial, or a glass dial. In all such cases legibility is the main factor sought and the bright metal is far plainer for hands and chapters than anything that may be substituted for them.

In tower clocks the rule is to have one foot of diameter of the dial for every ten feet of height. Thus a clock situated one hundred feet above the ground level should have a ten foot dial. On very large dials this rule is deviated from a little, but not much. All dials, except those of tower clocks, should be fastened to the movement, rather than to the case. This is particularly true where a seconds hand, with the small opening for the seconds hand sleeve, makes any twisting or warping of the case and consequent shifting of the dial liable to rub the dial against the sleeve at the seconds hand and thus interfere with the timekeeping.

The writer has in mind a case in which a large number of fine clocks were installed in a new brick and stone building. They were finely finished and no sooner had they been hung on the damp walls than the cases commenced to swell and twist. It was necessary three times to send a man to move the dials which had been attached to these clocks. As there were about thirty clocks it will be seen that this was expensive. After the walls had dried out the cases began to go back to the positions in which they were originally, as the mcisture evaporated from the cases, and the dials had consequently to be moved through another series. All told it took something like a week's work for one man to shift these dials half a dozen times during the first nine months of their installation. If these dials had been fastened on pillars on the movements, the shrinking and swelling of the cases would not have affected them.

It is for this reason that dials are invariably fastened on the movements of all high class clocks.

The characters on clock dials are still very largely Roman, the numerals being known as chapters. Attempts have been recently made to substitute Arabic figures and in such cases the Arabic figures remain upright throughout the series, while the chapters invariably point the foot of the Roman numeral toward the center of the dial. This makes the Roman numerals from IIII to VIII upside down, while in the Arabic numerals this inversion does not cccur.

The proportions generally sanctioned by usage have been found, after measuring clock dials, all the way from two to eighteen inches, and may be given in the following terms: With a radius of 26 mm. the minute circle is 1½ mm. The margin between minute circles and chapters is 1 mm. The chapters are 8½ mm. The width of the thick stems of the letters are ¾ mm. The width of an X is 4 mm. and the slanting of X's and V's is twenty degrees from a radius of the dial. The letters should be proportioned as follows: The breadth of an I and a space should equal one-half the breadth of an X, that is, if the X is one-half inch broad, the I will be three-sixteenths inch broad and the space between letters one-sixteenth inch, thus making the I plus one space equal to one-quarter inch or half the breadth of an X. The V's should be the same breadth as the X's. After the letters have been laid off in pencil, outline them with a ruling pen and fill in with a small camel's hair brush, using gloss black paint thinned to the proper consistency to work well in the ruling pen. Using the ruling pen to outline the letters gives sharp straight edges, which would be impossible with a brush in the hands of an inexperienced person.

For tower clocks the chapters and minutes together will take up one-third of the radius of the dial; the figures two-thirds of this, or two-ninths of the radius, and the minutes two-thirds of the remaining one-ninth of the radius, with every fifth minute more strongly marked than the rest.

We often hear stories concerning the IIII in place of IV. The story usually told is that Louis XIV of France was inspecting a clock made for him by a celebrated watchmaker of that day and remarked that the IV was an error. It should be IIII. There was no disputing the King and so the watchmaker took away the dial and had the IIII engraved in place of IV, and that it has thus remained in defiance of all tradition.

Mr. A. L. Gordon, of the Seth Thomas Clock Co., has the following to say concerning this story and thus furnishes the only plausible explanation we have ever seen for the continuance of this manifest error in the Roman numeral of the dial:

"That the attempt has been made to use the IV for the fourth hour on clock dials, any one making a study of them may observe. The dials on the Big Ben clock in the tower of the Parliament buildings, London, which may be said to be the most celebrated clock in the world, have the IV mark, and the dial on the Herald building in New York City also has it.

"That the IIII mark has come to stay all must admit, and if so there must be a good and sufficient reason. Art writers tell us that pictures must have a balance in the placing and prominence of the several subjects. Most conventional forms are equally balanced about a center line or a central point. Of the latter class the well known trefoil is a common example.

"A clock or watch dial with Roman numerals has three points where the numerals are heavier, at the IIII, VIII and XII. Fortunately these heavier numerals come at points equally spaced about the center of the dial and about a center line perpendicular to the dial. Of these three heavy numerals the lighter of them comes at the top and it is especially necessary that the other two, which are placed at opposite points in relation to the center line, should be balanced as nearly as possible. As the VIII is the heavier and cannot be changed, the balancing figure must be made to correspond as nearly as possible, and if marked as IV, it will not do so nearly as effectively as if the usual IIII is used."

It is comparatively an easy matter to make a metal dial either of zinc, copper or brass, by laying out the dial as indicated above with Roman chapters and numerals, after first varnishing the metal with asphaltum. This may be drawn upon with needle points which cut through the asphaltum and make a firmly defined line on the metal. It is best to lay out your dial in lead pencil and then take a metal straight edge and a needle point and trace through on the pencil marks. Mistakes may be painted out with asphaltum, so that the job becomes easy. After this has been done a comparatively dull graver may be used to cut or scrape away the asphaltum where the metal is to be etched and then the plate may be laid in a tray, a solution of chloride of iron poured on and rocking the tray will rapidly eat away the metal, forming sunken lines wherever the copper or brass is not protected by the asphaltum. This furnishes a rough surface on the etched portions, which enables the filling to stick much better than if it were smooth. In tracing the circles a pair of heavy, stiff, carpenters' compasses will serve where the watchmaker has not a lathe large enough to swing the dial. In all such cases it is best to start with a prick-punched center, tracing the minute circles and the serifs of the chapters with the compasses and then do your further division and marking by lead pencil, followed with the needle and then by the acids. It should be done before the holes are bored for the minute and seconds centers, as you then have an exact center to mark from and can go back to it many times.

This will be necessary in dividing the minute or seconds circle by hand (without an index on the lathe), as one of the tests of true division consists in having all marks lined up with a straight edge placed across the center. Thus IX and III should be in line with the center; VI and XII; X and IIII; I and VII, etc. It will readily be seen that for such purposes of reference the center should not be punched too large.

If it is desirable to ornament the dial, the desired ornament may be drawn on in the plain surface through the asphaltum and etched at the same time as the chapters and degrees. Or chapters and ornament may be drawn, pierced with a saw, engraved, filed up and backed up with a plain plate of another color. Gold ornament and silver background looks well.

Practically all the clocks having seconds hands carry that hand in such a position as to partially obscure the XII, with the exception of watchmakers' regulators, and these, if they have separate hour, minute and seconds circles, are made large enough to occupy the space between the center and the minute circle, placing the hour circle between the center and the thirtieth minute; the seconds between the center and the sixtieth minute. The reason for this is that in the watchmakers' regulators the hours are almost a matter of indifference; minutes are seldom referred to; the real comparison in watch regulation comes on the seconds hand. For this reason the seconds hand is made as large as possible and the chapters being placed on the hour circle by themselves, the seconds circle may occupy almost the entire distance between the center of the dial and the minute circle. They are placed one above the other because in regulators the time train is nearly always a straight-line train, which brings the seconds arbor vertically over the center arbor, and consequently the centers of the dials must be placed on a vertical line.

When the engraving has been properly done on a flat dial it is desirable to fill it with black in order to make it legible. There are several methods by which this may be done. The most durable is to make a black enamel and if it is a valuable clock the movement is generally worth a fine dial. The following formula will furnish a good black enamel:

Siliceous sand	12 parts
Calcined borax	20 parts
Glass of antimony	4 parts
Saltpetre	1 part
Chalk	2 parts
Peroxide of Manganese	5½ parts
Fine Saxony Cobalt	2 parts

The enamel is ground into coarse particles like sand, and the incised lines filled with it, after which the brass or copper plate is heated red hot to fuse the enamel. Two or three firings may be necessary to completely fill the lines; after filling they are stoned off level with the surface of the dial. Jeweler's enamel may be purchased of material dealers and used for the dials.

Black asphaltum mixed with a little wax or pitch, or even watchmakers' cement, used to fasten staffs and pinions into a lathe for turning, is also used on these dials and with a sufficient proportion of wax or pitch it prevents shrinking and forms a very satisfactory dial with the single exception that it cannot be cleaned with benzine or hot potash, which will dissolve the enamel. Shoemakers' heel ball is also used for repair jobs. In order to make either of these stick, the brass or copper plate is heated up so as to "hiss" as will a laundry flat iron when touched with a wetted finger, and a cement stick is rubbed over the letters to fill them; the excess of filling can be scraped off with an ivory scraper when at the right temperature—a little below the boiling point of water. Such filled letters can be lacquered over by going very quickly over the work so as not to dissolve the shellac in the cement.

Another way is to fill the letters with black lacquer. For quick repairs this is probably as good as any. Many of the old grandfather clocks have been filled in with a putty made with copal varnish and some black pigment. All putties shrink in drying and consequently crack and finally fall out. The wax and pitch are not subject to these disadvantages. If the plates are to be polished, polishing should precede the filling in of the letters, else the work may have to be done all over again. Black sealing wax and alcohol are also used, applied as a paint with a fine brush.

If the dial is to be silvered or gilt the blacking should be done first, and if to be electroplated the blacking should be what is known as the "platers' resist," which is composed chiefly of asphaltum and pitch dissolved in turpentine. It is also called "stopping-off" varnish, and has large use in the plating establishments to prevent deposition of metal where it is not desired.

The repairer who gets many grandfather clocks will often find that it is necessary to repaint the dial, generally because of a too vigorous scrubbing, or because of cracks or scaling, which the owner may dislike. It is always best, however, to be cautious in such matters, as many people value such a clock chiefly on account of its visible evidences of age and such cracks form generally a large proportion of such evidence. Therefore it is best never to touch an antique dial unless the owner desires it.

Such dials are usually sheet-iron, and tolerably smooth, so the metal will need but a few coats of paint to prepare it. For ground coats, take good, ordinary white-lead or zinc white, ground with oil, and if it has much oil mixed with it pour it off and add spirits of turpentine and Japan dryer—a teaspoonful of dryer for every half pint of paint.. The test for the paint having the right amount of oil left in it is, it should dry without any gloss. Rub every coat you apply with fine sand-paper, after it is perfectly dry, before applying the next coat of paint. For the final coat, lay the dial flat and go over it with French zinc-white. This coat dries very slow, and for a person not used to such work, is hard to manage. The next best (and for ordinary clock or watch making *the best*) for the last pure white coat is to take a double tube of Windsor & Newton's Kremnitz white, thinned with a little turpentine. Such tubes as artists use are the kind. Apply this last white coat with a flat, camel's hair brush. The tube-white should have turpentine enough added to cause it to flow freely, and sink flat and smooth after the brush. The letters or figures should be painted with ivory-black, which is also a tube color. This black is mixed with a little Japan, rubbing-varnish and turpentine, and the lettering is done with a small, sign writer's pencil. Any flowers or ornaments are painted on at the same time; and after they are dry the dial should be varnished with

Mastic or Damar varnish or white shellac. All kinds of coach (Copal) varnish are too yellow.

Painted dials on zinc will blister and crack off if subjected to extremes of heat and cold, unless they are painted with zinc white instead of lead for all white coats. The reason is the great difference in expansion between lead paint and metallic zinc. This case is similar to that of using an iron oxide to paint iron work of bridges, ships, etc., where other oxides will chip and scale off.

The metal dials on these old clocks were silvered by hand. When you get such a dial, discolored and tarnished, it can be cleaned in cyanide and resilvered, without sending it to an electroplater, by the following formula:

Dissolve a stick of nitrate of silver in half a pint of rain water; add two or three tablespoonfuls of common salt, which will at once precipitate the silver in the form of a thick, white curd, called chloride of silver. Let the chloride settle until the liquid is clear; pour off the water, taking care not to lose any chloride; add more water, thoroughly stir and again pour off, repeating till no trace of salt or acid can be perceived by the taste. After draining off the water add to the chloride about two heaped tablespoonfuls each of salt and cream of tartar, and mix thoroughly into a paste, which, when not in use, must not be exposed to the light. To silver a surface of engraved brass, wash the surface clean with a stiff brush and soap. Heat it enough to melt black sealing wax, which rub on with a stick of wax until the engraving is entirely filled, care being taken not to burn the wax. With a piece of flat pumice-stone, and some pulverized pumice-stone and plenty of water, grind off the wax until the brass is exposed in every part, the stoning being constantly in one direction. Finish by laying an even and straight grain across the brass with blue or water of Ayr stone. Take a small quantity of pulverized pumice-stone on the hand, and slightly rub in the same direction, which tends to make an even grain; the hands must be entirely free from soap or grease. Rinse the brass thoroughly, and before it dries, lay it on a clean board, and gently rub the surface with fine salt, using a small wad of clean muslin. When the surface is thoroughly covered with salt, put upon the wad of cloth, done up with a smooth surface, a sufficient quantity of the paste, say to a dial three inches in diameter a piece of the size of a marble, which rub evenly and quickly over the entire surface. The brass will assume a greyish, streaked appearance; add quickly to the cloth cream of tartar moistened with water into a thin paste; continue rubbing until all is evenly whitened. Rinse quickly under a copious stream of water; and in order to dry it rapidly, dip into water as hot as can be borne by the hands, and when heated, holding the brass by the edges, shake off as much of the water as possible, and remove any remaining drops with clean, dry cloth. The brass should then be heated gently over an alcohol lamp, until the wax glistens without melting, when it may be covered with a thin coat of spirit varnish, laid on with a broad camel's hair brush. The varnish or lacquer must be quite light-colored—diluted to a pale straw color.

It is now possible to buy silver plating solutions which can be used without battery and they will produce the same effect as the formula just given. If they happen to be in stock for the repairing of jewelry they may be used in cleaning the dials, but as this is liable to fall into the hands of many who are far from such conveniences, we furnish the original recipe, which can be executed anywhere the materials can be obtained.

If the dial is of brass, very good effects have been produced by stopping off portions of the dial in an ornamental pattern before silvering, and then lacquering after removing the resist. But for a plain black and brass dial a dip of strong sulphuric acid two parts, red fuming nitrous acid one part, and water one part, mixed in the open air and dipped or flowed over the dial, forms what is known as the platers' bright dip. After dipping the article should at once be rinsed in hot water and dried, and lacquered at once with a lacquer of light gold color. This makes a very neat and durable finish.

The satin effect may be obtained on a dial by prolonging the acid dip and otherwise proceeding as before. Many of these dials were of zinc and all that applies to brass or copper may be also executed in zinc, but in plating it will be found necessary to plate two or three times, as the single coating will apparently disappear into the zinc unless it is given a heavy deposit of copper in a plating bath. Where it is desired to obtain a bright gold color, the gold plating solutions now sold for the coloring of jewelry may also be used on either of these metals. For the reasons given above, however, they are not very successful on a zinc base.

Many of the cheap clocks have paper dials glued on a zinc plate and when the dial is soiled the repairer cleans them up by pasting another dial on top of the original. These dials are made on what is known as lithographic label paper; that is paper which is waterproof on one side, so that it will not shrink or swell when dampened. In addition to the lithograph coating they are generally given a varnish of celluloid by the clock manufacturers, thus making them practically waterproof. They are very cheap and the repairer will find that he will obtain in prestige from such new dials far more than they cost.

Tarnished metal dials are best cleaned by a dip of cyanide of potassium, of about the same strength as that used for cleaning silver. If the tarnished parts have been gilded, however, the cyanide should be excessively weak. Mining men use a cyanide solution for the recovery of gold, which is only two-tenths of one per cent cyanide, and this will collect all the gold from ore that runs from $10 to $15 to the ton, the pulp in such cases being left in the solution from seventy to ninety hours. The ordinary cyanide dip for the jeweler is one ounce to thirty-two of water, while the miner's solution is two-tenths of an ounce to one hundred ounces of water. You can see that with the strong cyanide solution the gilt surface will all be taken off unless very rapid dipping is strictly followed by thorough washing.

A novelty which keeps periodically coming to the front, say about once every ten years, is the luminous dial. This is done by painting the dial with phosphorus or a phosphorescent powder. Then when it is placed in the light it will absorb light and give it off in the dark until the evaporation of the phosphorus.

The composition and manufacture of this phosphorescent powder is effected in the following manner: Take 100 parts by weight of carbonate of lime and phosphate of lime, produced by calcination of sea-shells, especially those of the tridacna and cuttlefish bone, and 100 parts by weight of lime, rendered chemically pure by calcination. These ingredients are well mixed together, after which 25 parts of calcinated sea salt are added thereto, sulphur being afterward incorporated therewith to the extent of from 25 to 50 per cent of the entire mass, and a coloring matter is applied to the composition, which coloring matter consists of from 3 to 7 per cent of the entire mass of a powder composed of a mono-sulphide of calcium, barium, strontium, magnesium or other substance which has the property of becoming luminous in the dark, after having been impregnated with light. After these ingredients are well mixed, the composition is ready for use. Its application to clock dials is made either by incorporating suitable varnish therewith, such as copal, and applying the mixture with a brush to the surface of the dial, or by the production of a dial which has a self-luminous property, imparted to it during its manufacture. This is effected in the following manner: From 5 to 20 per cent of the composition obtained and formed as above described, is incorporated with the glass while it is in a fused state, after which the glass so prepared is molded or blown into the shape or article required. Another process consists of sprinkling a quantity of the composition over the glass article while hot, and in a semi-plastic state, by either of which processes a self-luminous property will be imparted to the article so treated.

Where enamel dials are chipped the cracks may be hidden by first pressing the cracks very slightly open and washing out. Then work in a colorless cement to fill the crack, allow to dry and stone down. Where holes have been left by the chipping, melt equal parts of scraped pure white wax and zinc white and let it cool. Warm the dial slightly and press the cold wax into the defective places and scrape off with a sharp knife and it will leave a white and lustrous surface. If too hard add wax; if too soft add some zinc white.

VARNISH FOR DIALS, ETC.—A handsome varnish for the dials of clocks, watches, etc., may be prepared by dissolving bleached shellac in the purest and best alcohol. It offers the same resistance to atmospheric influence that common shellac does. In selecting bleached shellac for this purpose be careful to get that which will dissolve in alcohol, as some of it being bleached with strong alkalies, is thereby rendered insoluble in alcohol. The shellac when dissolved should be of a clear light amber color in the bottle and this will be invisible on white paper when dry.

Colorless celluloid lacquer, known to jewelers as "silver lacquer" on account of its being used to prevent tarnish on finished hollow ware, also makes a good varnish to apply to dials, either metallic or painted. It is best to have it thin, flow it on the dial and then level the dial to dry.

Success in the repairing of a broken enameled clock dial will greatly depend upon the practical skill of the operator as well as of a knowledge of the process. If it is only desired to repair a chipped place on a dial, a fusible enamel of the right tint should be procured from a dealer in watchmakers' materials, which, with ordinary care, may be fused on the chipped place on the dial so as to give it a workmanlike appearance when finished off. The place to receive the enamel should be well cleaned, and the moist enamel spread over the place in a thin, even layer; and, after allowing it to dry, the dial may be held over a spirit lamp until the new enamel begins to fuse, when it may be smoothed down with a knife. The dial, after this operation, is left to cool, when any excess of enamel may be removed by means of a corundum file, and subsequently polished with putty powder (oxide of tin). The ingredients of enamel, after being fused into a mass, are allowed to cool, then crushed to powder and well washed to get rid of inpurities, and the resulting fine powder forms the raw material for enameling. It is applied to the object to be enameled in a plastic condition, and is reduced to enamel by the aid of heat, being first thoroughly dried by gentle heat, and then fused by a stronger one. The following is a good white enamel for dials:

Silver sand, 3 ounces; red lead, 3¾ ounces; oxide of tin, 2½ ounces; saltpeter, ½ ounce; borax, 2 ounces, flint glass, 1 ounce; manganese peroxide, 2 grains. The basis of nearly all enamels is an easily fusible colorless glass, to which the required opacity and tints are given by the addition of various metallic oxides, and these, on being fused together, form the different kinds of vitreous substances used by enamel workers as the raw material in the art of enameling.

The hands of timekeepers are worthy of more attention than is frequently bestowed upon them by watch and clockmakers. Their shape and general arrangement, and the neatness of their execution is often taken by the general public as an index to the character of the entire mechanism that moves them; and some are apt to suppose that when care is not bestowed on the parts of the time-piece which are most seen, much care cannot be expected to have been exercised on the parts of the watch or clock which are invisible to the general view. Although we are not prepared to fully endorse the opinion that when the hands of timepieces are imperfect in their execution, or in their general arrangement, all the mechanism must of necessity be imperfect also; still we think that in many instances there is room for improving the hands of timepieces, and we desire to direct more attention to this subject by the workmen.

In the general arrangement of the hands of watches and clocks, distinctness of observation should be the great point aimed at, and everything that has a tendency to lead to confusion should be carefully avoided. Clocks that have a number of hands radiating from one center, and moving round one circle—as for instance, center seconds, days of the month, equation of time, alarms and hands for other purposes—may show a good deal of mechanical skill on the part of the designer and maker of the timepiece; but so many hands moving together around one circle, although they may be of different colors, causes confusion, and requires considerable effort to make out what the different hands point to in a dim light, and this confusion is frequently increased by the necessity for a counterpoise being attached to some of the hands. As a rule timekeepers should be so arranged that never more than the hour and minute hand should move from one center on the dial. There may be special occasions when it is necessary or convenient to have center seconds to large dials; but these occasions are rare, and we are talking about the hands of timekeepers in every-day use for the ordinary purposes of life, and also for scientific uses. In astronomical clocks and watchmakers' regulators we find the hour, minute and second hands moving on separate circles on the same dial; and the chief reason for this arrangement is to prevent mistakes in reading the time. In chronometers, especially those measuring sidereal time, the hour hand is frequently suppressed, and the hours are indicated by a star wheel, or ring, with figures engraved on it, that show through a hole in the dial.

Hour and minute hands should be shaped so that the one can be easily distinguished from the other without any effort on the part of the observer. Probably a straight minute hand, a little swelled near the point, and a spade hour hand, are the shapes best adapted for this purpose, especially if the hands have to be looked at from a distance. There are occasions, however, when a spade hand cannot be used with propriety. In small watches and clocks having ornamental cases, hands of other designs are desirable, but whatever be the pattern used, or whatever color the hands may be made, it should ever be remembered that while a design in harmony with the case is perfectly admissible, the sole use of hands is to mark the time distinctly and readily.

The difference in the length of the hour and minute hands is also an important point in rendering the one easily distinguished from the other. The extreme point of the hour hand should extend so as to just cover the edge of the inside end of the numerals and the extreme point of the minute hand should cover about two-thirds of the length of the minute divisions. Hands made of this length will be found to mark the hours and minutes with great plainness, and the rule will be found to work well in dials of all sizes. As a general rule, the extreme points of the hands should be narrow. The point of the hour hand should never be broader than the thickest stroke of any of the numerals, and the extreme point of the minute hand never broader than the breadth of the minute lines; and in small work it is well to file the ends of the hands to a fine point. The ends of minute hands should in every instance be bent into a short, graceful curve pointing toward the dial, and as close to it as will just allow the point of the hand to be free. The minute hands of marine chronometers are invariably bent in this manner, and the hands of these instruments are usually models of neatness and distinctness.

Balancing hands by means of a counterpoise is a subject which requires some attention in order to effect the perfect poise of the hand without detracting anything from its distinctness. In watch work, and even in ordinary clock work, it seldom happens that any of the hands except the seconds require to be balanced, and then there is only one hand moving round the same circle, as is the case with seconds hands in general. We have become so accustomed to looking at seconds hands with projecting tails that we are apt to regard the appearance of the hands to be incomplete without the usual tail; but we must remember that the primary object in view in having a tail to a seconds hand is to counterpoise it, not to improve the looks of the hand itself. Poising becomes an actual necessity for a hand placed on so sensitive a part as the fourth wheel of a watch, or on the scape wheel of a fine clock. When only one hand moves in the same circle, like a seconds hand, the counterpoise may be effected by means of a projecting tail without in any way detracting from a distant reading of the hands, providing the tail is not made too long, and it is made of such a pattern that the one end can easily be distinguished from the other. In minute and hour hands, however, it is different. These two hands move round the same circle, and a counterpoise on the minute hand is liable at a distance to be mistaken for the hour hand.

The minute hands of large timepieces frequently require to be balanced, especially if the dial be large in comparison to the size of the movement; and in very large or tower clocks, whatever may be the size of the movement, it becomes an absolute necessity to balance the hands. In our opinion, tails should never be made on minute hands, when they can be avoided, and in cases where tails cannot be dispensed with, they should invariably be colored the same as the ground of the dial. In almost every instance, however, minute hands may be balanced in the inside, as is usual with tower clocks. A great many clocks used for railway and similar purposes in Europe have their minute hands balanced in this manner, and the plan works admirably; for in

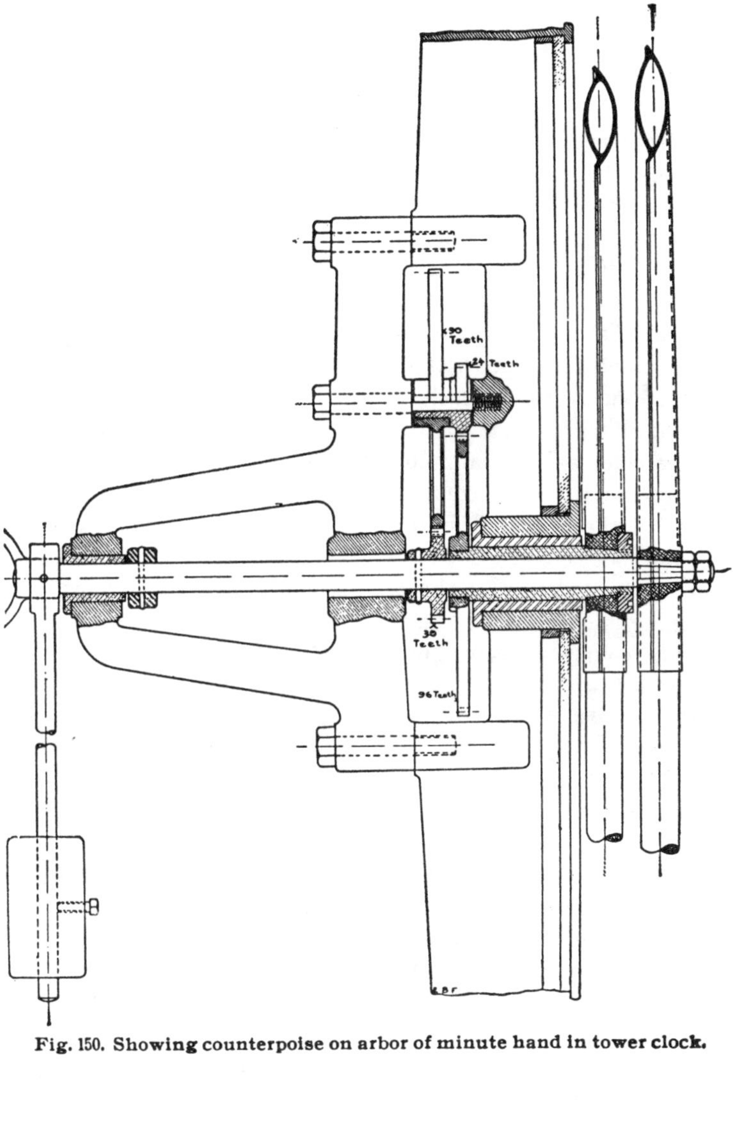

Fig. 150. Showing counterpoise on arbor of minute hand in tower clock.

addition to rendering the hands more distinct, the clocks require less power to keep them going than when the hands are balanced from the outside.

Tower clock hands are generally made of copper, elliptical in section, being made up of two circular segments brazed together at the edges, with internal diaphragms to stiffen them. The minute hand is straight and perfectly plain, with a blunt point. At the center of the dial the width of the minute hand is one-thirteenth of its length, tapering to about half as much at the point.

The hour hand is about the same width, ending just short of the dial figure and terminating in a palm or ornament. The external counterpoises are one-third the length of the minute hand, and of such a shape that they will not be confounded with either of the hands; a cylinder, painted the same color as the dial, and loaded with lead, makes a good counterpoise. This counterpoise may be partly on the inside of the dial if it is desired to keep it invisible, but it should not be omitted, as it saves a good deal of power, prevents the twisting of the arbors, and also assists in overcoming the action of the wind on the hands. Two-thirds of the counterpoise weight may be inside, as shown in Fig. 150.

To Blue a Clock Hand or a Spring.—To blue a piece of steel that is of some length, a clock hand for example, clockmakers place it either on ignited charcoal, with a hole in the center for the socket, and whitened over its surface, as this indicates a degree of heat that is approximately uniform, or on a curved bluing tray perforated with holes large enough to admit the socket. The center will become violet or blue sooner than the rest, and as soon as it assumes the requisite tint, the hand must be removed, holding it with tweezers by the socket, or by the aid of a large sized arbor passed through it; the lower side of the hand is then placed on the edge of the charcoal or bluing tray, and removed by gradually sliding it off toward the point, more or less slowly, according to the progress made with the coloring; with a little practice, the workman will soon be enabled to secure a uniform blue throughout the length and even, if necessary, to retouch parts that have not assumed a sufficiently deep tint.

Instead of a bluing tray, a small mass of iron, with a slightly rounded surface and heated to a suitable temperature, can be employed; but the color must not form too rapidly, and this is liable to occur if the temperature of the mass is excessive. Nor should this temperature be unevenly distributed.

A spring, after being whitened, can be blued in the same way. Having fixed one end, it is stretched by a weight attached to the other end, and the hot iron is then passed along it at such a speed that a uniform color is secured. Of course, the hot iron might be fixed and the spring passed over it. A lamp may be used, but its employment involves more attention and dexterity.

CHAPTER XXIII.

CLOCK CASING AND CASE REPAIRS.

PRECISION CLOCK CASES.—The casing of a precision clock is only secondary in importance to the compensation of its pendulum. The best construction of an efficient case can be ascertained only by most careful study of the conditions under which the clock is expected to be a standard timekeeper, and often the entire high accuracy sought by refined construction is sacrificed by an inefficient case and mounting.

The objects of casing a precision clock are as follows

a. To protect the mechanism from the effects of dust and dirt.

b. To avoid changes of temperature and barometric pressure.

c. To provide an enclosed space in which the gas medium in which the pendulum swings shall have any chemical constitution, of any hygroscopic condition.

d. There must be provided ready means of seeing and changing the condition of the pendulum, electric apparatus, movement, etc., without disturbing the case except locally.

Now if we hold the above considerations in view we can readily see that cast iron, wood and glass, with joints of wash leather (which is kept soft by a wax cement which does not become rancid with age), are the preferable materials.

The advantages of using cast iron for the pillar or body of the case are that it can be cast in such a shape as to require very little finishing afterwards, and that only such as planing parallel surfaces in iron planing machines. It makes a stiff column for mounting the pendulum when it rests upon a masonry foundation from below. Plates of glass can be clamped against the planed surfaces of iron piers (by putting waxed wash leather between the glass and the iron) so as to make air-tight joints without difficulty.

The mass of iron symmetrically surrounding the steel pendulum is the safest protection the clock can have against casual magnetic disturbances. In the language of electricians it "shields" the pendulum.

Suppose, then, we adopt as the first type of precision clock case which our present knowledge suggests, that of an iron cylinder or rectangular box resting on a masonry pier, and which has a table top to which the massive pendulum bracket is firmly bolted. This type admits of the weights being dropped in small cylinders outside of the cast iron cylinder or box. These weight cylinders, of course, end in the table top of the clock case above and in the projecting base of the flange of the clock case below.

With this construction it is a simple matter to cover the movement with a glass case, preferably made rectangular, with glass sides, ends and top, with metal cemented joints. The metal bottom edges of this rectangular box can be ground to fit the plane surface of the top of the clock case. Then, by covering the bottom edges with such a wax as was used in making the glass plates fit the iron case in front or back, we can secure an air-tight joint at the junction of the rectangular top glass case with iron case. In practice the wax to be used may be made by melting together and stirring equal parts of vaseline and beeswax. The proportions may be varied to give a different consistency of wax, and it may be painted on with a brush after warming over a small flame.

If the clock case will be exposed to a comparatively high temperature, say 95° F., then the beeswax can be 3 parts to 1 of vaseline. The good quality of this cement wax is that it does not change with age, or at least for several years, is very clean, and can be wiped off completely with kerosene, or turpentine, or benzine. In all joints meant to be air-tight, the use of rubber packing is to be avoided. It answers well enough at the start, but after several months it is sure to crack and leak air.

By an air-tight joint I do not mean a joint which will not leak air under any pressure which may be applied. It is not necessary that our pendulum should vibrate in a vacuum; all we want is that the pressure inside the clock case should be uniform; that it should not vary with the barometer outside. In actual practice we find it best to have the pressure inside the case as nearly as possible equal to the average atmospheric pressure outside. Now, if the barometer in a given locality never sinks below 27.5 inches, it is not necessary that the vacuum in the clock case be less than that represented by 29.5 inches of mercury pressure. So, too, if it were desirable to have the pressure inside the case greater than that outside, owing to some special form of joint which made the clock case less liable to leak out than to leak in, it might be that an inside pressure would be efficient at 31 inches of mercury. By not having the inside pressure vary but slightly from the outside, the actual pressure of air will not exceed one inch of mercury, or, say, ½ pound pressure to the square inch. This is a pressure which causes quite an insignificant strain upon any joint.

There are objections, however, to the use of air in an enclosed space for precision clocks and so the attempt has been made to use hydrogen. Air is, comparatively speaking, heavy. It is 14½ times as heavy as hydrogen gas, for instance. The pendulum, therefore, in moving through its arc has to push aside 14 times as much weight as it would have to in case it were surrounded by hydrogen. Then what might be called the "case friction" is greater than if we used hydrogen. By "case friction" I mean resistance and a disturbance to the pendulum depending on the effect of the currents of air produced by driving the air before the pendulum against the sides and front of the case. It is a well-established observation that small, cramped cases disturb the clock's rate more than large, roomy ones. This is because the air, having no room to go before the pendulum, is cushioned up against the side of the case at each pendulum swing, and acts as a resisting spring against the swing of the pendulum. By the time the pendulum has reached the end of its vibration the air has escaped upwards and downwards perhaps so that it no longer has its spring power to restore the loss of energy to the pendulum. This "case friction" is most pernicious in its action when associated with free falling weights in the clock case. Clock weights should always fall in separate compartments, and never in such a manner that they can affect the space in which the pendulum swings.

But this is a digression to explain the term "case friction" in its use in horology.

Precision clocks, almost without exception, have electric break-circuit attachments within the case. Most of these break-circuits are constructed so that there is a small spark every time the circuit is broken. The effect of such a spark in air is to convert a small portion of the air in the immediate neighborhood of the spark into nitrous acid gas. After several months there might be a considerable quantity of this gas in the case, with the certain result of rusting the nicer parts of the escapement.

Many attempts have been made to run a clock in an almost complete vacuum of air; but the volume to be exhausted is so large, and the leakage is so sure to occur after a time, that the attempt is now pretty generally abandoned. It will be inferred from what has preceded that a full atmosphere of hydrogen would only offer one-fourteenth the resistance to the pendulum that air would, and all the disturbances arising from the surrounding mediums would be only one-fourteenth for hydrogen of that which we would expect for air. Every consideration, therefore points to the use of hydrogen as the medium with which to fill our clock cases. It is inert, it forms no compounds under the influence of the electric spark, the case friction is no greater than would exist if we made an air vacuum of only about 1 inch of mercury, and hydrogen gas may be readily prepared. The method from dilute sulphuric acid and scrap zinc is the handiest, and it will be found described in almost any chemistry textbook or encyclopedia. Should the horologist wish to know something of the chemistry of the process, without pervious study, he will find it described in very simple language in any primary chemistry. The practical details of filling a clock case with hydrogen gas I have not yet worked out. It is evident that since hydrogen is 14½ times lighter than air, that by attaching a small tube to the source of hydrogen and to the top of the clock case, and another small outlet tube at the bottom of the clock case, that by gravity alone the hydrogen would fill the upper part of the case and drive the air before it out at the bottom. The hydrogen should be dry. To insure this it should pass through a tube containing quicklime, which, if it is a foot long and two inches in diameter, will be sufficient. No burning light or electric spark must be put into the case while filling, because the mixture of hydrogen with the air is very explosive when ignited. Great care must be used in making all joints when attempting to maintain an atmosphere of hydrogen as it leaks readily through the pores of wood iron and all joints. It is, therefore, better to treat the case friction as a constant element and simply keep it constant.

The above discussion has not considered the temperature question. It is important that the changes of temperature in a clock case should be as slow as possible and as small as possible. Professor Rogers, of the Harvard College Observatory, has shown that such bars as are used in pendulum rods of clocks are often several hours in taking up air temperatures many degrees different from that in which they were swinging. We have at the top of the pendulum a thin spring for suspension whose temperature decides its molecular friction; then we have the pendulum rod, and lastly the large bob, all of which take up any new temperature with different degrees of slowness. Now obviously no compensation can be made to act unless the temperatures are the same for all parts of the pendulum, and vary at the same rate. A number of years ago, there was a long discussion as to the temperature at the top and bottom of clock cases. It was shown that this regularly amounted to several degrees in the best clocks. It was to lessen this difference that at the Harvard College Observatory the Bonds built a deep well in the cellar, purposing to put the clock at its bottom. The idea was a good one, and were it not for the difficulty in getting at clocks in wells, and keeping water out, it would doubtless find favor where the utmost accuracy is desired.

A better plan is to run the clock at a high temperature, say 95° to 100° F. The oil is more liquid, the temperature can be more easily maintained, it can all take place in lighted, dry rooms, and the means for doing this we shall now consider.

Our iron case must now be housed in another outside case, which had better be of wood, with glass windows for seeing the clock face. A single thickness of wood would conduct heat too rapidly. It must therefore be made of two thicknesses, with an air space between. If the air space is left unfilled, the circulation of the air soon causes the inner wooden layer to be of the same temperature as the outer. It is necessary to prevent this circulation of air therefore by means of some substance which is a non-conductor of heat and which will prevent the air from circulating. The very best thing to be used in this connection is cotton batting, which has been picked out until it is as light and fibrous as possible. Then if the doors and windows of the wooden case are made of two thicknesses of extra thick glass, and are firmly clamped, by screws through their sashes or some other means, to the frame of the case, we have the best form possible for our completed case of the type I have described. It now remains to provide a layer of hot water pipes inside the clock room, heated by circulating hot water from the outside. The flame under the

Fig. 151. Section through clock room of the Waltham Watch Company

water tank outside, whether of gas or kerosine, to be automatically raised or lowered by any such thermostat arrangements as are in common use with chicken incubators, when the temperature varies from the point desired. Experience teaches that the volume of water had better be considerable, if there is considerable difference in the annual variations of temperature according to the seasons. Thus in Massachusetts or Illinois the temperature is likely to vary from — 30° F. to + 110° F., and the heating arrangements must be suitable to take care of this variation.

The Waltham Watch Company's clock room is an excellent example of the means taken to secure uniformity of temperature and absence of vibration.

The clock room, which is located in the basement of one of the buildings, is built with a double shell of hollow tile brick. The outer shell rests upon the floor of the basement, and its ceiling is within two or three inches of the basement ceiling. The inner shell is 10 feet square and 8 feet in height, measured from the level of the cellar floor. There is an 18-inch space between the walls of the inner and outer shell and a 9-inch space between the two ceilings. On the front of the building the walls are three feet apart to accommodate the various scientific instruments, such as the chronograph, barometer, thermostat, level-tester, etc. The inner house is carried down four feet below the floor of the basement, and rests upon a foundation of gravel. The walls of the inner house below the floor level consist of two thicknesses of brick with an air space between, and the whole of the excavated portion is lined, sides and bottom, with sheet lead, carefully soldered to render it watertight. At the bottom of the excavation is a layer of 12 inches of sand, and upon this are built up three solid brick piers, measuring 3 feet 6 inches square in plan by 3 feet in height, which form the foundation for the three pyramidal piers that carry the three clocks. The interior walls and ceilings and the piers for the clocks are finished in white glazed tiling. The object of the lead lining, of course, is to thoroughly exclude moisture, while the bed of sand serves to absorb all waves of vibration that are communicated through the ground from the various moving machinery throughout the works. At the level of the basement floor a light grating provides a platform for the use of the clock attendants.

Although the placing of the clock room in the cellar and the provision of a complete air space around the inner room would, in itself, afford excellent insulation against external changes of temperature, the inner room is further safeguarded by placing in the outer 18-inch space between the two walls a lamp which is electrically connected to, and controlled by, a thermostat. The thermostat consists of a composite strip of rubber and metal, which is held by a clamp at its upper end and curves to right or left under temperature changes, opening or closing, by contact points at the lower end of the thermostat, the electrical circuit which regulates the flame of the lamp. The thermostat is set so as to maintain the space between the two shells at a temperature which shall insure a constant temperature of 71 degrees in the inner clock house. This it does with such success that there is less than half a degree of daily variation.

The two clocks that stand side by side in the clock room serve to keep civil time, that is to say, the local time at the works. The clock to the right carries a twelve-hour dial and is known as the mean-time clock. By means of electrical connections it sends time signals throughout the whole works, so that each operative at his bench may time his watch to seconds. The other clock, known as the astronomical clock, carries a twenty-four-hour dial, and may be connected to the works, if desired. These two clocks serve as a check one upon the other. They were made at the works and they have run in periods of over two months with a variation of less than 0.3 of a second, or 1-259,000 part of a day. The third clock, which stands to the rear of the other two, is the sidereal clock. It is used in connection with the observatory work, and serves to keep sidereal or star time.

The rate, as observed at the Waltham works, rarely exceeds one-tenth of a second per day. That is to say, the sidereal clock will vary only one second in ten days, or three seconds in a month. The variation, as found, is corrected by adding or subtracting weights to or from the pendulum, the weights used being small disks, generally of aluminum.

Summing up, then, we find that the great accuracy obtained in this clock room is due to the careful elimination of the various elements that would exercise a disturbing influence. Changes of temperature are reduced to a minimum by insulation of the clock house within an air space, in which the temperature is automatically maintained at an even rate. Changes of humidity are controlled by the specially designed walls, by the lead sheathing of the foundation pit, by the preservation of an even temperature, and by placing boxes of hygroscopic material within the inner chamber. Errors due to vibration are eliminated by placing the clocks on massive masonry piers which stand upon a bed of sand as a shock-absorbing medium.

The astronomical clock is inclosed in a barometric case, fitted with an air pump, by which the air may be exhausted and the pendulum and other moving parts relieved from barometric disturbances. For it must be understood that variation in barometric pressure means a variation in the density of the air, and that the speed of the pendulum must necessarily be affected by such changes of density.

Restoring Old Cases.—Very often the watchmaker gets a clock which he knows will be vastly improved by varnish, but not knowing how to take off the old varnish he simply gives it a little sand paper or rubs it off with oil and lets it go at that. Varnishing such a clock thinly with equal parts of boiled oil and turpentine and allowing it to dry will often restore the transparency of the varnish; if uneven results are obtained a second coat may be necessary. Many of these old clocks have not been varnished for so many years that the covering of the wood looks like a cheap brown paint. To remove this in the ordinary way means endless labor, and if the case is inlaid with colored patterns of veneers, which are partly loosened by the glue drying out, the repairer is afraid to touch it for fear he will only make matters worse in the attempt to better them.

In the case of an old clock of inlaid marquetry, if the pieces of veneer have become partly loosened, the first thing to do is to make a thin, fresh glue. Work the glue under the veneer and then clamp it down tightly with a piece of oiled paper, or waxed paper, laid between the glue and the board used to clamp with and the whole firmly set down tight with screws or screw clamps. To make waxed paper dissolve parafine wax in benzine and flow or brush on the paper and let dry. After the glue has hardened comes the work of removing the varnish. To do this you will need some varnish remover, which can either be bought at the paint store, or made as follows:

Varnish Remover.—In doing such work the trick is to make sure that nothing put on the case will injure it, as a clock one hundred years old cannot be replaced. Therefore, if you are suspicious as to the varnish removers you can purchase, and do not want to take chances, you may make one of wood alcohol and benzole, or coal tar naphtha. Be sure you do not get petroleum naphtha, which is common gasoline. The coal tar naphtha is a wood product. The wood alcohol is also a wood product and the varnishes used upon furniture are vegetable gums, so that it will readily be seen that you are putting nothing on the antique with which it was not associated in its natural state. Equal parts of benzole and wood alcohol will dissolve gums instantaneously, so that if the oil has dried out of the varnish so much that the varnish has become opaque and only the rosins are left, the application of this fluid with a brush will cause instant solution, making the gums boil up and form a loose crust upon the surface of the wood, as the liquid evaporates, which it does very rapidly.

Varnishes containing shellac and some other gums are rather hard to dissolve and where an obstinate varnish is encountered it may be well to use wax in the varnish remover. This is done by shaving or chopping some parafine wax, dissolving it in the benzole, and when it is clear and transparent, add the wood alcohol. Upon the addition of the alcohol the wax immediately curdles so that the fluid becomes milky. In this condition it is readily brushed upon any surface and when the wax strikes the air it congeals and forms a crust which holds the liquid underneath and enables it to do its work instead of evaporating.

The wax also serves the purpose of allowing the workman to see just where he is putting his fluid and of holding it in position upon vertical surfaces or ceilings, round moldings, carved work and other places from which it will quickly run off. Only enough wax should be added to make it spread readily with the brush and after soaking it will be an easy matter to take a painter's putty knife, a case knife, or a scraper and laying it nearly flat on the wood remove all the varnish at one operation, wiping off the knife as fast as it becomes too full. After the bulk of the varnish is off some of the fluid, without the wax, may be used upon a cloth to go over and smooth up by removing the spots and stripes of varnish left by the knife, or in moldings, etc., where the knife cannot be applied, and we have our bare wood, which, after drying and sand papering, is ready for a fresh coat of XXX coach varnish, which should dry in 24 hours and harden in a week.

A very little work and practice in this will enable the workman to rapidly and cheaply clean up and repair antiques in such a way that it will add greatly to his reputation.

To restore the gloss of polished wood it is not always the best plan to employ true furniture polish. The majority of the so-called polishes for wood are based on a mixture of boiled linseed oil and shellac varnish, made by dissolving shellac in alcohol in the proportion of four ounces of shellac to a pint of alcohol. A little of the dissolved shellac is poured on to a canton-flannel rag, a few drops of the boiled linseed oil are placed on the cloth, and the wood to be polished is rubbed vigorously. About half an ounce of camphor gum dissolved with the shellac in the alcohol will greatly facilitate the operation of polishing.

A soft woolen rag, moistened with olive oil and vigorously rubbed on dull varnished surfaces, like old clock cases, will brighten the surface wonderfully. Some workmen add a few drops of a strong solution of camphor gum in alcohol to the olive oil.

The polishing of cases is accomplished by applying several coats of the best coach painters' rubbing varnish, when, after perfect drying, the surface is rubbed with a felt or a canton-flannel rag, folded flat, using water and the finest pulverized pumic stone. This operation smooths the surfaces. The final polishing of such work is done by rubbing with rotten stone and olive oil with the smooth side of canton flannel. To remove the last traces of smear caused by the oil, an old, soft linen cloth and rye flour is used. Of course, fine work like we see on new cases of fine quality is not likely to be produced by one who is unaccustomed to it; a man must serve a good, long apprenticeship in the varnish finishing business before he is competent for it; and even then some polishers fail to obtain the fine results achieved by others. The great danger is that the rubber will cut through the varnish and expose the bare wood on edges, corners and even in spots on plane surfaces, before he has removed the lumps and streaks of varnish on adjacent portions of the work. Whenever the varnish is flat and smooth in any spot, you must stop rubbing there.

Black wood clocks which have become smoked and dull should have the cases rubbed with boiled oil and turpentine on a piece of soft woolen rag; afterwards polish off with a dry rag. If the gloss has been destroyed it will have to be varnished. Flow the varnish well on and use 1½-inch brush and be careful to get the varnish on even and so as not to trickle. This is easy if you are careful to keep the varnish thin and do not go over the varnish a second time after spreading it on. Thin with turpentine and put very little on the case; it is already smooth and a mere film will give the gloss. For white filling on the engraving on black cases use Chinese white or get a good white enamel at a paint store.

Gilding on wood cases is done by mixing a little yellow dry color with thin glue and painting the cases with the mixture; the color lets you see what you are doing. When the glue has dried until it is "tacky," lay gold leaf on the painted portions and smooth down with cotton. If you have any holes do not attempt to patch them. It is easier and quicker to put on another sheet of gold leaf over the first one. After the gold is dry, it may be burnished with a bloodstone or smooth steel burnisher, or it may be left dead. Finish with colorless lacquer, very thin and smooth.

Imitation gold leaf, known to the trade as Dutch Metal, may be substituted for the gold leaf, if the latter is thought to be too expensive, but in such cases be sure to have the metal well covered with the lacquer, as unless this is done it will blacken in two or three years—sometimes in two or three months.

Bronze powder may be applied to the glue size with a tuft of cotton and well rubbed in until flat and smooth; then lacquer and dry. Never put on bronze paint, for the following reason: If we examine the bronze under a microscope we shall find that it is composed of flat scales like fish scales; if mixed as a paint they will be found lying at all angles in the painted work—many standing on edge. Such scales reflect the light away from the eye and make the work look dull and rough. If we rub these dry scales in gently on the sticky size, we will lay them all down flat and smooth, so that the work will glisten all over with an even color. Always lacquer bronzed work—yellow lacquer being the best—and put on plenty of lacquer.

Metal ornaments, when discolored, should be removed from the case, dipped in boiling lye to remove the lacquer, scratch brushed, dipped in ammonia to brighten, rinsed in hot water and dried in sawdust. They may then be lacquered with a gold lacquer, or plated in one of the gold plating solutions sold by dealers for plating without a battery and then lacquered, if bright. If they are of oxidized finish cleaning and lacquering is generally all that is necessary.

Oxidized metal cases, if badly discolored, should be sent to an electroplater to be refinished, as the production of smooth and even finishes on such cases, requires more skill than the clock repairer possesses, and he therefore could not do a good job, even if he had the necessary materials and formulae.

Marble cases are made of slabs, cemented together. Many workmen use plaster of paris by merely mixing it with water, though we rather think it better to use glue in the mixing, as plaster so mixed will not set as quickly as that mixed with water. After the case is cemented with the plaster, the workman can go over the joint with a brush and water colors, and with a little care should be able to turn out a job in which the joint will not be noticeable. Another cement much used for marble is composed of the white of an egg mixed with freshly slaked lime, but it has the disadvantage of setting very quickly.

Marble case makers use a cement composed of tallow, brick dust, and resin melted together, and it sets as hard as stone at ordinary temperatures.

It often happens that the marble case of a mantel clock is injured by some accident and its corners are generally the first to suffer. If the break is not so great as to warrant a new case or a new part the repairer may make the case a little smaller or file until the edges are reproduced, after which the polish is restored. Proceed as follows:

Take off from the damaged part as much as is necessary by means of a file, taking care however, not to alter the original shape of the case. Now grind off the piece worked with the file with a suitable piece of pumice stone and water and continue the grinding next with a water stone until all the scratches have disappeared, paying special attention to the corners and contours. After this has been done take a hard ball of linen, moisten it, and strew over it either tripoli or fine emery and proceed to polish the case with this. Finish the polishing with another linen ball, using on it still finer emery and rouge. Now dry the case and finish the polishing with a mixture of beeswax and oil of turpentine. This method may be employed for all kinds of marble, or onyx and alabaster cases.

In cases where the fractures are very deep, so that the object cannot be made much smaller without ruining the shape, the damaged parts may be filled with a cement, prepared from finely powdered marble dust and a little isinglass and water, or fish glue will answer very well. Stir this into a thick paste, which fill into the deep places and permit to dry; after drying, correct the shape and polish as described.

If the pieces which have been broken off are at hand they may be cemented in place again. Wet the pieces with a solution of water and silicate of potash, insert them in place and let them dry for forty-eight hours. If the case is made of white marble use the white of an egg and a little Vienna lime, or common lime will answer.

To Polish Marble Clock Cases.—It frequently becomes the duty of the repairer to restore and polish marble clock cases, and we would recommend him to make a thin paste of the best beeswax and spirits of turpentine, clean the case well from dust, etc., then slightly cover it with the paste, and with a handful of clean cotton, rub it well, using abundant friction, finish off with a clean old linen rag, which will produce a brilliant black polish. For light colored marble cases, mix quicklime with strong soda water, and cover the marble with a thick coating. Clean off after twenty-four hours, and polish well with fine putty powder.

To Remove Oil Spots From Marble.—Oil spots, if not too old, are easily removed from marble by repeatedly covering them with a paste of calcined magnesia and benzine, and brushing off the magnesia after the dissipation of the oil; this may have to be repeated several times. Another recipe reads as follows: Slaked lime is mixed with a strong soap solution, to the consistency of cream; this is placed upon the oil spot, and repeated until it has disappeared. In place of this mixture, another one may be used, consisting of an ox gall, 125 grains of soapmaker's waste lye and 62½ grams of turpentine, with pipe clay, to the consistency of dough.

Cutting Clock Glasses.—You will sometimes want a new glass for a clock. I get a lot of old 5x7 negatives and scald the film off in plain hot water, rinse well and dry. Now I lay my clock bezel on a piece of paper and trace around with a pencil, inside measure. Now remove the bezel and trace another circle around the outside of this circle about one-eighth inch. Now, lay the paper on a good, solid, smooth surface, glass on top, and with a common wheel glass-cutter follow around the outside line, free handed, understand. The paper with marked circle on is under the glass, and you can see right through the glass where to follow with the cutter. Now cut the margins of glass so as to roughly break out to one-half inch of your circle cut, running the cuts out on the side, then carefully break out.

CHAPTER XXIV.

SOME HINTS ON MAKING A REGULATOR.

Of all the instruments used by a watchmaker in the prosecution of his business, there is probably none more important than his regulator. Its purpose is to divide time into seconds, and it is the standard by which the practical results of his labors are tested; the guide which all the other time-keepers in his possession are made to follow and the arbitrator which settles all disputes regarding the performance of his watches.

No regulator has yet been constructed that contains within itself every element for producing absolutely accurate time-keeping. At intervals they must all be corrected from some external source, such as comparison with another time-keeper, the error of which is known, or by the motion of the heavenly bodies, when instruments for that purpose are available. Before beginning to make a regulator, the prudent watchmaker will first reflect on the various plans of constructing all the various details of an accurate time-keeper, and select the plan which, in his opinion, or in the opinion of those whom he may consult on the subject, will best accomplish the object he has in view.

In former years a regulator case was made with the sole object of accommodating the requirements of the regulator, and every detail in the construction of the case was made subservient to the necessities of the clock. The plain, well-made cases of former years are now almost discarded for those of more pretentious design. If the general change in the public taste demands so much display, there can be no objection. It is perfectly harmless to the clock, if the designers and makers of the cases would only remember that narrow waists or narrow necks on a case, although part of an elegant design, do not afford the necessary room for the weight and freedom of the pendulum; that the doors and other openings in the case must be constructed with a view to exclude dust; and that the back should be made of thick, well-seasoned hardwood, such as oak or maple, so as to afford the means of obtaining as firm a support for the pendulum as possible.

When a regulator case is known to have been made by an inexperienced person, which sometimes happens, or when we already have a case, it is always the safest course for those who make the clock to examine the case personally and see the exact accommodation there is for the clock. Sometimes, when we know beforehand, we can, without violating any principle, vary the construction a little, so as to make the weight clear the woodwork of the inside of the case, and in other respects complete the regulator in a more workmanlike manner by making the necessary alterations in the clock at the beginning of its construction, instead of after it has been once finished agreeably to some stereotyped arrangement.

The arrangement of the mechanism of an ordinary regulator is a simple operation compared with some other horological instruments of a more complex character. We are not limited in room to the same extent as in a watch, and the parts being few in number a regulator is more easily planned than timekeepers having striking or automatic mechanism for other purposes combined with them; yet it often happens that the inexperienced make serious blunders in planning a regulator, and, as the clock approaches completion, many errors make themselves visible, which might have been avoided by the exercise of a little more forethought. It may be that, when the dial is being engraved, the circles do not come in the right position, or the weight comes too close to the pendulum, or the case, or the cord comes against a pillar, or other faults of greater or less importance appear, all of which might have been obviated by taking a more comprehensive view of the subject before beginning to make the clock. The best way to do this is to draw a plan and side and front elevations to a scale.

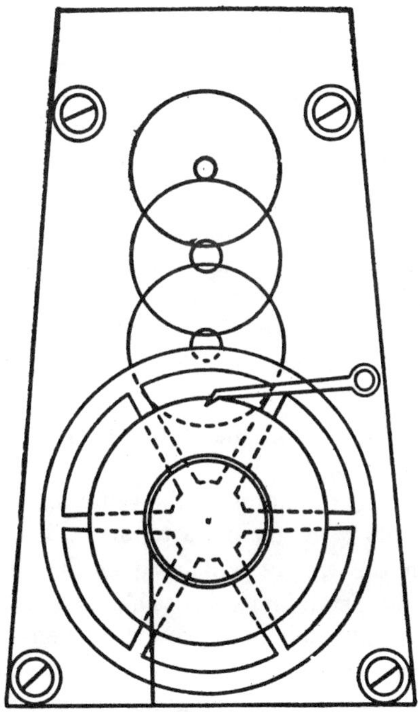

Fig. 152

The position which the barrel and great wheel should occupy is worthy of serious consideration. In most of the cheap regulators, as well as in a few of a more expensive order, the barrel is placed in a direct line below the center wheel, as is shown in Fig. 152. This arrangement admits of a very compact movement, and it also allows the weight to hang exactly in the center of the case, which some think looks better than when it hangs at the side, especially when there is a glass door in the body of the case. But while a weight hanging in the center of a case may be more pleasing to the eye than when it hangs at the side, this is an instance where looks can, with great propriety, be sacrificed for utility, because when the weight hangs in the center it comes too close to the pendulum, and is very liable to disturb its motion. In proof of this statement, let any reader who has a regulator with a light pendulum and a comparatively large weight hanging in front of it, closely watch the length of the arc the pendulum vibrates when the weight is newly wound up and when it is down opposite the pendulum ball, and he will observe that the length of vibration of the pendulum varies from five to fifteen minutes of arc, according to the position in which the weight is placed; that the pendulum will vibrate larger arcs when the weight is above or below the ball than when it is opposite it; and if the clock has a tendency to stop from any cause, that it will generally do so more readily when the weight is opposite the pendulum ball than when it is in any other position. For this reason I would dispense with the symetrical looks of the weight hanging in the center of the case, which, after all, is only a matter of taste, and construct the movement so that the weight will hang at the side, and as far away from the pendulum as possible.

Fig. 153 is intended to represent the effect which placing the barrel at either side has on throwing the weight away from the pendulum. A is the center wheel; B and C are the great wheels and barrels with weights hanging from them; D is the pendulum. It will be noticed by the diagram that the weight at the left of the pendulum is exactly the diameter of the barrel farther away from the pendulum than the weight on the right. On close inspection it will also be observed that on the barrel C the force of the weight is applied between the axis of the barrel and the teeth of the wheel, while on the barrel B the axis of the barrel lies between the point where the force is applied and the point where the teeth act on the pinion; consequently a little more of the effective force of the weight is consumed by the extra amount of pressure and friction on the pivots of the barrel B than there is in C.

Notwithstanding this disadvantage, I would for a regulator recommend the barrel to be placed at the left side of

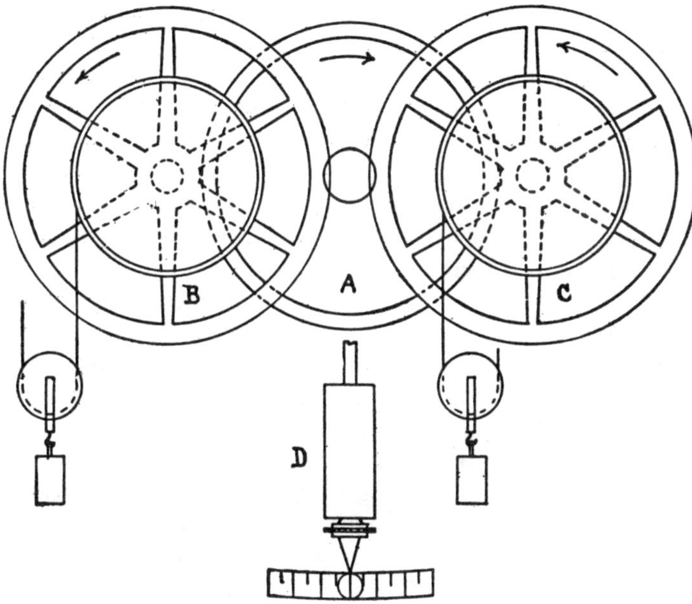

Fig. 153

the center wheel, because the weight may thereby be led a sufficient distance from the pendulum in a simple manner. If we place the barrel at the right, and thereby secure the greatest effective force of the weight, and then lead the weight to the side by a pulley, we will lose a great deal more by the friction of the pulley than we gain by the proper application of the weight.

In a regulator with a Graham escapement but little force is required to keep it going, and there is usually accommodation for an abundance of power; therefore we cannot use a little of this superabundant available force to better advantage than by placing the barrel at the left side of the clock, and thereby throw the weight a sufficient distance from the pendulum in the simplest manner.

The escapement we assume to be the old dead beat, as for time-keeping it is equal to a gravity escapement while possessing advantages undesirable to sacrifice for a doubtful improvement. The advantages it possesses over any form of gravity escapement are: it has fewer pieces and not so many wheels; it takes very much less power to drive; is not liable to fail in action while winding, if the maintaining power should be rather weak; while for counting seconds and estimating fractions, its clear, definite, and equable beat has great superiority over the complication of noises made by a gravity escapement.

Full directions for making this and other escapements have already been given, but in a regulator there are some considerations which will not be encountered in connection with the escapements of ordinary clocks, where fine time-keeping is not expected. We have previously stated that the center of suspension of the pendulum should be exactly in line with the axis of the escapement and we will now endeavor to state plainly how important this is in a fine clock and the reasons for it. Mr. Charles Frodsham, the noted English chronometer maker, has conducted a series of careful experiments and the results were communicated in a report to the British Horological Society, as follows:

When we talk of detached escapements, or any escapement applied to a pendulum, it is necessary to bear in mind that there is always one-third at the least of the pendulum's vibration during which the arc of escapement is intimately mixed up with the vibration, either in locking, unlocking, or in giving impulse; therefore, whatever inherent faults any escapement may possess are constantly mixed up in the result; the words "detached escapement" can hardly be applied when the entire arc of vibration is only two degrees; or, in other words, what part of the vibration is left without the influence of the escapement?—at most one degree. In chronometers the arc of vibration is from ten to fifteen times greater than the arc of escapement.

The dead-beat escapement has been accused of interfering with the natural isochronism of the pendulum by its extreme friction on the circular rests, crutch, and difficulty of unlocking, etc., all of which we shall show is only so when improperly made.

When the dead-beat escapement has been mathematically constructed, and is strictly correct in all its bearings, its vibrations are found to be isochronous for arcs of different extent from 0.75 of a degree to 2.50 degrees; injurious friction does not then exist; the run up on the locking has no influence, nor is there any friction at the crutch; oil is not absolutely necessary, except at the pivots; and there is no unlocking resistance nor any inclination to repel or attract the wheel at its lockings.

The general mode of making this escapement is very defective and indefinite, and entirely destroys the naturally isochronous vibration of the pendulum.

The following is the usual rate of the same pendulum's performance in the different arcs of vibration with an escapement as generally constructed after empirical rules:

Arc of vibration 3°	rate per diem 9.0 seconds.
Arc of vibration 2½°	rate per diem 6.0 seconds.
Arc of vibration 2°	rate per diem 3.5 seconds.
Arc of vibration 1½°	rate per diem 1.5 seconds.
Arc of vibration 1°	rate per diem 0.0 seconds.

Thus for a change of vibration of 1°, we have a daily error of 3.5. No change of suspending spring will alter inherent mechanical errors destructive of the laws of motion. With clocks made in the usual manner, whether you apply a long or short spring, strong or weak, broad or narrow, you will not remove one fraction of the error; so the sooner the fallacy of relying upon the suspending spring to cure mechanical errors is exploded the better.

That the suspending spring plays a most important part must be admitted, since, when suspended by a spring, a pendulum is kept in motion by a few grains only, whereas, if supported on ordinary pivots, 200 lbs. weight would not drive it 2′ beyond its arc of escapement, so great would be the friction at the point of suspension.

The conditions on which alone the vibrations of the pendulum will be isochronous are the following:

1. That the pendulum be at time with and without the clock, in which state it is isochronous "suspended by a spring."

2. That the crutch and pallets shall each travel at the same precise angular velocity as the pendulum, which can only happen when the arc each is to describe is in direct proportion to its distance from the center of motion, that is, from the pallet axis.

3. That the angular force communicated by the crutch to the pendulum shall be equal on both sides of the quiescent point; or, in other words, that the lead of each pallet shall be of the same precise amount.

4. That any number of degrees marked by the crutch or pallets shall correspond with the same number of degrees shown by the lead of the pendulum, as marked by the index on the degree plate.

5. That the various vibrations of the pendulum be driven by a motive weight in strict accordance with the theoretical law; that is, if a 5-lb. weight cause the pendulum to double its arc of escapement of 1°, and consequently drive it 2°, all the intermediate arcs of vibration shall in practice accord with the theory of increasing or diminishing their arcs in the ratio of the square roots of the motive weight.

To accomplish the foregoing conditions, there is but one fixed point or line of distance between the axis of the escape wheel and that of the pallet, and that depends upon the number of teeth embraced by the pallets and only one point in which the pallet axis can be placed from which the several lines of the escapement can be correctly traced and properly constructed with equal angles, and equal rectangular lockings on both sides, so that each part travels with the same degree of angular velocity, which are the three essential points of the escapement.

Much difference of opinion has been expressed upon the construction of the pallets, as to whether the lockings or circular rests should be at equal distances from the pallet axis, with arms and impulse planes of unequal length, or at unequal distances from the pallet axis, with arms and impulse planes of equal length. In the latter case the locking on one side is three degrees above, and on the other three degrees below the rectangle, whereas in the former the tooth on both sides reposes at right angles to the line of pressure; but the length of the impulse planes is unequal. When an escapement is correctly made upon either plan, the results are very similar.

It is possible to obtain equal angles by a false center of motion or pallet axis; but then the arcs of repose will not be equal. This, however, is not of so much consequence as that of having destroyed the conditions Nos. 2, 3, 4; for even at correct centers, if the angles are not drawn off correctly by the protractor, and precisely equal to each other, the isochronous vibrations of the pendulum will be destroyed, and unequal arcs will no longer be performed in equal times; the quiescent point is not the center of the vibration, except when the driving forces are equal on both sides of the natural quiescent point of the pendulum at rest.

Now this is the very pith of the subject, and which few would be inclined to look for with any hope of finding it it the solution of this important question, the isochronism of the pendulum.

One would naturally suppose that unequal arcs on the two sides of the vertical lines would not seriously affect the rate of the clock, but would be equal and contrary, and consequently a balance of errors, and so they probably are for the same fixed vibration, but not for any other; because different angles are driven with different velocities, the short angle has a quicker rate of motion than the long. Five pounds motive weight will multiply three times the pendulum's vibration over an arc of escapement of 0.75°; but the same pendulum, with an arc of escapement of 1°, would require 11.20 lbs. to treble its vibration; the times of the vibration vary in the same ratio as the sum of the squares of the differences of the angles of each pallet, compared with the spaces passed over.

From this it will be seen that the exact bending point of the pendulum spring should be opposite the axis of the fork arbor when regulating the clock and this may have to be determined by trial, raising or lowering the plates by screws in the arms of the suspending brackets until the proper position is found, when the movement may be clamped firmly in position by the binding screws, see Fig. 158.

On common clocks the crutch is simply riveted on its collet and bent as required to set the clock in beat, but for a first-class clock a more refined arrangement is usually adopted. There are other plans, but perhaps none so thoroughly sound and convenient as the following. The crutch itself is made of a piece of flat steel cut away so as to leave a round boss at the bottom for the fork, and a round boss at the top to fit on a collet on the pallet arbor, a part projecting above to be embraced between a pair of opposing screws. On the collet is fixed a thin brass plate with two lugs projecting backwards from the frame, these lugs being drilled and tapped to receive the opposing screws in a line. The boss of the crutch lies flat against this plate, and is held up to it by a removable collet. The collet may be pinned across or fitted keyhole fashion, in either case so as to hold the crutch firmly, allowing it to move with a little stiffness under the influence of the screws. With this arrangement the adjustment to beat may be made with the utmost delicacy by slacking one screw and advancing the other, taking care that in the end they are well set home so as to make the crutch practically all one piece with the arbor. Milled heads are most convenient for these screws, and being placed at the top they are easily got at. The crutch should always be fitted with a fork to embrace the pendulum rod, as this ensures the impulse being given directly through the center, and with the same object the acting sides of the fork should be truly square to the frame. A slot in the pendulum rod with a pin acting in it is never so sure of being correct, as, although the surfaces may be rounded, it is very unlikely that the points of contact will be truly in the plane of the axis of the rod. The slightest error in this respect will tend to cause wobbling of the bob, although, to avoid this, great attention must also be given to the suspension spring, the pin on which it hangs, and the pin and the hole at the top of the pendulum rod. All these points must be in a true line, and the spring symmetrical on both sides of the line in order that the impulse may be given exactly opposite the center of the mass, otherwise wobbling must occur, although perhaps of an amount so small as to be difficult of detection, and this is not a matter of small importance, as it has an effect on the rate which could be mathematically demonstrated.

The frames of many regulators are made too large and heavy. In some cases there may be good reasons for making them large and heavy, but in most instances, and especially when the pendulum is not suspended from the movement, it would be much better to make the frames lighter than we frequently find them. Very large frames present a massive appearance, and convey an idea of strength altogether out of proportion to the work a regulator is required to perform. They are more difficult and more expensive to make than lighter ones, and after they are made they are more troublesome to handle, and the pivots of the pinions are in greater danger of being broken when the clock is being put together than when they are moderately light.

In a clock such as we have under consideration, where the frame is not to be used as a support for the pendulum, but simply to contain the various parts which constitute the movement, the thickness of the frames may with propriety be determined on the basis of the diameter of the majority of the pivots which work into the holes of the frames. The length of the bearing surface of a pivot will, according to circumstances, vary from one to two and a half times the diameter of the pivot. The majority of the pivots of our regulator will not be more than .05 or .06 of an inch in diameter; consequently a frame 0.15 of an inch thick will allow a sufficient length of bearing for the greater portion of the pivots, and will also allow for countersinks to be made for the purpose of holding the oil. If thin plates are used one or two of the larger pivots should be run in bushes placed in the frame, as described in Fig. 155.

The length and breadth of the frame, and also its shape, should be determined solely on the basis of utility. There can be no better shape for the purpose cf a regulator than a plain oblong, without any attempt whatever at ornament. For our regulator a frame nine inches long and seven inches broad will allow ample accommodation for everything, as may be seen on referring to Fig. 157.

The plates are made of various alloys: cast-brass, nickel-silver, and hard-rolled sheet-brass. It is difficult to make plates of cast-brass which would be even, free from specks, etc., but cast plates may very well be made of ornamental patterns and bushings of brass rod inserted, or they may be jeweled as shown in Figs. 154, 155, 156. Nickel, or

German silver, makes a fine plate, but it is difficult to drill the small holes through plates of four-tenths of an inch in thickness, on account of the peculiar toughness of the metal, so that bushings are necessary. The best material where the holes are to be in the plates is fine, hard-rolled sheet brass; it should have about 4 oz. of lead to the 100 lbs., which will make it "chip free," as clockmakers term it, rendering it easy to drill; the metal is so fine and condensed to that extent by rolling, that the holes can be made with the greatest degree of perfection. The many improvements in tools and machinery have effected great changes and improvements in clock-making. It once was quite a difficult task to drill the small holes in the plates with the ordinary drills and lathes; now we lay the plates after they are soldered together at the edges (which is preferable to pinning), on the table of an upright drill, and with one of the modern twist-drills the task is rendered a very easy one. After the pivot-holes are drilled, we run through from each side a round broach, finished lengthwise and hardened, which acts as a fine reamer, straightening and polishing the holes exquisitely. A little oil should be used on the reamer to prevent sticking. The method of fitting up the pivot-holes invented by LeRoy, a French clockmaker of some note, is shown in Fig. 154. It is a sectional view of the plate at the pivot-hole. It will be observed that instead of countersinking for the oil, the reverse is the case. A is a hardened steel plate counterbored into the clock plate B, and held in its place by the screws. There should be a small space between the steel plate and the crown of the arch for the oil. After the clock has been put together it is laid down on its face

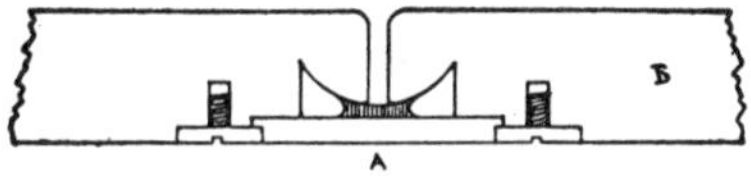

Fig. 154

or side, a drop of oil is put to the pivot end, and the steel plate immediately put on; and the oil will at once assume the shape of the shaded spot in the drawing, being held in the position at the center of the pivot by capillary attraction, until it is exhausted by the pivots; the steel plates also govern the end play of the pinions. The pivot ends being allowed to touch the plates occasionally, the shoulders of the pinions are turned away into a curve, and, of course, do not bear against the plate, as in most clocks.

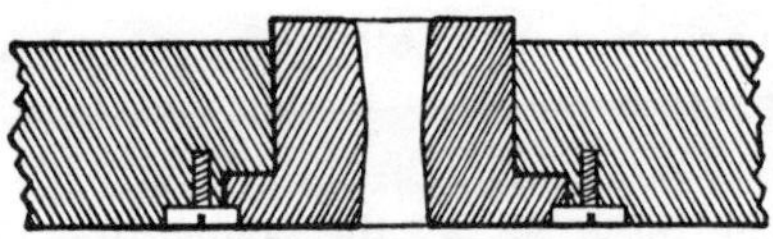

Fig. 155

Glass plates may be used instead of steel, or rose cut thin garnets, or sapphires, with the flat sides smoothly polished, may be bought of material dealers and set in bezels like a cap jewel. They are very hard and smooth for the pivot

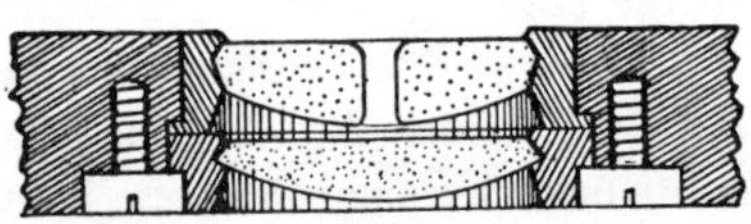

Fig. 156

ends, and the state of the oil at the pivots can be seen at any time. Clocks fitted up in this manner have been running many years without oiling.

When fitted up in this way the plates may be thicker. We have made the clock plates about four-tenths of an inch in thickness, which allows of counterboring, and admits of long bearings for the barrel arbor, which are so liable to be worn down in the holes by the weights; and the pivots of the pinions, by being a little longer, do not materially increase the friction.

In first-class clocks, when all the materials are as hard as possible, the wheels and pinions high numbered, the teeth, pinions, pivots, and holes smooth, true, and well polished, the amount of wear is very slight, especially if the driving weight has no useless excess. Yet there are advantages in having some parts jeweled, such as the pallets and the four escapement holes. The cost of such jeweling is not an objection, while the diminished friction of the smooth, hard surfaces is worth the extra outlay. The holes can be set in the bushes described in Fig. 156, the end stones being cheap semi-precious stones, either rose cut or round.

For jeweling the pallets, dovetailed slots may be made so that the stones will be of a wedge shape; there is no need for cutting the slots right through as in lever watch pallets. The stones will be held more firmly if shaped as wedges lying on a bed of the steel and exposing only the circular resting curve and the driving face. The slots can be filed out and the stones ground on a copper lap to fit, fixed with shellac and pressed firmly home while warm. The grinding and polishing of the acting surfaces are done exactly as described for hard steel, only using diamond powder instead of emery. The best stones are pale milky sapphires, such as are useless as gems, this kind of stone being the hardest.

The holes may be much shorter when jeweled, as the amount of bearing surface required with stones is less than with brass; this results in less adhesion through the oil, and less variation of force through its changes of consistency. The 'scape wheel may also be thinner with similar results, and less weight to be moved besides. So the advantages of jeweling are worth consideration.

It is important to finish the wheels and pinions before drilling any holes in the plates and then to definitely locate the holes after trial in the depthing tool.

For the clockmaker's use the next in value to the wheel-cutting engine is a strong and rigid depthing tool, for it is by means of this instrument that the proper center distances of wheels and pinions can be ascertained, and all errors in sizes of wheels and pinions, and shapes of teeth, are at once detected before the holes are drilled in the plates. In fact, this tool becomes for the moment the clock itself; and if the workman will consider that as the wheels and pinions perform in the tool for the little time he is testing them, so they will continue to run during the life of the clock, he will not be too hasty in allowing wheels to go as correct when a hundredth of an inch larger or smaller, and another test, would, perhaps, make the pitching perfect.

There are various kinds of depthing tools in use, but many of them are objectionable for the reason that the centers are so long that the marking points on their outer ends, are too far from the point where the pitching or depthing is being tested, and the slightest error in the parallelism of these centers is, of course, multiplied by the distance, so that it may be a serious difference. Having experienced some trouble from this cause, we made an instrument with very short centers, on the principle that the marking points, or centers, should be as near the testing place as possible. We succeeded in making one with a difference of only three-fourths of an inch, which was so exact that we had no further trouble. It was made on the Sector plan, but upright, so that the work under inspection, whether wheels and pinions, or escapements, could be observed closely, and with a glass, if necessary.

It is very important that the posts or pillars and side-plates of clocks should be made and put together in the most thorough manner; the posts should be turned exact to length and have large shoulders, turned true, so that the plates, when put together without screws, should fit accurately, for if they do not, when the screws are driven, some of the pivots will be cramped. We prefer iron for the posts, it being stiffer, and better retaining the screw threads in the ends, which in brass are liable to strip unless long

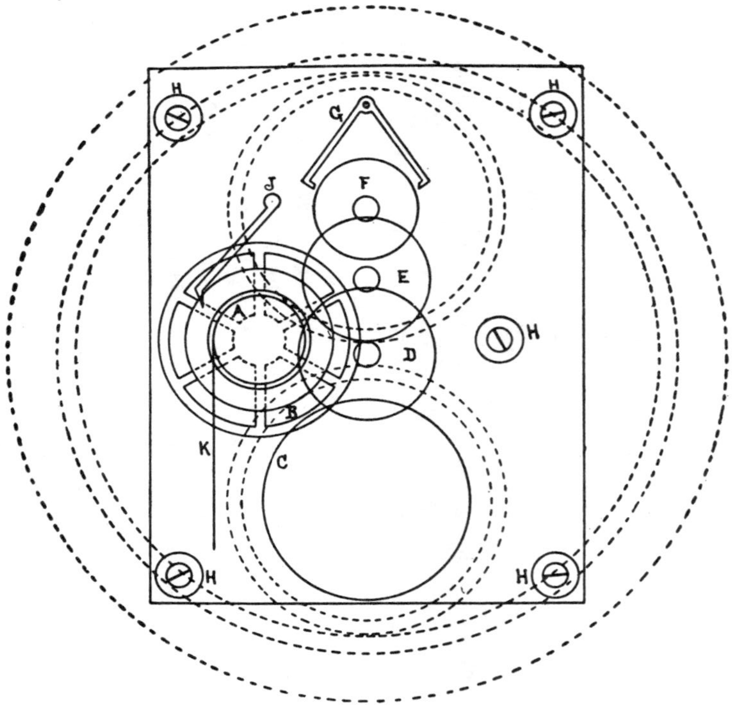

Fig. 157

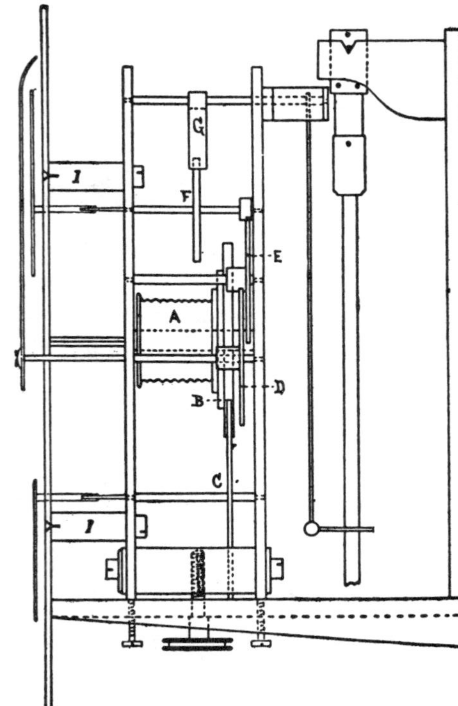

Fig. 158

and deep holes are tapped. Steel pillars should be blued after being finely finished, thus presenting a pleasing contrast. The plate screws should also be of steel, with large flat heads, turned up true, and having a washer next to the plate. Brass pillars are favored by many and are easier turned in a small lathe, but they should be much larger than the steel ones.

When the pillars are made of brass round rod of proper diameter is the best stock. If this cannot be procured, a pattern is turned from wood, and a little larger in every respect than the pillar is desired to be. If there is to be any ornament put on the pillar, it is never made on the pattern, because it makes it more difficult to cast, and besides, the ornamentation would all be spoiled in the hammering. The pattern must be turned smooth, and the finer it is the better will be the casting. After the casting is received the

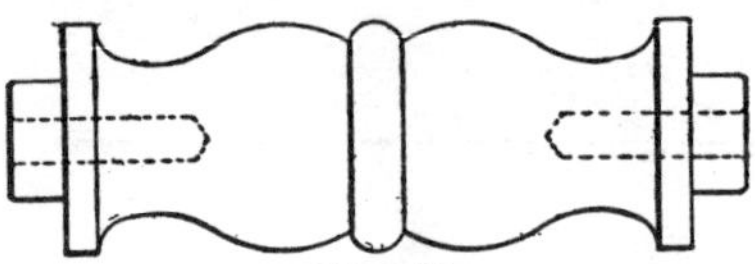

Fig. 159

first thing to be done is to hammer the brass, and then center the holes, because it will be seen from Fig. 159 that there are holes for screws at each end of the pillar. Holes of about .20 of an inch are then bored in the ends of the pillars, and should be deep, because deep holes do no harm and greatly facilitate the tapping for the screws. After the holes are tapped, run in a bottoming tap and then countersink them a little, to prevent the pillar from going out of truth in the turning. It will depend a great deal on the conveniences which belong to the lathe the pillars are turned in as to how they will be held in the lathe and turned. If the holes in the ends of the pillars have been bored and tapped true, and if the lathe has no kind of a chuck or face plate with dogs, suitable for holding rods, the best way is to catch a piece of stout steel wire in the chuck and turn it true, cut a true screw on it, and on this screw one end of the pillar, and run the other end in a male center. However, if the screws are not all perfectly true, and the centers of the lathe not perfectly in line, this plan will not work well, and it will be necessary to catch a carrier on to the pillar and turn it between two male centers.

The dial feet are precisely the same as the pillars, only smaller. These dial feet are intended to be fastened in the frame by a screw, the same as the pillars; but it will be observed that the screw which is intended to hold the dial on the pillar is smaller. The dial feet will be turned in precisely the same manner as the pillars. For finishing the plain surfaces of the pillars and dial feet, an old 6 or 7-inch smooth file makes a good tool The end of the file is ground flat, square or slightly rounded, and perfectly smooth. The smoother the cutting surface the smoother the work done by it will be. It is difficult to convey the idea to the inexperienced how to use this tool successfully. In the first place, a good lathe is necessary, or at least one that allows the work to run free without any shake. In the second place, the tool must be ground perfectly square, that is, it is not to be ground at an angle like an ordinary cutting tool. Then the rest of the lathe must be smooth on the top, and the operator must have confidence in himself, because if he thinks that he cannot turn perfectly smooth, it will be a long time before he is able to do it. A tool for turning the rounded part of the pillar, if a pattern of this style is decided on, is made by boring a hole, the size of the desired curve, in an old file, or in a piece of flat steel, and smoothing the hole with a broach and then filing away the steel. The shoulders should be smooth and flat, or a very little undercut, and the ends of the pillars should be rounded as is shown in Fig. 159, because rounded points assist greatly in making the frames go on to the pillars sure and easy, and greatly lessen the danger of breaking a pivot when the clock is being put together.

When a washer is used the points of the pillars project half the thickness of the washer through the frames, the hole in the washer being large enough to go on to the points of the pillars.

Figure 160 is an outline of the cock required for the pallet arbor, and the only cock that will be required for the regulator. It is customary, in some instances, to use a cock for the scape-wheel and also for the hour-wheel arbors,

Fig. 160

but for the scape-wheel arbor I consider that a cock should never be used when it can be avoided. The idea of using a cock for the scape-wheel arbor is to bring the shoulder of the pivot near to the dial and thereby make the small pivot that carries the seconds hand so much shorter; and so far this is good, but then the distance between the shoulders of the arbor being greater, when a cock is used the arbor is more liable to spring and cause the scape-wheel to impart an irregular force to the pendulum through the pallets. This is the reason why I prefer not to use a cock except when the design of the case is such that long dial feet are necessary, and renders the use of a cock indispensable. In the present instance, however, the dial feet are no longer than is just necessary to allow for a winding square on the barrel arbor, and therefore a cock for the scape wheel is superfluous. It is better to use a long light socket for the seconds hand than put a cock on the scape-wheel arbor in ordinary cases. Except for the purpose of uniformity a cock on the hour wheel is always superfluous, although its presence is comparatively harmless. The front pivot of the hour-wheel axis can always be left thick and strong enough should the design of the case require the dial feet to be extra long.

For the pallet arbor, however, a cock is always necessary, and it should always be made high enough to allow the back fork to be brought as near to the pendulum as possible, so as to prevent any possibility of its twisting when the power is being communicated from the pallets to the pendulum. This cock should be made about the same thickness as the frames, and about half an inch broad. Make the pattern out of a piece of hard wood, either in one solid piece or by fastening a number of pieces together. The pattern should be made a little heavier than the cock is required to be when finished, and it should also be made slightly bevelled to allow it to be easily drawn from the sand when preparing the mould for casting. After it is cast the brass should be hammered carefully, and then filed square, flat, and smooth.

Screws are better and cheaper when purchased, but they may be made of steel or brass rod by any workman who is provided with a set of fine taps and dies. If purchased they should be hardened, polished and blued before using them in the regulator. The threads of screws vary in proportion to the size of the screw and the material from which it is made. A screw with from 32 to 40 turns to the inch, and a thread of the same shape as the fine dies for sale in the tool shops make, is well adapted for the large screws in a regulator. However, it is not threads of the screws I desire to call attention to so much, although it must be admitted that the threads are of primary importance. It is the shape of the heads and the points which is too often neglected.

A thread, or a thread and a half, cut down on the point of a screw, will allow it to enter easier than when the point is flat, round, or shaped like a center. This is not a new idea for making the points of screws, but the plan is either not known to many, or it is not practiced to the extent it ought to be.

The shape of the head of a screw should also always be based on utility, and the shape that will admit of a slit into it that will wear well should be selected. A round head ought never to be used, because a head of this shape does not present the same amount of surface to the screwdriver that a square head does. It is the extreme end of the slit that is most effective, and in round-headed screws this part is cut away and the value of the head for wearing by the use of the screwdriver is the same as if the head of the screw was so much smaller. A chamfered head may suit the tastes of some people better than a perfectly flat head, but in a head of this shape the slit must be cut deeper than in a square head, because the chamfered part of the head is of little or no use for the screwdriver to act against. The slits should always be cut carefully in the center of the head and the sides of the slit filed perfectly flat with a thin file and the slight burr filed off the edge to prevent the top of the head getting bruised by the action of the screwdriver. The shape of the slit which is best adapted for wearing is one slightly tapered, with a round bottom. The round bottom gives greater strength to the head, and prevents the heads of small screws from splitting.

I have dwelt at some length on these little details because a proper attention to them goes a long way in the making of a clock in a workmanlike manner, and it is desirable that the practical details should be as minute as possible.

The construction of the barrel is a subject which requires a greater amount of consideration than is sometimes bestowed upon it. We often meet with regulator barrels which have considerable more brass put into them than is necessary. The value of this extra metal is of little or no consequence. It is the unnecessary pressure the weight of it causes on the barrel pivots, and the consequent increase of friction, which is objectionable. For this reason the weight of the barrel, as well as the weight of every other part of the clock that moves on pivots, should be made no heavier than is absolutely necessary to secure the required amount of strength. In every instance, except when the diameter is required to be very small, the barrel should be made of a piece of thin brass tubing with two ends of cast brass fastened into it.

Figure 161 is a sectional view of the ends of a barrel; the diagram on the right is the end where the great wheels rest against, and the one on the left is the other end. The insides of both these ends are precisely the same, but the outsides differ a little. It will be observed that there is a

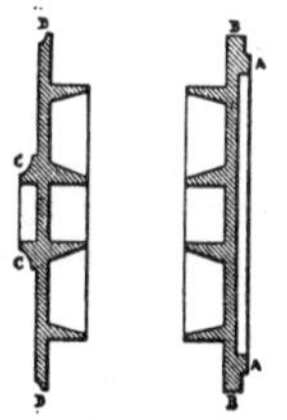

Fig. 161

little projection near the hole on the outside of the front end. This projection is left with the view of making the hole in the center longer, and thereby causing this end to take a firmer hold on the barrel arbor. The back end, or the end that the great wheels rest against, and where the ratchet teeth are cut, is shaped precisely like the diagram on the right of Fig. 161. If you cannot get brass plate of sufficient thickness for the ends of the barrel they must be cast.

The patterns for these barrel ends should be made without any hole in the center, and in every way heavier and thicker than they are to be when finished, because it is difficult to obtain good and solid castings when the patterns are made thin, although it is by no means impossible to make them so. Like all brass castings used for the clockmaker's purpose, they should be carefully hammered, and, although these pieces are of an irregular shape, they can be easily hammered regularly with the aid of narrow-faced hammers or punches, and with the exercise of a little patience. After hammering, the castings should be placed on a face plate in the lathe, and the tube which is to form the top part of the barrel fitted easy and without shake on to the flanges and the other parts of the castings turned down to the required thickness, and a hole a little less than 0.3 of an inch diameter bored in the center of each before it is removed from the face plate. The tube which is to form the top of the barrel should be no heavier than is just necessary to cut a groove for the cord, and for this regulator it should be 1.5 inch diameter outside measurement, 1.5 inch long, and turned perfectly true on the ends.

The hole in the front end of the barrel, which is the end nearest to the dial, should be broached a little from the inside, and the other end broached a little larger from the outside. The reason for broaching the holes in this manner is to cause the thickest part of the barrel arbor to be at the place where the great wheels work, because, in making a barrel for a regulator, it will generally be found that the arbor requires to be thickest in this particular place. The arbor should be made from a piece of fine cast steel a little more than 0.3 of an inch thick, and not less than four inches long. It is always well to have the steel long enough. This steel should be carefully centered and turned true, and of the same size and taper as the holes in the barrel ends. It is not necessary that the barrel arbor should be hardened and tempered, except on special occasions. In most cases it will last as long as any other part of the clock if it is left soft, and it is much easier to make when soft. Before fitting the arbor to the barrel ends it is well to place the ends into the tube that is to form the top of the barrel, because a better fit can be made in this way than when each is fitted separately. When the arbor has been fitted, a good and convenient way of fastening it together is, to use soft solder. It can be easily heated to the required degree of heat with the blow-pipe. A very little solder is sufficient for the purpose, and if the joints have been well fitted the solder will not show when the work is finished. Care should be taken to notice that the solder adheres to the arbors properly. Perhaps it would be well to mention here that, should the clockmaker not have access to a cutting engine with conveniences attached to it for cutting the barrel ratchet after the barrel has been put together, the ratchet should be cut first.

When the different pieces which constitute a barrel have been fastened together the brass work has next to be turned true, and the grooves cut for the cord to run in. It is best not to turn anything off the arbor till the grooves are cut, because they are usually cut smoother when the arbor is strong. The most important points to notice when turning a barrel is to be sure that the top is of equal diameter from the one end to the other, and that the bearing where the great wheels rest against are perfectly true, because, if the top of a barrel is of unequal thickness, the weight will pull with unequal force as it runs down, and if the bearing on the end be out of truth the great wheels will also be very liable to get out of truth, as their position on the barrel is altered by winding the clock up.

The shape of the outside of the barrel ends, as is represented in Fig. 161, will be found to be good and serviceable. AA is the bearing for the great wheels to rest against; BB is where the ratchet teeth are to be cut. There must be a little turned off the face of BB, as is shown in the diagram, so as to prevent the great wheel from rubbing on the teeth. The space between AA and the barrel arbor is turned smooth.

Although it is by no means an absolute necessity to have a groove cut in the top of the barrel, yet it is extremely desirable that there should be one, so that the cord may always be guided with certainty as the clock is wound up. It has long been a disputed question whether the cord should be fastened at the front end of the barrel and wind towards the back, or whether it should be fastened at the back and wind towards the front. I am not aware that there is any violation of principle, so far as the regularity of the power is concerned, whether the cord runs one way or the other. I understand it to be solely a question of keeping the weight clear of the case and the pendulum ball. In ordinary constructed regulator cases this object will be best attained by cutting the screw so that the cord can be fastened at the front of the barrel and wind towards the back; because in making it in this way, the weight is the length of the barrel farther away from the front of the case when it is wound up, and about the same distance farther away from the pendulum ball when it is nearly run down, than if the cord was fastened at the back end of the barrel and wound towards the front. The cutting of the groove is usually done in an ordinary screw cutting lathe.

In making the pivots on a barrel it is the usual custom to make the back pivot smaller than the front one but, with all due respect for this time-honored custom, I would direct a little attention to the philosophy of continuing to make the barrel pivots of a regulator in this manner. Friction varies with pressure; a large pivot has a greater amount of friction than a smaller one, because the pressure on the sliding surface of the revolving body is farther away from the center of motion in one case than in the other. In regulators where the barrel pivots are of a different size, the effective force of the weight will vary slightly according as the weight is fully wound up or nearly run down. In one instance the pressure of the weight is more directly on the large pivot than it is on the smaller one; and in the other instance the pressure is more directly on the small pivot than it is on the larger one, and when the weight is half wound up, or half run down, the pressure is equal on both pivots.

In the center pinion and in some of the other arbors of a clock, it is sometimes necessary to make one pivot considerably larger than the other; but in these cases the difference in the size of the pivots does not affect the regularity of the transmission of the power, because the pressure that turns the wheel is always at the same point. In a regulator barrel, however, the pressure of the cord and weight shifts gradually from one end of the barrel to the other, as the clock runs down, and when the pivots are of unequal thickness the power is transmitted nearly as irregular as if the top of the barrel was slightly conical and both pivots of the same size. For the above reason, I think, that it will be plain to all that in a fine clock both of the barrel pivots should be made of an equal diameter. The front pivot should be made no larger than is absolutely necessary for a winding square, and when we take the fact into consideration that a fine clock with a Graham escapement requires considerable less power to keep it in motion than an eight-day marine chronometer does, we may safely conclude that the winding squares of many regulators of the Graham class might be made smaller. A pivot about 0.2 of an inch will secure a sufficient amount of strength. For the reasons mentioned above, the back pivot should be exactly the same diameter, and although the effects of friction will be slightly greater when both pivots are of an equal size, still the force of the weight will be transmitted more regularly, which is the object aimed at. Where the plates are bushed a length of two to three diameters is long enough for the pivot holes.

The stop works, maintaining powers and general arrangement of the great wheel, ratchets and clicks, have been so fully described and illustrated on pages 282 to 290, Figs. 83 to 87, that it would be useless duplication to repeat them here, and the reader is therefore referred to those pages, for full particulars. This is also the case with the purely mechanical operations of cutting the wheels and pinions, hardening, polishing, staking, etc.; all have been fully treated; but there are some further considerations which may be mentioned here. The practical value of making pinions with very high numbers is very much overrated. I know of two clocks situated in the same building that are compared every other day by transit observation. They both have Graham escapements and mercurial pendulums, and are equally well fitted up, and as far as the eye can detect, they are about equally well made in all the essential points, with only this difference: one clock has pinions of eight, and the other pinions of sixteen leaves, yet for two years one clock ran about equally as well as the other. In fact, if there was any difference, it was in favor of the clock with the eight-leaved pinions. In giving this example, I must not be understood to be placing little value on high-numbered pinions. I know that in some instances they can be used to advantage. The idea that I want to illustrate at present is, that it is not in this direction that we are to search for the means of improving the rates of regulators.

A pinion as low as eleven leaves can be made so that the action of the tooth will begin at or beyond the line of centers; but as eleven is an inconvenient number to use in clock-work, we may with great propriety decide upon twelve as being a sufficient number of leaves for all the pinions used in a regulator having a Graham escapement.

In arranging the size of the wheels in a regulator, the diameters of the center and third wheels are determined by the distance between the center of the minute and the center of the seconds hand circle on the dial. As the dials of regulators are usually engraved after the dial plates have been fitted, and as the position of the holes in the dial for the center and scape wheel pivots to come through determines the size of the seconds circle, it may be well to mention here that, for a twelve-inch dial, two and a half inches is a good distance for the center of the minute circle to be from the center of the seconds circle. Consequently the center and third wheels must be made of such a diameter as will raise the scape wheel arbor two and a half inches from the center arbor, and the other wheels must be made proportionably larger, according to the number of teeth they contain.

We all know what a difficult matter it is to make a cutter that will cut a tooth of the proper shape; but when the cutter is once made and carefully used, we also know that it will cut or finish a great number of wheels without injury. For this reason, those who are contemplating making only one, or at most but a few regulators, will find the work will be greatly simplified by making the wheels of a diameter proportionate to the number of teeth they contain, and for all practical purposes the cutter that cuts or finishes the teeth of one wheel will be sufficiently accurate for the others. If we make all the pinions with the same number of leaves they will also all be nearly of the same diameter, and may be cut, or rather the cutting operation may without any great impropriety be finished with one cutter.

An opinion prevails among a certain class of workmen that the teeth of the great wheel and leaves of the center pinion should be made larger and stronger than the other wheels and pinions, because there is a greater strain upon them than on the other. However reasonable this idea may seem, a little consideration will show that in the case of a regulator, with a Graham escapement, where so little motive power is required to keep it in motion, an arrangement of this nature is altogether unnecessary. The smallest teeth ever used in any class of regulators are strong enough for the great wheel; and if there be a greater amount of strain on the teeth of the great wheel in comparison with the teeth of the third wheel, for example, then make the great wheel itself proportionately thicker, as is usually done, according to the extra amount of strain that it is to bear. The teeth of wheels and the leaves of pinions wear more from imperfect construction than from any want of a sufficient amount of metal in them.

If we assume the distance between the center of the minute and the center of the seconds circle to be 2½ inches, and also assume that the clock will have a seconds pendulum, and all the pinions have 12 leaves, and the barrel make one turn in 12 hours, then the following is the diameter the wheels will require to be, so that the teeth may all be cut with one cutter, and also the number of teeth for each wheel:

Great wheel 144 teeth. Diameter 3.40 inches for the pitch circumference.

Hour wheel 144 teeth. Diameter 3.40 inches for the pitch circumference.

Center wheel, 96 teeth. Diameter 2.26 inches for the pitch circumference.

Third wheel 90 teeth. Diameter 2.11 inches for the pitch circumference.

Scape wheel 30 teeth. Diameter 1.75 inches for the pitch circumference.

The number of arms or crosses to be put in a wheel is usually decided by the taste of the person making the clock. There is, however, another view of the subject, which I would like to mention. With the same weight of metal a wheel will be stronger with six arms than with four or five, and as lightness, combined with strength, should be the object aimed at in making wheels, I prefer six arms to four or five for the wheels of a regulator.

Figs. 157 and 158 are front and side elevations of the proposed regulator movement, showing the size and position of the wheels, the size of the frames, the positions of the pillars, dial feet, etc. The dotted large circular lines on Fig. 157 show the position the hour, minutes, and seconds circles will occupy on the dial. According to the ordinary rules of drawing, the dotted lines would infer that the movement is in front of the dial, and perhaps it may be necessary to explain that in the present instance these lines are made dotted solely with the view of making the diagram more distinct, and are not intended to represent the dial to be at the back of the movement. A is the barrel, B is the great wheel, which turns once in twelve hours; C is the hour wheel, which works into the great wheel, and also turns once in twelve hours; D is the center wheel, which turns once in an hour, and carries the minute hand; E is the third wheel, and F is the scape wheel, which turns once in a minute and carries the seconds hand; G is the pallets; H the pillars, and I is the dial feet; J is the maintaining power click, and K shows the position of the cord. Neither the hour or great wheels project over the edge of the frame, and it will be observed that a clock of this arrangement is remarkable for its simplicity, having only four wheels and three pinions, with the addition of the scape wheel and the barrel ratchets. There are no motion or dial wheels, the wheel C turning once in 12 hours, carrying the hour hand. The size and shape of the frames and the position of the pillars, allows the dial feet to be placed so that the screws which hold the dial will appear in symmetrical positions on the dial.

Formerly the term "astronomical" was applied to clocks which indicated the motions and times of the earth, moon, and other celestial bodies, but at present we may take it as indicating such as are used in astronomical observatories. In all essential particulars they are the same as first class watchmakers' regulators, the most obvious departure being that the hour hand is made to revolve only once a day, the dial being divided into twenty-four hours. This only requires an intermediate wheel and pinion in the motion work, and, assuming the hour hand to be driven from the center arbor, there will be the usual hour and minute wheels and cannon pinion. The most suitable ratio for these are $1/4$ and $1/6 = 1/24$, and, as any numbers, being multiples, may be used, they may as well be selected so as to be cut with the same tools as the wheels of the train. Two pinions of 20 and wheels of 80 and 120 suit very well; $20 \div 80$ and $20 \div 120 = 20/80 \times 20/120 = 400/9600 = 1/24$, and the hands will both go in the same direction.

Some astronomical clocks show mean solar, and others sidereal time; this requires no structural alteration, merely a little shortening of the pendulum in the latter case, which can be done with the regulating nut.

LIST OF ILLUSTRATIONS.

INDEX.

A

C

D

E

S

T